T0300961

Electromagnetics for High-Speed Analog and Digital Communication Circuits

Modern communications technology demands smaller, faster, and more efficient circuits, the design of which requires a good understanding of circuit theory and electromagnetics. This book reviews the fundamentals of electromagnetism as applied to passive and active circuit elements, highlighting the various effects and potential problems in designing a new circuit. The author begins with a review of the basics: the origin of resistance, capacitance, and inductance, from a circuit and field perspective; then progresses to more advanced topics such as passive device design and layout, resonant circuits, impedance matching, high-speed switching circuits, and parasitic coupling and isolation techniques. Using examples and applications in RF and microwave systems, the author describes transmission lines, transformers, and distributed circuits. State-of-the-art developments in Si-based broadband analog, RF, microwave, and mm-wave circuits are also covered. With up-to-date results, techniques, practical examples, many illustrations, and worked examples, this book will be valuable to advanced undergraduate and graduate students of electrical engineering and practitioners in the IC design industry. Further resources for this title are available at www.cambridge.org/9780521853507.

ALI M. NIKNEJAD obtained his Ph.D. in 2000 from the University of California, Berkeley, where he is currently an associate professor in the EECS department. He is a faculty director at the Berkeley Wireless Research Center (BWRC) and the co-director of the BSIM Research Group. Before his appointment at Berkeley, Niknejad worked for several years in industry designing CMOS and SiGe ICs. He has also served as an associate editor of the *IEEE Journal of Solid-State Circuits*, and was a co-recipient of the Jack Raper Award for Outstanding Technology Directions Paper at ISSCC 2004.

Electromagnetics for High-Speed Analog and Digital Communication Circuits

ALI M. NIKNEJAD

CAMBRIDGE
UNIVERSITY PRESS

University Printing House, Cambridge CB2 8BS, United Kingdom

One Liberty Plaza, 20th Floor, New York, NY 10006, USA

477 Williamstown Road, Port Melbourne, VIC 3207, Australia

314-321, 3rd Floor, Plot 3, Splendor Forum, Jasola District Centre, New Delhi - 110025, India

79 Anson Road, #06-04/06, Singapore 079906

Cambridge University Press is part of the University of Cambridge.

It furthers the University's mission by disseminating knowledge in the pursuit of education, learning and research at the highest international levels of excellence.

www.cambridge.org
Information on this title: www.cambridge.org/9780521853507

First published 2007
Reprinted with corrections 2008

A catalogue record for this publication is available from the British Library

ISBN 978-0-521-85350-7 Hardback

Contents

Preface

Why another EM book? There are virtually thousands of books written on this subject and yet I felt the urge to write another one.

The idea for this book germinated in my mind on a long and uneventful drive from Berkeley to San Diego. I had just completed my first year of graduate school at Berkeley and had started a research project on analyzing spiral inductors. It occurred to me that studying electromagnetics as a circuit designer was a lot easier than studying it as an undergraduate at UCLA. Even though I took many EM courses during my undergraduate education, very little of it actually stuck with me. Much like all those foreign languages we learn in high school or college, without any practice, we quickly lose our skills. When we find ourselves at that critical moment in a foreign country, our language skills fail us. While EM is the foundation of much of electrical engineering, somehow it's treated as a foreign tongue, spoken only by the few learned folks in the the field. But learning EM should not be like learning Greek or Latin!

That summer I spent many weekends in San Diego visiting my family. During these trips I'd take my EM books down to the beach and study. I'd plant myself on the beach at La Jolla or Del Mar and work my way through my undergraduate EM text. This time around, things were making a lot more sense, since I had an urgent need to actually learn electromagnetics. But I observed that having a circuits background was somewhat equivalent to speaking a related derived tongue. I realized that many people out there also missed the boat on learning EM, since they learned it without any background, desire, or need to learn it. But many of those same people, after taking a lot of high-frequency electronics courses, feel they need to relearn this important subject. If you're one of those people, this book is written for you!

When I was an undergraduate student, EM courses were a required part of every EE student's education. No matter how painful, you had to work your way through two or three courses. But today the situation has changed dramatically. Many schools have made this an optional course and, much to our horror, many students simply skip it! Even though they do take EM as part of their physics education, the emphasis is on fundamentals, with no coverage of important engineering topics such as transmission lines or waveguides. Today, more than ever, this seems like a tragedy. High-speed digital, RF, and microwave circuits abound, necessitating the training of engineers in the art and science of electronics, electromagnetics, communication circuits, antennas, propagation, etc.

With the availability of high-speed 64-bit microprocessors, server farms, Gb/s networks, and mass storage, many practical problems are now computationally tractable. Workers in the field of high-speed electronics are increasingly turning to commercial electromagnetic solvers to tackle difficult problems. As powerful as EM solvers are today, it still takes a lot of skill to set up and run a problem. And at the end of a long five hour simulation, can you trust the results? Did you actually set up the problem correctly? Are the boundary conditions appropriate? Is the field accuracy high enough? These are difficult questions and can only be answered by observing the currents, voltages, and electric and magnetic fields with a trained eye.

The focus of this book is the application of electromagnetics to circuit design. In contrast to classical analog integrated circuit design, passive components play an integral role in the design of RF, microwave, and broadband systems. Most books dedicate a section or at best a chapter to this all important topic.

The book begins with the fundamentals – the origins of resistance, capacitance, and inductance. We spend a great deal of time reviewing these fundamental passive elements from a circuit and field perspective. With this solid foundation, the book progresses to more advanced applications. A chapter on passive device design and layout reviews state-of-the-art layout techniques for the realization of passive devices in an integrated circuit environment. Important circuit applications such as resonant circuits and impedance matching are covered extensively with an emphasis on the inner workings of the circuitry (rather than a cookbook approach) in order to uncover important insights into the insertion loss of these circuits. Next, the book moves to active two-port circuits and reviews the co-design of amplifiers with passive components. Two-port circuit theory is used extensively to understand optimal power gain, stability, activity, and unilateral gain. Transmission lines, transformers, and distributed circuits form the core of the advanced circuit applications of passive elements. These topics are taught in a coherent fashion with many important examples and applications to RF and microwave systems. The time-domain perspective is covered in a chapter on high-speed switching circuits, with a detailed discussion of the transient waveforms on transmission lines and transmission line dispersion. Parasitic coupling and isolation techniques are the topic of an entire chapter, including discussion of package, board, and substrate coupling. An introduction to the analysis and design of passive microwave circuits is also covered, serving as a bridge to an advanced microwave textbook.

Acknowledgments

I would like to thank all the people who have helped me write this book. Much of this material was inspired by teaching courses at Berkeley and so I thank all the students who read the original lecture notes and provided feedback in EECS 105, 117, 142, 217, and 242 (thanks to Ke Lu for detailed feedback). This book would not be as interesting (assuming you find it so) without real circuit applications drawn from literature and from our own research projects. Thanks to my colleagues and collaborators at Berkeley who have created a rich and stimulating research environment. In particular, thanks to my BWRC colleagues, Robert Brodersen, Jan Rabaey, Bora Nikolic, Robert Meyer, Paul Wright, and John Wawrzynek. And thanks to Professor Chenming Hu for inviting me to be a part of the world-famous BSIM team. Thanks to Jane Xi for her hard work and dedication to the BSIM team. Special thanks goes to the graduate student researchers. In particular, thanks to Sohrab Emami and Chinh Doan who were key players in starting the Berkeley 60 GHz project and OGRE. Many of the high-frequency examples come from our experience with this project. Thanks to Professor Andrea Bevilacqua (University of Padova, Italy) for a stimulating research collaboration on UWB. Thanks to Axel Berny and his love of oscillators.

Though I take responsibility for any errors in the book, I have my graduate students to thank for the countless errors they were able to find by reading through early drafts of the manuscript. Thanks to Ehsan Adabi, Bagher (Ali) Afshar, Mounir Bohsali, Yuen Hui Chee, Wei-Hung Chen, Debo Chowdhury, Mohan Dunga, Gang Liu, Peter Haldi, Babak Heydari, and Nuntachai Poobuapheun. They provided detailed feedback on various chapters of the book.

Also thanks to my friends and colleagues for reviewing the book. In particular I'm grateful to Dr. Manolis Terrovitis, Eric Hoffman, Professor Hui Wu, and Professor Hossein Hashemi for taking the time to review the book and provide feedback.

Finally, thanks to the folks who supported our research during the past four years. Special thanks to DARPA and the TEAM project, in particular thanks to Barry Perlman and Dan Radack for your support of university research. Thanks to BWRC member companies, in particular ST Microelectronics, Agilent Technologies, Infineon, Conexant Systems, Cadence, and Qualcomm. Thanks to Analog Devices, Broadcom, Berkeley Design Automation, and National Semiconductor for your support through the UC MICRO and UC Discovery programs. And thanks to SRC and member companies for supporting research of compact modeling at Berkeley. Thanks in particular to Jim Hutchby of SRC, Keith Green of Texas Instruments, Weidong Liu of Synopsys, Judy An of AMD, Josef Watts and Jack Pekarik of IBM, and Ben Gu of Freescale.

1 Introduction

1.1 Motivation

The history of electronics has been inextricably linked with the growth of the communications industry. Electronic communication served as a major enabling technology for the industrial revolution. When scientists and engineers learned to control electricity and magnetism, it did not take long for people to realize that the electromagnetic force would enable long-range communication. Even though the basic science of Maxwell's equations was well understood, it took much longer for practical applications to fully exploit all the fantastic possibilities such as radio, television, and personal wireless communication.

At first only crude wires carrying telegraph signals were rolled out sending Morse code,[1] digital signals at speeds limited by human operators. In this regard it is ironic that digital communication predates analog communication. Telegraph wires were laid alongside train tracks, making long-range communication and transportation a practical reality. Sending signals faster and further ignited the imagination of engineers of the time and forced them to study carefully and understand the electromagnetic force of nature. Today we are again re-learning and inventing new digital and analog communication systems that are once again compelling us to return to the very fundamental science of electricity and magnetism.

The topic of this book is the high-frequency electromagnetic properties of passive and active devices. For the most part, passive devices are resistors, capacitors, transformers, and inductors, while active devices are transistors. Most applications we draw from are high-frequency circuits. For example, radio frequency (RF) circuits and high-speed digital circuits both depend on a firm understanding of passive devices and the environment in which they operate.

Circuit theory developed as an abstraction to electromagnetics. Circuit theory is in effect the limit of electromagnetics for a circuit with negligible dimension. This allows spatial variations and time delay to be ignored in the analysis of the circuit. As such, it allowed practicing engineers to forego solving Maxwell's equations and replaced them with simple concepts such as KCL and KVL. Even differential equations were eliminated and replaced with algebraic equations by employing Laplace transforms. The power and popularity of circuit theory was due to its simplicity and abstraction. It allowed generations of engineers to

---

[1] Or as Paul Nahin suggests in [41] we should more correctly call this "Vail" code.

solve difficult problems with simple and yet powerful tools. In effect, it allowed generations of engineers to forego reading a book such as this one.

So why read another book on electromagnetics? Why bother learning all this seemingly complicated theory when your ultimate goal is to build circuits and systems for communication and information management?

We live today at the intersection of several interesting technologies and applications. Integrated circuit technology has enabled active devices to operate at increasingly higher frequencies, turning low-cost Si technology into a seemingly universal panacea for a wide array of applications. CMOS digital circuits are switching at increasingly higher rates, pushing multi GHz operation. Si CMOS, bipolar, and SiGe technology have also enabled a new class of low-cost RF and microwave devices, with ubiquitous deployment of cellular phones in the 800 MHz–2 GHz spectrum, and high-speed wireless LAN in the 2–5 GHz bands. There seems to be very little in the way of enabling Si technology to exploit the bandwidths up to the limits of the device technology. In a present-day digital 130 nm CMOS process, for instance, circuits are viable up to 60 GHz [55] [13].

At the same time, wired communication is pushing the limits. Gigabit Ethernet and high-speed USB cables are now an everyday reality, and people are already pursuing a 10 Gb/s solution. Optical communication is of course at the forefront, with data rates in the 40 Gb/s range now commercially viable and at relatively low cost.

The simultaneous improvement in active device technology, miniaturization, and a host of new applications are the driving force of today's engineering. As integrated circuits encompass more functionality, many traditionally off-chip components are pushed into the IC or package, blurring the line between active devices and circuits and passive devices and electromagnetics. This is the topic of this book.

Technology enhancements

The limitations in frequency and thus speed of operation is usually set by the active device technology. One common figure of merit for a technology is, the unity-gain frequency f_T, the frequency at which the short-circuit current gain of the device crosses unity. Another important figure of merit is f_{max} the maximum frequency of oscillation, or equivalently the frequency where the maximum power gain of a transistor drops to unity. Since f_{max} is a strong function of layout and parasitics in a process, it is less often employed. In contrast, the f_T depends mostly on the dimensions of the transistor and the transconductance

$$f_T = \frac{1}{2\pi} \frac{g_m}{C_\pi + C_\mu} \tag{1.1}$$

It can be shown [50] that the device f_T is inversely related to the transistor dimensions. For a long-channel MOSFET the key scaling parameter is the channel length L

$$f_T \approx \frac{1}{2\pi} \frac{3\mu(V_{gs} - V_t)}{2L^2} \tag{1.2}$$

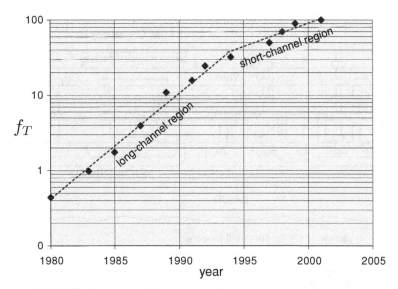

Figure 1.1 The improvements in device unity–gain frequency f_T over the past two decades due to device scaling.

while in the limit for short channel transistors the scaling changes to L^{-1} since the current is limited by velocity saturation

$$I_{ds,sat} = W Q_i v_{sat} = W C_{ox}(V_{gs} - V_t)v_{sat} \qquad (1.3)$$

resulting in

$$f_T \propto \frac{v_{sat}}{L} \qquad (1.4)$$

For a bipolar junction transistor (BJT) the critical dimension is the base width. In the limit that base transit limits the frequency of operation

$$f_T \approx \frac{1}{2\pi} \frac{1}{\tau_B} \propto \frac{1}{W_B^2} \qquad (1.5)$$

As integrated circuit manufacturing technology has improved exponentially in the past three decades, so has the f_T of the device, giving circuit designers increasingly faster devices. A plot of the device f_T over the years for a MOSFET device is shown in Fig. 1.1, and the exponential growth in technological advancements can be seen clearly.

It is important to note that this improvement in performance only applies to the intrinsic device. Early circuits were in fact limited by the intrinsic transistor and not the parasitic routing and off-chip environment. As circuit technology has advanced, though, the situation has reversed and now the limitation is set by the parasitics of the chip and board environment, as well as the performance of the passive devices. This is why the material of this book is now particularly relevant. It can be shown that a good approximation to the CMOS device f_{max} is given by [42]

$$f_{max} \approx \frac{f_T}{2\sqrt{R_g(g_m C_{gd}/C_{gg}) + (R_g + r_{ch} + R_s)g_{ds}}} \qquad (1.6)$$

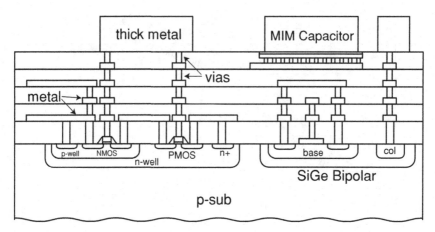

Figure 1.2 Cross section of a SiGe BiCMOS process.

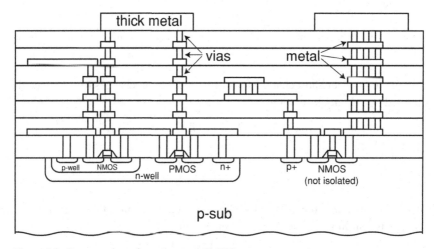

Figure 1.3 Cross section of an advanced CMOS process.

where the device performance is a strong function of the loss, such as the drain/source resistance R_s, R_d, and the gate resistance R_g. These parasitics are in large part determined by layout and the process technology.

While early integrated circuit technologies were limited to a few types of different active devices and a few layers of aluminum interconnect metal, present-day process technology has a rich array of devices and metal routing. In an advanced Si process, shown in Fig. 1.2, high-performance SiGe HBT devices are complemented by MOS and PN-junction varactors, metal-insulator-metal (MIM) high-density and high-quality capacitors, and thick-metal for low-loss interconnect and inductors/transformers. Even a digital CMOS process, as shown in Fig. 1.3, has many advanced capabilities. In addition to several flavors of MOS active devices (fast thin oxide, thick oxide, high/low V_T), there are also enhanced isolation structures and triple-well (deep n-well) devices, and many layers of interconnect that allow construction of high-quality, high-density capacitors and reasonably high-quality inductors.

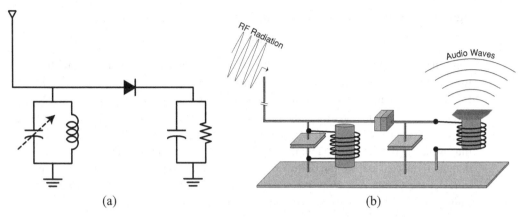

Figure 1.4 (a) A simple AM receiver circuit. The resistor represents a high-input impedance earphone. (b) A physical realization of the simple AM receiver circuit.

Radio and wireless communication

Early radio systems were essentially all passive. To see this look into the back of an old radio where a few active devices (vacuum tubes or transistors) are surrounded by tens to hundreds of passive devices. Consider the circuit diagram of a very simple AM receiver shown in Fig. 1.4a. The antenna drives a resonant tank tuned to the center frequency of the transmitting station. This signal is fed into a peak detector that follows the peak of the RF signal. The low-pass filter time constant is only fast enough to follow the low-frequency audio signal (generically the baseband signal) and yet too slow to follow the RF, thus removing the RF signal and retaining the low-frequency audio. This received signal is usually too weak to drive a speaker but can be heard through a sensitive headphone. A simple audio amplifier can be used to strengthen the signal.

It is interesting to note that this AM receiver can be physically realized by merely using contacts between a few different pieces of metal and semiconductors. This is shown in Fig. 1.4b. The resonant tank is simply a piece of wire wound into a coil which contacts with the capacitor, two metal plates in close proximity. The diode can be realized as the junction of a metal and semiconductor. Finally, to convert electric energy into sound we can use another large inductor coil and use the time-varying magnetic force to move a paper thin cone driven by a magnetic core. Magnetic materials have been known since ancient times and therefore since the metal age we have had the capability to build radio receivers! In fact, it is not surprising that radios often crop up accidentally.[2]

Most modern radios operate based on an architecture invented by Edwin Armstrong. The block diagram of such a system, called a super-heterodyne receiver, is shown in Fig. 1.7. This receiver incorporates a local oscillator (LO), a block that primarily converts DC power into RF power at the oscillation frequency. A mixer takes the product of this signal and the

---

[2] For instance my old answering machine also picked up the radio. Sometimes you could hear it as you were waiting for the tape recorder to rewind. At least this was a desirable parasitic radio.

signal received by the antenna. Recall the following trigonometric identity[3]

$$2 \cos(\omega_{LO} t) \cos(\omega_{RF} t) = \cos((\omega_{LO} + \omega_{RF})t) + \cos((\omega_{LO} - \omega_{RF})t) \qquad (1.7)$$

Note that the product of the received RF signal and the local oscillator signal produces two new signals, one centered at the difference frequency and one centered at the sum frequency. If we put a bandpass filter at one of these frequencies, call it the intermediate frequency, IF, we can electronically tune the radio by simply changing the LO frequency. This is accomplished by using a frequency synthesizer (a PLL or phase locked loop), and thus we avoid building a variable filter common to the early radios. The important point is that the IF is fixed and we can build a very selective filter to pinpoint our desired signal and to reject everything else. Why not simply set LO equal to RF to move everything to DC? This is in fact the direct-conversion or zero-IF architecture. It has some shortcomings such as problems with DC offset,[4] but its main advantage is that it lowers the complexity of the RF section of a typical radio.

At the heart of the frequency synthesizer is the voltage controlled oscillator (VCO). The VCO is an oscillator where the output frequency is a function of a control voltage or current.[5] To build a VCO we need a way to change the center frequency of a resonant tank. The resonant tank is simply an inductor in series or in parallel with a capacitor. One typical realization is to use varactor, a variable capacitor. A reversed biased diode serves this purpose nicely, as the depletion region width, and thus the small-signal capacitance, is a function of the reverse bias. It seems that a super-heterodyne receiver has simply moved the variable resonant tank from the antenna front end to a variable resonant tank in the VCO! Have we gained anything? Yes, because the frequency of the VCO can be controlled precisely in a feedback loop (using an accurate frequency reference such as a crystal), eliminating any problems associated with absolute tolerances in components in addition to drift and temperature variation.

The radio has once again emerged as a critical application of passive devices spawned by the growth and popularity of wireless telephones, in particular the cellular phone. By limiting the transmitter powers and taking advantage of spatial diversity (re-using the same frequency band for communication for points far removed – for non-adjacent cell sites), a few hundred radio channels can be used to provide wireless communication to millions of people. Modern cell phones employ complicated radio receivers and transmitters (transceivers) employing hundreds and thousands of passive devices. Early cell phones used simple architectures such as the super-heterodyne receiver but the demand for low-cost and small footprints has prompted a re-investigation of radio architectures.

The layout of a modern 2.4 GHz transceivers for 802.11b wireless LAN (WLAN) is shown in Fig. 1.5 [7]. The IC is implemented in a 0.25μ CMOS process and employs several integrated passive devices such as spiral inductors, capacitors, and resistors. The spiral inductors comprise a large fraction of the chip area. The next chip shown in Fig. 1.6

---

3   I recall asking my trig teacher about the practical application of the subject. After scratching her head and pondering the question, her response was that architects use trig to estimate the height of buildings! A much better answer would have been this equation.

4   And $1/f$ noise in MOS technology.

5   This makes a nice AM to FM modulator, as well.

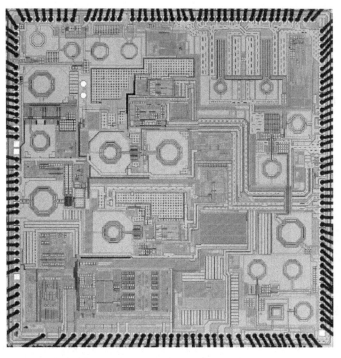

Figure 1.5 A 2.4 GHz CMOS 802.11b Wireless LAN Transceiver [7]. (Copyright 2003, IEEE)

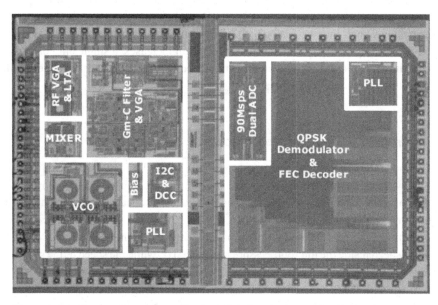

Figure 1.6 A direct-conversion satellite broadband tuner-demodulator SOC [17] operates from 1–2 GHz. (Copyright 2003, IEEE)

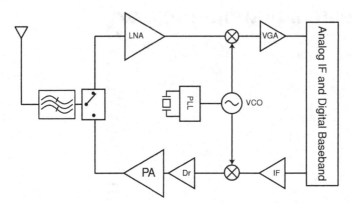

Figure 1.7 The block diagram of an Armstrong super-heterodyne transceiver.

[17] is an integrated direct-conversion satellite broadband tuner-demodulator "system-on-a-chip" (SOC). The chip is implemented in a $0.18\,\mu$ CMOS process and employs MIM capacitors and spiral inductors. It operates in the 1–2 GHz band, requiring broadband operation and high linearity. Notice that the digital baseband has been integrated on to a single chip along with the sensitive analog and RF blocks. This brings about several important challenges in the design due to the parasitic coupling between the various blocks. A triple-well process and lead-less package technology are used to maximize the isolation.

In general, integrating an entire transceiver on to a single chip has many challenges. The power amplifier (PA) or PA driver can injection lock the VCO through the package and substrate, causing a spurious modulation. Digital circuitry can couple seemingly random switching signals into the analog path, effectively increasing the noise floor of the sensitive RF and analog blocks. As the level of integration increases, a single chip or package may contain several systems in operation simultaneously, requiring further understanding and modeling of the coupling mechanisms.

Computers and data communication

Computers and data communication, particularly the Internet, have given rise to a new tidal wave in the information revolution. The speed of computers has improved drastically due to technological improvements in transistor, microprocessor, memory, and system bus architectures. Computer circuits move and process discrete time signals at a frequency determined by the system clock. For instance, in the current generation of computers the clock speed inside the microprocessor is several GHz, while the speed of the system bus and memory lag behind by a factor of 2–3. This is because inside the microprocessor everything is small and dense and signals travel short distances in the presence of small parasitics (mainly capacitance). Off-chip, though, the system bus environment is characterized by much longer distances and much larger parasitics, such as non-ideal dispersive transmission lines along the board traces. Modern computer networks, like gigahertz Ethernet LAN, also operate at high frequencies over wires, necessitating a complete understanding of distributed transmission line effects. These topics are covered in Chapters 9 and 12.

High-speed wireless data communication is the focus of much research and development. The next and future generations of cellular technology will bring the Internet from our homes and offices into virtually every location on earth. Wireless LAN systems enable short-range high-speed data communication without the expensive network infrastructure. A physical network infrastructure requires time-consuming distribution of cables to every office in a building. A wireless system can be up and running in minutes or hours as opposed to days or months.[6]

In such systems cost and size will force many external passive components on to the chip environment, where knowledge of parasitic coupling and loss is critical in a successful low-cost implementation. In this book we spend a great deal of time discussing inductors, capacitors, transformers, and other key passive elements realized in the on-chip environment.

Microwave systems

Microwave systems employ higher frequencies where the wavelength $\lambda = c/f$ is of the order of centimeters or millimeters. Thus the lumped circuit approach fails since these structures are a significant fraction of a wavelength and spatial variation begins to play as important a role as time variation. Such systems were first employed in World War II for radar systems.[7] In a radar system, the small wavelength allows us to construct a highly directional antenna to focus a beam of radiation in a given direction. By observing the reflection, we can compute the time-of-flight and hence the distance to an object. By also observing the Doppler frequency shift, we can compute the speed of the object.

Perhaps the greatest difficulty in designing microwave systems below 10 GHz is that the operating frequency is in an intermediate band where lumped element circuit techniques do not strictly apply and microwave methodology results in prohibitively large circuits. At 3 GHz, the wavelength is 10 cm in air and about 5 cm in silicon dioxide, while an integrated circuit has dimensions of the order of millimeters, thus precluding distributed elements such as quarter-wave transmission lines. But using advances pseudo-lumped passive devices such as inductors, transformers, and capacitors, microwave ICs can be realized with minimal off-chip components.

Many early microwave systems were designed for military applications where size and cost were of less concern in comparison to the quality and reliability. This led to many experimental and trial-and-error design approaches. Difficult system specifications were met by using the best available technology, and often expensive and exotic processes were employed to fabricate high-speed transistors. New microwave systems, in contrast, need to be mass produced and cost and size are the main concerns. Fortunately high-volume process

---

[6] I seem to recall that it took a year for a network upgrade to occur in Cory Hall at Berkeley!
[7] It is ironic that the EEs of the time lacked the necessary skills to build such systems and the project was handed off to the physicists at the MIT Radiation Lab.

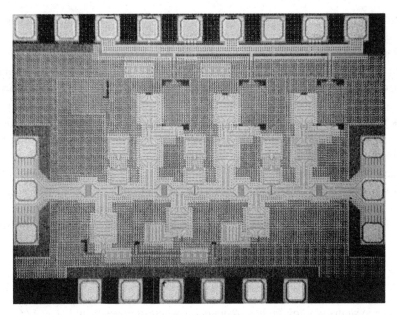

Figure 1.8 A three-stage 60 GHz CMOS LNA implemented in a digital 130 nm process.

technology using silicon is now readily available. The speed is now sufficient to displace many specialized technologies. Since high-volume microwave systems are primarily being designed by circuit engineers as opposed to microwave engineers, the lack of knowledge of electromagnetics and distributed circuits can be an impediment to successful integration and implementation.

Higher-frequency bands offer new opportunities to exploit sparsely used spectrum. The 60 GHz "oxygen absorption" band is a prime example, providing 7 GHz of unlicensed bandwidth in the US. An example of a 60 GHz multi-stage low-noise amplifier (LNA) is shown in Fig. 1.8. Here transmission lines play a key role as inductors, interconnect, and resonators. A 60 GHz single-transistor mixer, shown in Fig. 1.9, employs a hybrid coupler (see Section 15.7) to combine the RF and LO signal. Spiral inductors are also employed in the IF stages. Both of these chips were fabricated in a digital 130 nm CMOS process. Another CMOS microwave circuit is shown in Fig. 1.10. This is a circular standing-wave 10 GHz oscillator, employing integrated transmission lines in the resonator [11]. There is a beautiful connection between this oscillator and the orbit of an electron in a hydrogen atom. Similar to the wave function of an electron, the electromagnetic mode must satisfy the periodic boundary condition, and this determines the possible resonant modes of the structure.

Optical communication

Fiber-optic communication systems allow large amounts of data to be transmitted great distances with relatively little attenuation. At optical frequencies, metals are too lossy for long-haul communication without amplification, and so the energy is confined inside a thin fiber of glass by total internal reflection. The flexibility and low cost of this material has displaced more traditional waveguides made of rigid or semi-rigid and expensive materials.

Figure 1.9 A single-transistor 60 GHz CMOS mixer implemented in a digital 130 nm process.

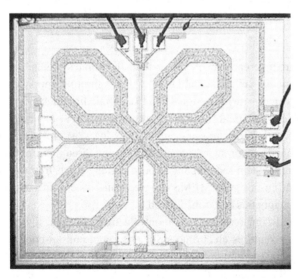

Figure 1.10 A circular standing-wave oscillator operates at 10 GHz [11]. (Copyright 2003, IEEE)

The main bottleneck in optical systems, though, is the electronics. While new optical amplifiers may displace electronic regeneration techniques, routing digital signals on fiber lines requires electronic circuits. Much research effort is dedicated to completely displacing electronic routing through mechanical means. This may seem like a backward step

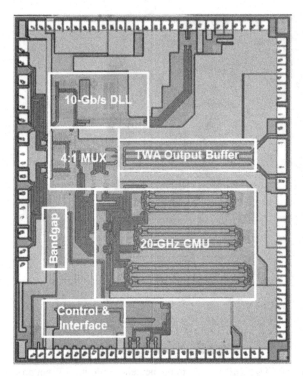

Figure 1.11 A 40 Gb/s SONET MUX/CMU implemented in SiGe technology. The integrated 20 GHz CMU employs coupled oscillators and transmission line resonators [56]. (Copyright 2003, IEEE)

since mechanical systems are inherently much slower than electronic systems. But it is the throughput of electronic systems, not the switching rate, that is limiting optical communication systems. Transistors can only process signals up to a certain limit due to intrinsic f_T and extrinsic parasitics. Micro-electrical-mechanical systems (MEMS) technology allow micromechanical systems to reside on the same substrate as low-cost electronic circuits. A simple MEMS-based hinged mirror switch does not limit the throughput of an optical system, whereas an electronic system requires high-speed digital signals to travel through transistors and parasitics. The co-design of MEMS and electronic systems in a chip requires an understanding of electromagnetics and transduction.

Even if routing of optical signals is done without electronic means, the generation of high-speed optical signals requires electronic systems to pool and un-pool dozens of lower-speed signals on to a fast common serial line. This requires accurate timing and synchronization, such as voltage-controlled oscillators phase locked to reference signals. Optical and high-speed digital circuits require routing of high-speed signals in the presence of board and transmission line parasitics, as well as the design of passive elements for frequency synthesis. An example optical transceiver chip incorporating the multiplexers (MUX) and clock-multiplication unit (CMU) is shown in Fig. 1.11, where transmission lines are used in a coupled oscillator to generate a low-phase noise high-frequency oscillation.

1.2 System in Package (SiP): chip and package co-design

Present design methodology draws a clear line between the chip and the package. This clean interface between the chip and the package simplifies design considerably, since the two can be analyzed independently. As already noted, the package is a major source of parasitic energy coupling in current transceivers and the careful evaluation of the package is a critical step in the design. At higher frequencies, though, the line between the package and chip will blur even further. In fact, the package could be a crucial and beneficial part of the system, as many key passive components could reside in the package. Package and chip co-design will require new design techniques and a more intimate understanding of the electromagnetic effects.

The simultaneous simulation of the interaction of the board environment, the package, and the on-chip passives is presently beyond the scope of present-day simulation tools. One problem is the vastly different scales employed in board environment (centimeters), package (sub-cm), and chip environment (sub-micron geometries). While future tools may overcome this limitation, present-day engineers must have a firm grasp of fundamental engineering to analyze and design such structures. For instance, instead of simulating the entire system, key coupling issues can be dealt with only by simulating critical sub-portions of the system. Identification of key elements and analysis of such systems are two of the goals of this book.

1.3 Future wireless communication systems

The conceptual transceiver shown in Fig. 1.12 naturally partitions into several sections, most notably the RF front-end section, the analog core, and the digital signal processing (DSP) unit. The arrows represent the flow and transfer of information among the various sections. The RF receive section amplifies the weak incoming RF signal and converts it into an intermediate frequency. The choice of IF frequency mostly determines the architecture of the system. The analog section performs further amplification and filtering and converts the signal into digital form. From here, the signal is usually buffered and driven off-chip into the DSP core where the final demodulation and detection are performed. Similarly, on the transmit side, information is encoded and modulated in the DSP core. The analog section converts the signal from digital discrete time to continuous time and up-converts the center frequency to the transmit band. In this transceiver each section communicates with every other section in a well-defined manner. Isolation between the various sections is achieved by physical separation and proper frequency planning to avoid interference.

There are several things to note about this system. Note that the entire system is physically much smaller than the wavelength. Thus, the quasi-static approximation is applicable and much of the internal circuitry can be considered as lumped elements. Also, due to the small size, there is no significant coupling between the external electromagnetic fields propagating in free space and the on-chip signals. The transduction of RF energy from free space to the volume of the conductors is achieved externally through an antenna element. This naturally partitions the radiation incident on the system from the transceiver. The RF energy flows

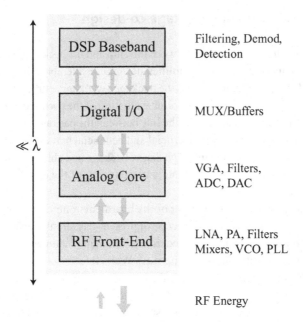

Figure 1.12 A block diagram of a present-day advanced wireless transceiver.

into the system along a well-defined path from the package interface between the board and the chip. Due to the physically large size and requirement for high quality, many passive devices in this system are realized externally, such as surface acoustic wave (SAW) filters, wire-wound high-quality inductors, and low-insertion loss diode switches. To save cost and power, the trend has been to integrate as many of the passive devices on-chip as possible. At higher frequencies this is increasingly a necessity due to the large parasitic impedances present in the board environment. MEMS technology may allow many off-chip components to be miniaturized and moved on chip.

To improve isolation in present-day transceivers, many options such as deep trenches, triple-well devices, or special technologies, such as SOI (Silicon On Insulator), are effective. The proper grounding and regulation of supplies are also important. While these techniques are effective at mitigating the coupling at lower microwave frequencies, they are much less effective at higher frequencies.

A hypothetical transceiver at a much higher frequency is shown in Fig. 1.13. Due to the large physical size of the chip and package relative to the wavelength, an array of antenna elements is integrated directly into the package. The DSP core is the heart of the system and there is constant feedback between the baseband and RF sections to optimize the power and throughput in the system. For instance, the phased antenna array can be used to boost the directivity of the receiver or transmitter to improve the SNR or to lower the power. Spatial diversity in the antenna elements can also boost the data throughput per fixed bandwidth beyond Shannon's limits.

We see that the package now plays a much larger role in this hypothetical transceiver. The package is physically much larger than the chip and contains an ever-increasing density

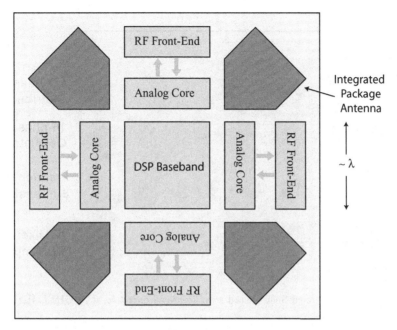

Figure 1.13 A block diagram of a hypothetical future wireless transceiver.

of signal lines in close proximity. If the package design is not performed simultaneously as an integral part of the system, the electromagnetic coupling can serve as a parasitic feedback path that can compromise the stability and lower the sensitivity of the system. These problems are naturally more severe at higher frequencies. An example of a high-frequency system with large levels of microwave integration is shown in Fig. 1.14 [24], where a bank of eight receivers operate in parallel to down-convert signals from an antenna array. The receiver can select the appropriate phase shift in each path to form a high-directivity antenna, or to improve the spatial diversity of the receiver. Proper routing of signals and isolation are key elements in the successful design of such a system.

1.4 Circuits and electromagnetic simulation

Circuits are really a part of electromagnetics. As we alluded to earlier, circuit theory can adequately replace electromagnetics when circuit dimensions are much smaller than the wavelength of electromagnetic fields. This allows all time variation to be discarded and difficult partial differential equations (PDE) resulting from Maxwell's equations to be reduced to ordinary differential equations (ODE).

Many tools are available for solving a system of non-linear ODEs resulting from applying KVL and KCL to nodes and loops in a circuit. Since Maxwell's equations are linear, the non-linearity is introduced primarily through the transistors and junctions in the circuit. The most common tool for solving ODEs in the circuit designer's arsenal is SPICE and its many variants. Frequency domain solvers (harmonic balance) and period steady-state (PSS)

3.5 mm

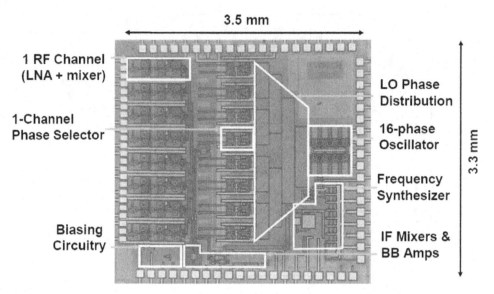

Figure 1.14 A fully integrated SiGe phased array front-end operating at 24 GHz [24]. (Copyright 2003, IEEE)

solvers are also available and are helpful when analyzing periodic time-variant circuits in the presence of noise.

PDE solvers, by contrast, are much slower and harder to use than ODE solvers such as SPICE. This is because a space of three to four dimensions (three spatial dimensions and one dimension for time) is involved. This has a large impact on traditional matrix solvers that require $O[n^{2-3}]$ time to solve a system of n unknowns. Time variation can be removed for a harmonic steady-state solution, but this even further complicates incorporating PDE solutions in an ODE solver. For instance, if you simulate an inductor with an electromagnetic (EM) solver, how do you take the frequency domain solution into a time-domain solver and efficiently compute the solution?

Electromagnetic simulation and analysis is a critical component in successfully designing active and passive components. This is missing from the circuit curriculum, since circuit designers usually do not deal with the innards of components such as transistors and inductors. Instead, they employ a lumped circuit model that captures the details. For instance, the BSIM models incorporated in Berkeley SPICE and commercial simulators model transistor device physics with hundreds of elements and parameters to describe a sub-micron transistor accurately. A transistor, though, is a very small element and most of the physics fit well into a lumped element model. A typical inductor, by contrast, is relatively simple, containing perhaps three elements to model the inductance, the resistive loss, and capacitance of the structure. This model serves as long as the inductor structure is very small relative to the wavelength. But as the frequency of operation increases, several shortcomings of the model are present. For example, the loss is a complicated function of frequency. Even if we only model skin effect, the resistive loss varies like $\sqrt{f}$. How do we put such a variation in resistance in a time-domain SPICE simulation? There is no easy solution.

In addition to SPICE, then, the circuit designer of today and of the future must be able to understand and model electromagnetic effects. In the past each block was partitioned into a lumped model (designed by device and EM engineers) and the circuit designer could ignore the complicated details. For modern circuits this is no longer possible as even short wires on chip have sufficient parasitic inductance, loss, and capacitance as to impact the performance of circuits. The situation is exacerbated by the increasingly higher densities at which circuits are packed on a single die, and electromagnetic coupling becomes the bottleneck in the achievable performance. As EM solution techniques expand with improvements in computational speed and capacity,[8] the use of an EM solver will become an integrated part of the design of circuits much as using SPICE is standard practice today.

---

[8] It is interesting to note that each generation of advance is fueled by new computation tools running on the previous generation of computers. This self-feeding cycle of advancement means that the current generation of tools is never good enough for the next generation. That's why we need engineers who understand the fundamentals and can partition difficult problems into digestible pieces. That's why engineers will never be replaced by computers!

2 Capacitance

2.1 Electrostatics review

Let's begin with the all important Gauss' Theorem. It is easy to show that the electric "flux" crossing any sphere surrounding a point source is constant and equal to the charge enclosed

$$\oint_S \mathbf{D} \cdot d\mathbf{S} = \int_0^{2\pi} \int_0^{\pi} \frac{q}{4\pi r^2} r^2 \sin\theta d\theta d\varphi = \frac{q}{4\pi} 2\pi \times 2 = q \qquad (2.1)$$

This result is due to the $1/r^2$ dependence of the electric field. Gauss' Law proves that for *any* surface (not just a sphere), the result is identical

$$\oint_S \mathbf{D} \cdot d\mathbf{S} = q \qquad (2.2)$$

Furthermore by superposition, the result applies to any distribution of charge

$$\oint_S \mathbf{D} \cdot d\mathbf{S} = \int_V \rho dV = q_{\text{inside}} \qquad (2.3)$$

First of all, does it makes sense? For instance, if we consider a region with no net charge, then the flux density crossing the surface is zero. This means that the flux lines entering the surface equal the flux lines leaving the surface. Since charge is the source/sink of electric fields, this does make intuitive sense. We can prove that the flux crossing an infinitesimal surface of any shape is the same as the flux crossing a radial cone. Notice that if the surface $d\mathbf{S}$ is tilted relative to the radial surface by an angle θ, its cross-sectional area is larger by a factor of $1/\cos\theta$. The flux is therefore a constant

$$d\Psi = \mathbf{D} \cdot d\mathbf{S} = D_r dS \cos\theta = D_r \frac{dS'}{\cos\theta} \cos\theta = D_r dS' \qquad (2.4)$$

For any problem with symmetry, it is easy to calculate the fields directly using Gauss' Law. Take a long (infinite) charged wire as shown in Fig. 2.1. If the charge density is constant with density given by λ C/m, then by symmetry the field is radial. Applying Gauss' Law to a small concentric cylinder surrounding the wire

$$\oint \mathbf{D} \cdot d\mathbf{S} = D_r 2\pi r d\ell \qquad (2.5)$$

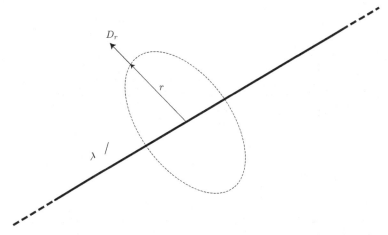

Figure 2.1 An infinitely long charged wire.

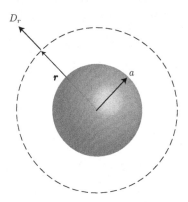

Figure 2.2 A cloud of charge of uniform density.

Since the charge inside the cylinder is simply $\lambda d\ell$

$$D_r 2\pi r d\ell = \lambda d\ell \tag{2.6}$$

$$D_r = \frac{\lambda}{2\pi r} \tag{2.7}$$

Another easy problem is a cloud of charge Q shown in Fig. 2.2. Since the charge density is uniform $\rho = Q/V$, and $V = \frac{4}{3}\pi a^3$

$$\oint \mathbf{D} \cdot \mathbf{dS} = \begin{cases} \rho \frac{4}{3}\pi r^3 = Q\left(\frac{r}{a}\right)^3 & r < a \\ Q & r > a \end{cases} \tag{2.8}$$

But $\oint \mathbf{D} \cdot \mathbf{dS} = 4\pi r^2 D_r$

$$D_r = \begin{cases} \frac{Qr}{4\pi a^3} & r < a \\ \frac{Q}{4\pi r^2} & r > a \end{cases} \tag{2.9}$$

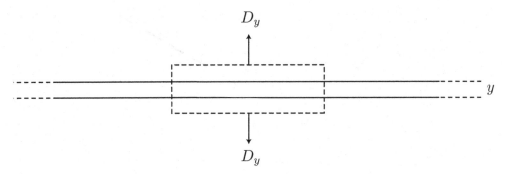

Figure 2.3 A plane charged uniformly with surface density ρ_s.

Consider an infinite plane that is charged uniformly with surface charge density ρ_s, shown in Fig. 2.3. By symmetry, the flux through the sides of a centered cylinder intersecting with the plane is zero and equal at the top and bottom. The flux crossing the top, for instance, is simply DdS, where D can only possess a $\hat{y}$ component by symmetry. The total flux is thus $2DdS$. Applying Gauss' Law

$$2DdS = \rho_s dS \tag{2.10}$$

The electric field is therefore

$$\mathbf{E} = \hat{y} \begin{cases} \frac{\rho_s}{2\epsilon_0} & y > 0 \\ \frac{-\rho_s}{2\epsilon_0} & y < 0 \end{cases} \tag{2.11}$$

We can reformulate Gauss' Law in differential form as follows. Applying the definition of divergence to the electric flux density D, we have

$$\lim_{V \to 0} \frac{\oint_S \mathbf{D} \cdot \mathbf{dS}}{V} = \frac{q_{\text{inside}}}{V} = \rho = \nabla \cdot \mathbf{D} \tag{2.12}$$

Therefore we have the important result that at any given point

$$\nabla \cdot \mathbf{D} = \rho(x, y, z) \tag{2.13}$$

This is the analog to the equivalent statement

$$\oint_S \mathbf{D} \cdot \mathbf{dS} = \int_V \rho dV \tag{2.14}$$

The *Divergence Theorem* is a direct proof of this relationship between the volume integral of the divergence and the surface integral (applies to any vector function $\mathbf{A}$)

$$\int_V \nabla \cdot \mathbf{A} dV = \oint_S \mathbf{A} \cdot \mathbf{dS} \tag{2.15}$$

Here V is any bounded volume and S is the closed surface on the boundary of the volume. Application of this theorem to the electric flux density D immediately gives us the differential form of Gauss' Law from the integral form.

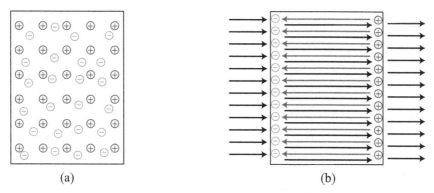

Figure 2.4 (a) An ideal conductor can be modeled as a sea of free electrons roaming about among ionized background molecules. (b) These electrons readily respond to an external field. The rearrangement of charge produces an internal field that perfectly balances (cancels) the applied field.

Perfect conductors

Perfect conductors are idealized materials with zero resistivity. Common metals such as gold, copper, and aluminum are examples of good conductors.

A fuzzy picture of a perfect conductor, shown in Fig. 2.4a, is a material with many mobile charges that easily respond to an external field. These carriers behave as if they are in a vacuum and they can move unimpeded through the metal. These mobile charges, though, cannot leave the material due to a large potential barrier (related to the work function of the material).

We may argue that the electric field is zero under static equilibrium based on the following argument. Under static equilibrium charges cannot move. So we might argue that if an external field is applied, then all the charges in a perfect conductor would move to the boundary where they would remain due to the potential barrier. But this argument ignores the fact that the external field can be canceled by an *internal* field due to the rearrangement of charges (see Fig. 2.4b).

Thus we have to invoke Gauss' Law to further prove that in fact there can be no net charge in the body of a conductor. Because if net charge exists in the body, Gauss' Law applied to a small sphere surrounding the charge would require a field in the body, which we already hypothesized to be zero. Therefore our argument is consistent. In conclusion, we are convinced that a perfect metal should have zero electric field in the body and no net charge in the body. Thus, we expect that if any net charge is to be found in the material, it would have to be on the boundary.

We could in fact define a perfect conductor as a material with zero electric field inside the material. This is an alternative way to define a perfect conductor without making any assumptions about conductivity (which we have not yet really explored). A perfect conductor is also an equipotential material under static conditions, or the potential is everywhere constant on the surface of a perfect conductor. This is easy to prove since if $E \equiv 0$ in the material, then $\int \mathbf{E} \cdot \mathbf{dl}$ is likewise zero between any two points in the material body.

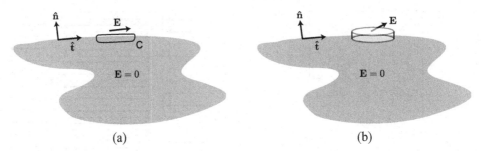

Figure 2.5 (a) The tangential boundary condition can be found by calculating the line integral of the field over a small path partially penetrating the conductor. (b) The normal boundary condition is discovered by calculating the flux of the field through a small cylinder ("pill box") partially embedded in the conductor.

Perfect conductor boundary conditions

Let us begin with *tangential* boundary conditions, shown in Fig. 2.5a. It is easy to show that the electric field must cross the surface of a perfect conductor at a normal angle. Since we have

$$\int_C \mathbf{E} \cdot \mathbf{dl} \equiv 0 \tag{2.16}$$

for any path C, choose a path that partially crosses into the conductor. Since $E = 0$ inside the conductor, the only contributions to the integral are the side walls and E_t, or the tangential component along the path.

In the limit, we can make the path smaller and smaller until it is tangent to the surface and imperceptibly penetrates the material so the side wall contributions vanish.

$$\int_C \mathbf{E} \cdot \mathbf{dl} = E_t \Delta \ell \equiv 0 \tag{2.17}$$

We are thus led to conclude that $E_t \equiv 0$ at the surface. Another argument is that since $\mathbf{E} = -\nabla \phi$ and furthermore since the surface is an equipotential, then clearly E_t is zero since there can be no change in potential along the surface.

Next consider the *normal* boundary conditions. Consider a "pill-box" hugging the surface of a perfect conductor as shown in Fig. 2.5b. The electric flux density is computed for this surface. As before, we will make the volume smaller and smaller until the sidewall contribution goes to zero. Since the bottom is still in the conductor and the field is zero, this term will not contribute to the integral either. The only remaining contribution is the normal component of the top surface

$$\oint \mathbf{D} \cdot \mathbf{dS} = D_n dS = Q_{\text{inside}} = dS\rho_s \tag{2.18}$$

We have thus shown that $D_n = \rho_s$

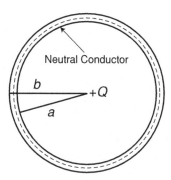

Figure 2.6 Charge placed inside a neutral conducting spherical shell.

Example 1

Consider a charge placed at the center of a spherical shell made of perfect conducting materials. Let us find the surface charge density on the inner and outer surface of the conductor.

Let us apply Gauss' Law for a sphere lying inside the conductor (shown with the dashed line in Fig. 2.6). Since $\mathbf{E} = 0$ on this surface, the charge inside must likewise be zero $Q_{\text{inside}} = 0$, which implies that there exists a uniform charge density (by symmetry) of $\rho_{\text{inner}} = -Q/S_i$ where $S_i = 4\pi a^2$.

If we now consider a larger sphere of radius $r > b$, since the sphere is neutral, the net charge is just the isolated charge Q. Thus

$$D_r 4\pi r^2 = Q + \underbrace{\rho_{\text{inside}} S_1}_{-Q} + \rho_{\text{outside}} S_2 = Q \tag{2.19}$$

Thus $\rho_{\text{outside}} = Q/S_2 = \frac{Q}{4\pi b^2}$

The fields have been sketched in Fig. 2.7. Radial symmetry is preserved and an induced negative and positive charge density appear at the shell surfaces.

In summary, the radial electric fields are given by (and plotted in Fig. 2.8)

$$\begin{aligned} D_r &= \frac{Q}{4\pi r^2} & r < a \\ D_r &= 0 & a \leq r \leq b \\ D_r &= \frac{Q}{4\pi r^2} & r > b \end{aligned} \tag{2.20}$$

Let us compute the potential from the fields. Take the reference point at ∞ to be zero and integrate along a radial path. For $r > b$

$$\phi = -\int_{\infty}^{r} E_r dr = \frac{Q}{4\pi \epsilon r} \tag{2.21}$$

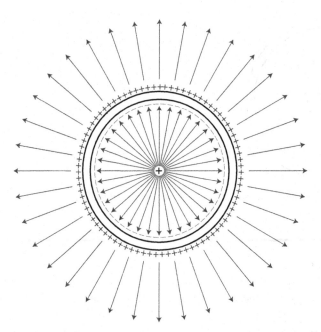

Figure 2.7 Induced charge and electric fields on a neutral spherical shell.

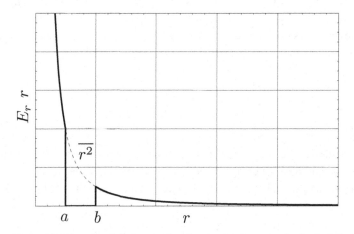

Figure 2.8 Plot of radial electric flux density for the neutral spherical shell.

Likewise for $a \leq r \leq b$, since E_r is zero for the path inside the conducting sphere, the potential remains constant at $\frac{Q}{4\pi\epsilon_0 b}$ until we exit the conductor. Finally, once outside the conductor for $r < a$ we continue integrating

$$\phi = \frac{Q}{4\pi\epsilon b} - \int_a^r \frac{Q}{4\pi\epsilon_0 r^2} dr \tag{2.22}$$

ϕ_r r

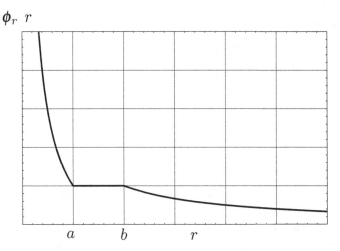

Figure 2.9 The electric potential for a charge placed at the center of a spherical shell.

Thus we have

$$\phi = \frac{Q}{4\pi \epsilon_0}\left(\frac{1}{b} + \frac{1}{r} - \frac{1}{a}\right)$$ (2.23)

A plot of the potential is shown in Fig. 2.9. Note that potential on the conductor is non-zero and its value depends on the amount of charge at the center.

Example 2

What if we now ground the shell from the previous example? That means that $\phi = 0$ on the surface of the conductor. The work done in moving a point charge from infinity to the surface of the shell is thus zero. Choose a radial path

$$\phi = -\int_{\infty}^{r} E_r dr = 0$$ (2.24)

Since the function E_r is monotonic, $E_r = 0$ everywhere outside the sphere! A grounded spherical shell acts like a good shield. If $E_r(b) = 0$, then $\rho_s = 0$ at the outer surface as well.

To find the charge on the inner sphere, we can use the same argument as before unchanged so $\rho_{inside} = -Q/S_1$. But now the material is not neutral! Where did the charge go? Imagine starting with the ungrounded case where the positive charge is induced on the outer surface. Then ground the sphere and notice that charge flows out of the sphere into ground thus "charging" the material negatively.

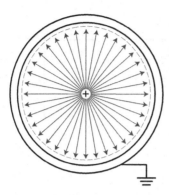

Figure 2.10 Induced charge and electric fields on a grounded spherical shell.

The fields have been sketched in Fig. 2.10. Radial symmetry is preserved and an induced negative charge density appears at inner shell surfaces. The field outside the shell is zero.

Poisson's equation (aka Fish equation)

We already have an equation for calculating $\mathbf{E}$ and ϕ directly from a charge distribution and vice versa. If we know $\mathbf{E}$, then we can find ρ by simply taking the divergence, $\rho = \epsilon \nabla \cdot \mathbf{E}$. Alternatively if we know ϕ

$$\rho = \epsilon \nabla \cdot \mathbf{E} = -\epsilon \nabla \cdot \nabla \phi \tag{2.25}$$

The operator $\nabla \cdot \nabla$ is a new operator and we call it the Laplacian ∇^2. Thus we have Poisson's equation $\nabla^2 \phi = -\frac{\rho}{\epsilon}$. In a charge-free region we have $\nabla^2 \phi = 0$. This is known as Laplace's equation. Let us untangle this beast of an operator

$$\nabla \cdot \nabla \phi = \left(\frac{\partial}{\partial x}\hat{\mathbf{x}} + \frac{\partial}{\partial y}\hat{\mathbf{y}} + \frac{\partial}{\partial z}\hat{\mathbf{z}} \right) \cdot \left(\frac{\partial \phi}{\partial x}\hat{\mathbf{x}} + \frac{\partial \phi}{\partial y}\hat{\mathbf{y}} + \frac{\partial \phi}{\partial z}\hat{\mathbf{z}} \right) \tag{2.26}$$

since all cross products involving mutual vectors die (e.g. $\hat{\mathbf{x}} \cdot \hat{\mathbf{y}} \equiv 0$), we have

$$\nabla^2 \phi = \frac{\partial^2 \phi}{\partial x^2} + \frac{\partial^2 \phi}{\partial y^2} + \frac{\partial^2 \phi}{\partial z^2} \tag{2.27}$$

Other coordinate systems are not so nice!

Dielectrics

So far we have only considered electrostatics in a vacuum and with ideal conductors. We now expand our discussion to include dielectric materials. Dielectric materials are often good insulators. A perfect insulator cannot conduct DC current because there are no mobile charges as all electrons (free carriers) in the material are bound to the nuclei. But there is a response to an external field due to the dielectric properties of the material.

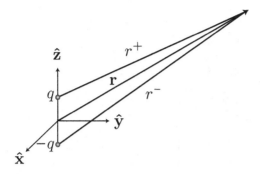

Figure 2.11 Calculation of the electric potential at a point **r** distant from the dipole.

Review of dipoles

An electric dipole consists of two equal and opposite charges, q, separated by a distance, d. With reference to Fig. 2.11, consider the potential due to an electric dipole at points far removed from the dipole $r \gg d$

$$\phi(\mathbf{r}) = \frac{q}{4\pi\epsilon r^+} + \frac{-q}{4\pi\epsilon r^-} = \frac{q}{4\pi\epsilon}\left(\frac{1}{r^+} - \frac{1}{r^-}\right) \tag{2.28}$$

We can simplify the above expression by the application of the Law of Cosines, since vector summation is fundamentally related to a triangle. Note we can think of adding two vectors $\mathbf{r}$ and $\hat{\mathbf{z}}d/2$ as forming $\mathbf{r}^+ = \mathbf{r} + \hat{\mathbf{z}}d/2$. The length of a vector is given by $|\mathbf{a}|^2 = \mathbf{a}\cdot\mathbf{a}$. We can therefore express r^+ in terms of r and d

$$r^+ = \sqrt{\mathbf{r}^+ \cdot \mathbf{r}^+} = \sqrt{\left(\mathbf{r} + \frac{d}{2}\hat{\mathbf{z}}\right)\cdot\left(\mathbf{r} + \frac{d}{2}\hat{\mathbf{z}}\right)} \tag{2.29}$$

or simplifying a bit and recalling that $r \gg d$

$$r^+ = r\sqrt{1 + \left(\frac{d}{2r}\right)^2 + \frac{d}{r}\cos\theta} \approx r\sqrt{1 + \frac{d}{r}\cos\theta} \tag{2.30}$$

So for the potential we have

$$\frac{1}{r^+} = \frac{1}{r}\frac{1}{\sqrt{1 + \frac{d}{r}\cos\theta}} \approx \frac{1}{r}\left(1 - \frac{d}{2r}\cos\theta\right) \tag{2.31}$$

and

$$\frac{1}{r^-} \approx \frac{1}{r}\left(1 + \frac{d}{2r}\cos\theta\right) \tag{2.32}$$

So that the potential is finally given

$$\phi(\mathbf{r}) = \frac{q}{4\pi\epsilon}\left(\frac{1}{r^+} - \frac{1}{r^-}\right) = \frac{-qd\cos\theta}{4\pi\epsilon r^2} \tag{2.33}$$

Unlike an isolated point charge, the potential drops like $1/r^2$ rather than $1/r$.

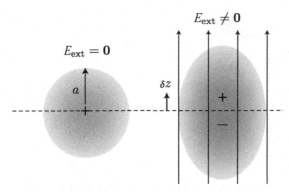

Figure 2.12 A spherically symmetric electron cloud has zero average dipole moment. But under the application of an external field, E_{ext}, the electron cloud is distorted (exaggerated for clarity) to produce a dipole moment.

The potential calculation was easy since we did not have to deal with vectors. The electric field is simply $\mathbf{E} = -\nabla \phi$. Since our answer is in spherical coordinates, and there is no φ variation due to symmetry, we have

$$\nabla \phi = \hat{\mathbf{r}} \frac{\partial \phi}{\partial r} + \hat{\boldsymbol{\theta}} \frac{1}{r} \frac{\partial \phi}{\partial \theta} = \hat{\mathbf{r}} \frac{2qd \cos \theta}{4\pi \epsilon r^3} + \hat{\boldsymbol{\theta}} \frac{1}{r} \frac{qd}{4\pi \epsilon r^2} \sin \theta \tag{2.34}$$

Thus the electric field is given by

$$\mathbf{E} = -\frac{qd}{4\pi \epsilon r^3} (\hat{\mathbf{r}} 2 \cos \theta + \hat{\boldsymbol{\theta}} \sin \theta) \tag{2.35}$$

We can define a vector dipole moment $\mathbf{p} = q\mathbf{d}$, where $\mathbf{d}$ is a vector connecting the positive and negative charge. In terms of $\mathbf{p}$, we can rewrite the potential in a coordinate independent manner

$$\phi(\mathbf{r}) = \frac{\hat{\mathbf{r}} \cdot \mathbf{p}}{4\pi \epsilon r^2} \tag{2.36}$$

For a spherically symmetric atom and electron cloud distribution, shown in Fig. 2.12, the time average dipole moment is zero. But the application of an external field can distort the electronic charge distribution to produce a net dipole.

It is easy to see why this happens. The electrons are bound to the nucleus through the internal force of the atom. This force is much stronger than our applied field and so the probability that an electron will obtain sufficient energy from the field to leave the atom is negligible. But the average electronic distribution is the minimum energy configuration. If the electrons displace slightly from their equilibrium positions, then the new configuration yields lower overall energy in the presence of a field.

It is important to note the displacement $\Delta z \ll a$. This is because the *internal* field of an atom is already very strong

$$E_{int} \sim \frac{e}{4\pi \epsilon a^2} \sim 10^{11} \text{V/m} \tag{2.37}$$

This is several orders of magnitude larger than any field we typically impart on to the atom externally. To first order, then, the response is linear with the field. This is because for a

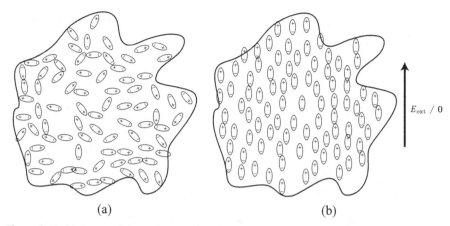

(a)   (b)

Figure 2.13 (a) A material consisting of molecules with net dipole moments has a net zero dipole moment due to the random orientation of the molecules. (b) The dipoles tend to align under the application of a sufficiently strong electric field.

small displacement, any function looks linear

$$f(x + \Delta x) \approx f(x) + \Delta x f'(x) \tag{2.38}$$

Atomic dipole moment

Based on the fact that $E_{ext} \ll E_{int}$, we expect that the induced dipole moment in the atom to be small

$$p = e \Delta z \tag{2.39}$$

But since there are a lot of atoms, the net effect can be large. The constant of proportionality is known as the electronic polarizability coefficient α_e

$$\mathbf{p} = \alpha_e \mathbf{E} \tag{2.40}$$

Typically $\alpha_e \sim 10^{-40} \mathrm{F} \cdot \mathrm{m}^2$.

Ionic polarizability

For more complex molecules, like CH_4, there can be larger-scale polarization. Instead of each individual atom getting polarized, the shared covalent bond electrons can rearrange and produce a larger polarization. In other cases, the molecules may have built-in polarization due to ionic or covalent bonds. Water is a good example of this. As shown in Fig. 2.13a, normally the dipole moments are pointing in random directions so the net polarization for $E_{ext} = 0$. But the application of an external field can align the dipoles, shown in Fig. 2.13b. The response can be huge due to the large number of molecules involved.

Unfortunately, the effect of thermal energy tends to re-randomize the orientation so we expect that on average only a small fraction of the dipoles are aligned. In fact, we can show that the probability that a molecule be aligned is related to the energy of a dipole in an external field. The energy is proportional to p^2 and so the fraction of polarized molecules depends on the ratio of p^2 and $3k_B T$

$$\alpha_0 = \frac{p^2}{3k_B T} \tag{2.41}$$

Typically $\alpha_0 \sim 10^{-38} Fm^2$

Total polarization

In general, the total polarization arises from different sources (induced and permanent). If the density of polarization is N per volume, the net polarization is

$$\mathbf{P} = N\alpha\mathbf{E} = \epsilon_0 \chi_e \mathbf{E} \tag{2.42}$$

χ_e is known as the electric susceptibility. We are now in a position to derive the potential due polarized matter. Let us rewrite the potential due to a single dipole

$$\phi(r) = \frac{1}{4\pi\epsilon_0} \frac{\mathbf{p} \cdot \hat{\mathbf{r}}}{r^2} = -\frac{1}{4\pi\epsilon_0} \mathbf{p} \cdot \nabla \frac{1}{r} \tag{2.43}$$

Notice that $\nabla \frac{1}{r} = \nabla r^{-1} = \hat{\mathbf{r}} \frac{\partial r^{-1}}{\partial r} = -\frac{\hat{\mathbf{r}}}{r^2}$. Integrating over all dipoles we have

$$\phi(\mathbf{r}) = \frac{1}{4\pi\epsilon_0} \int_V -\mathbf{P} \cdot \nabla \frac{1}{|\mathbf{r} - \mathbf{r}'|} dV \tag{2.44}$$

Note that $\nabla \mathbf{R}^{-1} = -\nabla' R^{-1}$, where $R = |r - r'|$. Using the identity $\nabla \cdot (a\mathbf{A}) = a\nabla \cdot \mathbf{A} + \mathbf{A} \cdot \nabla a$

$$\phi(\mathbf{r}) = \frac{1}{4\pi\epsilon_0} \left(\int_V \nabla' \cdot \frac{\mathbf{P}}{|\mathbf{r} - \mathbf{r}'|} dV - \int_V \frac{\nabla' \cdot \mathbf{P}}{|\mathbf{r} - \mathbf{r}'|} dV \right) \tag{2.45}$$

By Divergence Theorem we can recast this into

$$\phi(\mathbf{r}) = \frac{1}{4\pi\epsilon_0} \left(\oint_S \frac{\mathbf{P} \cdot \hat{\mathbf{n}}}{|\mathbf{r} - \mathbf{r}'|} dS - \int_V \frac{\nabla' \cdot \mathbf{P}}{|\mathbf{r} - \mathbf{r}'|} dV \right) \tag{2.46}$$

Effective volume and surface charge

Notice that these integrals look like an ordinary potential calculation for surface charge and volume charge. Define a surface charge density $\rho_s = \mathbf{P} \cdot \hat{\mathbf{n}}$ and a volume charge density $\rho_v = -\nabla' \cdot \mathbf{P}$. Then we have

$$\phi(\mathbf{r}) = \frac{1}{4\pi\epsilon_0} \left(\oint_S \frac{\rho_s}{|\mathbf{r} - \mathbf{r}'|} dS + \int_V \frac{\rho_v}{|\mathbf{r} - \mathbf{r}'|} dV \right) \tag{2.47}$$

There must be a good explanation for this coincidence! Here is the intuitive picture. If the material is uniform and the external field is uniform, then the polarization is also uniform and $\nabla' \cdot \mathbf{P} = 0$. The interpretation is that all the internally induced polarizations cancel out.

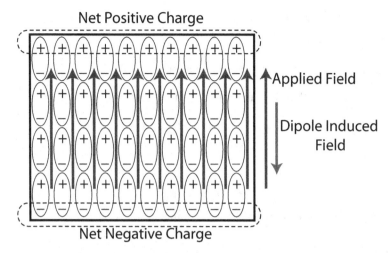

Figure 2.14 For a uniform material polarized by an external field, the internal dipole moments cancel. Only the dipoles on the boundary produce net charge.

As shown in Fig. 2.14, the dipoles on the boundary, though, remain unbalanced and thus produce a net charge. If the polarization is non-uniform, though, then $\nabla' \cdot \mathbf{P} \neq 0$ and a volume polarization arises.

Now back to Gauss' Law. Since the bound charges can create net charge, we must include their contribution in Gauss' Law

$$\nabla \cdot \epsilon_0 \mathbf{E} = \rho_{\text{free}} + \rho_{\text{bound}} \qquad (2.48)$$

where $\rho_{\text{bound}} = -\nabla \cdot \mathbf{P}$. Substituting in the above equation

$$\nabla \cdot \epsilon_0 \mathbf{E} + \nabla \cdot \mathbf{P} = \rho_{\text{free}} \qquad (2.49)$$

Grouping terms we have (where ρ is now understood to be free of charge)

$$\nabla \cdot (\epsilon_0 \mathbf{E} + \mathbf{P}) = \rho_{\text{free}} = \rho \qquad (2.50)$$

Electric flux density vector

We can re-define the electric flux density

$$\mathbf{D} = \epsilon_0 \mathbf{E} + \mathbf{P} \qquad (2.51)$$

so that it depends only on free charge

$$\nabla \cdot \mathbf{D} = \rho \qquad (2.52)$$

Since the polarization is proportional to $\mathbf{E}$

$$\mathbf{D} = \epsilon_0 \mathbf{E} + \epsilon_0 \chi_e \mathbf{E} = \epsilon_0 (1 + \chi_e) \mathbf{E} \qquad (2.53)$$

The dielectric constant captures the polarization effects in a macroscopic coefficient $\epsilon = \epsilon_r \epsilon_0 = \epsilon_0 (1 + \chi_e)$

Dielectric boundary conditions

Normal D

Applying Gauss' Law to the interface of two materials

$$\oint_S \mathbf{D} \cdot d\mathbf{S} = (\hat{\mathbf{n}} \cdot \mathbf{D}_1)\Delta S - (\hat{\mathbf{n}} \cdot \mathbf{D}_2)\Delta S = \rho_s \Delta S \qquad (2.54)$$

Thus the discontinuity in the normal field must be due to surface charge

$$\mathbf{D}_{1n} - \mathbf{D}_{2n} = \rho_s \qquad (2.55)$$

For dielectrics, there is no "free" surface charge and so $\rho_s = 0$

$$\hat{\mathbf{n}} \cdot \mathbf{D}_1 = \hat{\mathbf{n}} \cdot \mathbf{D}_2 \qquad (2.56)$$

or $D_{1n} = D_{2n}$. The normal electric field is discontinuous $\epsilon_1 E_{1n} = \epsilon_2 E_{2n}$.

Tangential E

Take the line integral around the interface. Since the field is static, the line integral along any closed path is identically zero

$$\int_C \mathbf{E} \cdot d\mathbf{l} = (E_{1t} - E_{2t})\Delta w = 0 \qquad (2.57)$$

This means that the tangential field is always continuous, $E_{1t} = E_{2t}$. Then the tangential component of D is *not* continuous

$$\frac{D_{1t}}{\epsilon_1} = \frac{D_{2t}}{\epsilon_2} \qquad (2.58)$$

2.2 Capacitance

Capacitance of a conductor is defined as

$$C \triangleq \frac{q}{\phi} \qquad (2.59)$$

where a larger C means that an object can store more charge at a fixed potential. It is not so different than the volume of a tank of water. The bigger the tank, the more water it can store. Hence it is a measure of capacity to store charge. As an example, consider an isolated spherical conductor of radius a. By Gauss' Law we know the radial field is given by ($r \geq a$)

$$D_r = \frac{q}{4\pi r^2} \qquad (2.60)$$

If we integrate this field to arrive at the potential, we have

$$\phi = -\int_\infty^a E_r dr = -\frac{q}{4\pi\epsilon} \int_\infty^a \frac{dr}{r^2} = \frac{q}{4\pi\epsilon} \left(\frac{1}{r}\right)\Big|_\infty^a = \frac{q}{4\pi\epsilon a} \qquad (2.61)$$

The capacitance is therefore

$$C = 4\pi\epsilon a \propto a \qquad (2.62)$$

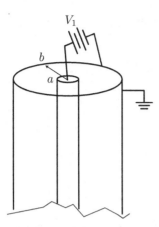

Figure 2.15 Setup for the calculation of the capacitance per unit length of a coaxial structure.

The larger sphere can hold more charge at a fixed potential. Equivalently, dumping a given charge on to a larger sphere will increase its potential less than a smaller sphere.

We can also define the capacitance between two objects as

$$C_{12} = \frac{q_1}{\phi_{12}} \tag{2.63}$$

where we fix the potential between the two objects at ϕ_{12} and measure the amount of charge transfer between the objects. In the above equation a charge of q_1 has been transfered from object 2 to 1 and therefore a charge of $q_2 = -q_1$ will reside on object 2. We can re-interpret the capacitance of a single object as the capacitance relative to a reference at infinity.

Example 3

Let us practice and calculate the capacitance per unit length for a coaxial structure shown in Fig. 2.15. We can solve Poisson's equation to find the potential in the region between the conductors[1]

$$\nabla^2 \phi = 0 \tag{2.64}$$

By symmetry ϕ is only a function of r and not a function of θ or z

$$\nabla \cdot \nabla \phi = \frac{1}{r} \frac{\partial}{\partial r} \left(r \frac{\partial \phi}{\partial r} \right) = 0 \tag{2.65}$$

$$r \frac{\partial \phi}{\partial r} = K_1 \tag{2.66}$$

$$\frac{\partial \phi}{\partial r} = \frac{K_1}{r} \tag{2.67}$$

The general solution is therefore $\phi(r) = K_1 \ln r + K_2$.

---

[1] Note the form of the Laplacian in cylindrical coordinates differs from rectangular coordinates.

This solution must satisfy the boundary conditions that $\phi(a) = V_1$ and $\phi(b) = 0$

$$\phi(r = a) = V_1 = K_1 \ln a + K_2 \tag{2.68}$$
$$\phi(r = b) = 0 = K_1 \ln b + K_2 \tag{2.69}$$

Solving these equations we have

$$\phi(r) = \frac{V_1 \ln r/b}{\ln a/b} \tag{2.70}$$

The field is given by $\mathbf{E} = \hat{\mathbf{r}} E_r = -\hat{\mathbf{r}} \frac{\partial \phi}{\partial r}$

$$\mathbf{E} = \frac{\hat{\mathbf{r}} V_1}{r \ln b/a} \tag{2.71}$$

Recall that $\mathbf{D} \cdot \hat{\mathbf{n}} = \rho_s$ on the surface of conductors. Therefore for a length of ℓ of the coaxial conductors

$$\frac{\epsilon V_1}{a \ln(b/a)} = \frac{q}{2\pi a \ell} \tag{2.72}$$

Solve for the charge to find the capacitance

$$q = \frac{2\pi\epsilon}{\ln \frac{b}{a}} \ell V_1 \tag{2.73}$$

The capacitance per unit length is therefore

$$C' = \frac{C}{\ell} = \frac{2\pi\epsilon}{\ln \frac{b}{a}} \tag{2.74}$$

Parallel plate capacitor

A very convenient idealization is a parallel plate capacitor. Here we take two conducting plates of area A and separate them by a distance t. If $t \ll \min(dx, dy)$, then the field is largely independent of x and y and only a function of z. This is true especially in regions inside the plates far from the edges of the capacitor. Now charge the upper plate to a voltage of V_0 relative to the bottom plate. For convenience take the reference of zero potential at the bottom plate.

If we neglect the edge effects, then the solution of the potential in the charge-free region between the plates is governed by a one-dimensional differential equation.

$$\nabla^2 \phi = \frac{\partial^2 \phi}{\partial z^2} = 0 \tag{2.75}$$

subject to the boundary conditions $\phi(0) = 0$ and $\phi(t) = V_0$. The solution is clearly a linear function (integrate the above equation twice) $\phi(z) = C_1 z + C_2$ and applying the boundary conditions to solve for C_1 and C_2 we arrive at

$$\phi(z) = \frac{V_0}{t} z \tag{2.76}$$

The electric field $E = -\nabla\phi$ can thus be computed by a simple derivative and is a constant within the capacitor

$$\mathbf{E} = -\hat{\mathbf{z}}\frac{V_0}{t} \tag{2.77}$$

By Gauss' Law, the normal component of the electric flux density is equal to the charge density on the plates

$$\mathbf{D} \cdot \hat{\mathbf{n}} = \rho \tag{2.78}$$

Since the field is in fact everywhere perpendicular to the conductor, we simply have

$$\rho_{\text{top}} = -\hat{\mathbf{z}} \cdot \mathbf{D} = \epsilon\frac{V_0}{t} \tag{2.79}$$

and

$$\rho_{\text{bot}} = \hat{\mathbf{z}} \cdot \mathbf{D} = -\epsilon\frac{V_0}{t} \tag{2.80}$$

To find the capacitance we simply note that we are interested in the coefficient $q = CV_0$. To get the total charge, we multiply the constant charge density by the area of the plate $q = \rho_{\text{top}}A = \frac{\epsilon A}{t}V_0 = CV_0$, so the capacitance is given by $C = \frac{\epsilon A}{t}$. This equation is intuitively satisfying. The capacitance goes up with A, since, for a fixed charge on the plates, the charge density drops and so does the potential giving a larger capacitance. Likewise, if we increase t, the capacitance drops since now there is less motivation for positive (negative) charge to flow on to the top (bottom) plate! The charges are more distant from their beloved negative (positive) charges.

For two conductors of any shape, the capacitance is defined as

$$C \triangleq \frac{Q}{\phi_{12}} \tag{2.81}$$

The potential difference ϕ_{12} is the line integral of $\mathbf{E}$ over *any* path from conductor 1 to conductor 2. By Gauss' Law, the positive charge on conductor 1 is equal to the electric flux crossing *any* surface enclosing the conductor. The capacitance is therefore written as

$$C = \frac{\oint_S \mathbf{D} \cdot d\mathbf{S}}{-\int_\Gamma \mathbf{E} \cdot d\mathbf{l}} \tag{2.82}$$

where S surrounds the top plate and Γ is a path from one plate to the other.

From circuit theory we know how to add capacitors in series or in parallel. Parallel caps are easier. If we connect two conductors and connect them to a potential V_0, the charge is simply the total charge

$$Q = Q_1 + Q_2 = C_1V_0 + C_2V_0 = (C_1 + C_2)V_0 \tag{2.83}$$

In general we have

$$C_{||} = C_1 + C_2 + C_3 + \dots \tag{2.84}$$

For series capacitors, note that the applied voltage is divided between the capacitors

$$V_0 = V_1 + V_2 = \frac{Q_1}{C_1} + \frac{Q_2}{C_2} \tag{2.85}$$

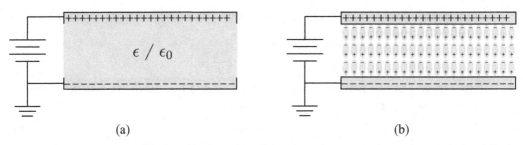

(a)   (b)

Figure 2.16 A parallel plate capacitor with a dielectric has larger capacitance since the induced dipole moments create an internal electric field which opposes the applied field.

The central observation is that the charge on each capacitor is the same, $|Q_1| = |Q_2|$. This is because the floating node must have zero net charge and thus $-Q_1 = Q_2$. The result can be easily generalized

$$\frac{1}{C_{\text{series}}} = \frac{1}{C_1} + \frac{1}{C_2} + \frac{1}{C_3} + \ldots \tag{2.86}$$

Parallel plate capacitor with dielectric

One important application for a dielectric is to insulate conductors from one another (coaxial transmission line) while providing mechanical support. In a capacitor there is the additional benefit that the capacitance increases. As shown in Fig. 2.16, the capacitance increases because the effect of the polarization is to produce a net charge density at the top and bottom of the plates that subtracts from the net charge on the conductors.

So for a fixed applied field (voltage), the parallel plate capacitor can support more charge with a dielectric. Another perspective is if we fix the amount of charge on the plates, then the field in the dielectric is reduced and hence the applied electric field is smaller.

Capacitance matrix

The concept of capacitance can be generalized to multiple conductors

$$q_1 = C_{11}V_1 + C_{12}V_2 + C_{13}V_3 + \ldots$$
$$q_2 = C_{21}V_1 + C_{22}V_2 + C_{23}V_3 + \ldots \tag{2.87}$$
$$\vdots$$

Each coefficient C_{ii} represents the self capacitance. It can be computed by applying $V_i = 1\text{V}$ to conductor i while grounding all other conductors. Then C_{ii} is simply the total charge on the conductor. Likewise, to find C_{ij}, we apply a voltage of $V_j = 1\text{V}$ to conductor j while grounding all other conductors. Then C_{ij} is again simply the total charge on conductor i.

When we say we "grounding" all other conductors, we mean that we connect them to a voltage source of zero volts. Recall that voltage is always defined relative to a reference. For instance, we may take one of the conductors as the "ground" reference and then measure all absolute voltages relative to this conductor. Otherwise we may also connect the conductors to a much larger body, one with infinite capacitance. Then charge can be freely removed or added to the "ground" without changing its potential. Think of using the ocean as the reference "ground." Then removing or adding several buckets of water will not alter the "level" (voltage) of the ocean.

We may also express the voltage on each conductor in terms of the total charge on each conductor in the system in the following manner

$$v_1 = P_{11}q_1 + P_{12}q_2 + P_{13}q_3 + \cdots$$
$$v_2 = P_{21}q_1 + P_{22}q_2 + P_{23}q_3 + \cdots \qquad (2.88)$$
$$\vdots$$

To find P_{ij}, we add a charge of 1C to conductor j and leave all other conductors neutral while observing the voltage at conductor i.

The definition of capacitance follows directly from the laws of electrostatics. Let us place N conductors in vacuum. Then the electric potential at any point in space is given by

$$\phi(r) = \int_V \frac{\rho(r')}{4\pi\epsilon|r-r'|} dV' = \sum_i \int_{V_i} \frac{\rho(r')}{4\pi\epsilon|r-r'|} dV' \qquad (2.89)$$

Notice that this equation scales linearly with the absolute amount of charge on conductor i. So we may perform the integral based solely on geometric calculations to obtain coefficient P_{ij}

$$\phi(r) = \sum_i q_i \int_{V_i} \frac{\rho(r')/q_i}{4\pi\epsilon|r-r'|} dV' \qquad (2.90)$$

or in matrix form, we may write $v = Pq$. If the matrix P is not singular, we may invert this equation to obtain $q = P^{-1}v$. We may be temped to call P_{ij}^{-1} a capacitance, but notice that these coefficients are in terms of the potential V_i relative to a common reference

$$q_1 = c_{11}V_1 + c_{12}V_2 + c_{13}V_3 + \cdots$$
$$q_2 = c_{21}V_1 + c_{22}V_2 + c_{23}V_3 + \cdots \qquad (2.91)$$
$$\vdots$$

To relate c_{ij} to C_{ij}, simply equate the total charges

$$q_1 = C_{11}V_1 + C_{12}V_{12} + C_{13}V_{13} = c_{11}V_1 + c_{12}V_2 + c_{13}V_3 + \cdots \qquad (2.92)$$

Since $V_{1i} = V_1 - V_j$, we have

$$C_{11} = c_{11} + c_{12} + c_{13} + \cdots \qquad (2.93)$$

Also $C_{ij} = -c_{ij}$, and so these matrices are related but not the same. Notice that $c_{ij} < 0$ is logical, since if we put a positive voltage on node j and observe the induced charge on node i, it should be negative. On the other hand $C_{ij} > 0$, since if we connect the positive terminal of a battery to node i and the negative node of the battery to node j, then the charge on node i should be positive. Capacitors in SPICE are always of the C_{ij} form, and hence positive.

Electrostatic energy of a capacitor

Consider the energy required to charge a capacitor. The amount of work released to move a charge dQ from the positive terminal to the negative terminal is $dU = V dQ$. This work must be stored in the capacitor

$$dU = V dQ = CV dV \tag{2.94}$$

Integrating over the voltage on the capacitor we have

$$U = \frac{1}{2}CV^2 \tag{2.95}$$

We say that this energy is stored in the field of the capacitor. This "field" business will become clear as we develop these ideas further. For a parallel plate capacitor, the field is constant and equal to $E = V/d$. If we substitute in the energy equation, we have

$$U = \frac{1}{2}CE^2d^2 \tag{2.96}$$

Substituting further with $C = \epsilon A/d$

$$U = \frac{1}{2}\frac{\epsilon A}{d}E^2d^2 = \frac{1}{2}\epsilon V E^2 \tag{2.97}$$

where $V = d \times A$ is the volume of the region in between the plates. Since the fields are confined to this volume, we may speculate that the energy density is also so confined.

In this particular case, we have

$$u = \frac{U}{V} = \frac{1}{2}\epsilon E^2 = \frac{1}{2}\mathbf{D} \cdot \mathbf{E} \tag{2.98}$$

We will show later that this is true in general for any electrostatic field.

Now consider the energy required to charge two conductors to voltages ϕ_1 and ϕ_2. Let us do the calculation in phases. First apply voltage ϕ_1 to conductor 1 but keep conductor 2 grounded. A charge $q_1 = c_{11}\phi_1$ flows on to conductor 1, whereas a charge $q_2 = c_{21}\phi_1$ flows on to conductor 2. Since conductor 2 is grounded, there is no energy required to add or remove charge from it. For conductor 1, though, the energy required is $\frac{1}{2}c_{11}\phi_1^2$. Now raise the voltage on conductor 2 from zero to ϕ_2. Additional work of $\frac{1}{2}c_{22}\phi_2^2$ is required. But an additional charge of $q_1 = c_{12}\phi_2$ also flows on to conductor 1. The work required to do this is $c_{12}\phi_1\phi_2$ (no integration is required since the potential is fixed at ϕ_1).

The total work is therefore the sum of the various terms

$$W = \frac{1}{2}c_{11}\phi_1^2 + \frac{1}{2}c_{22}\phi_2^2 + c_{12}\phi_1\phi_2 \tag{2.99}$$

But if we had reversed the order of charging the conductors, we would have arrived at the following result

$$W' = \frac{1}{2}c_{11}\phi_1^2 + \frac{1}{2}c_{22}\phi_2^2 + c_{21}\phi_2\phi_1 \tag{2.100}$$

But the energy of the system would surely be the same, or $W = W'$, which implies that $c_{12} = c_{21}$. Thus the capacitance matrix is symmetric. A symmetric matrix with non-zero diagonal is invertible, thus justifying why we could go freely from the P matrix to its inverse C matrix.

Let us derive this result more directly as follows. The energy increment is given by $dW = dQv$, where we add charge dQ at constant voltage v. On conductor 1 we have

$$q_1 = c_{11}V_1 + C_{12}V_2 + \ldots \tag{2.101}$$
$$dq_1 = c_{11}dV_1 + C_{12}dV_2 + \ldots \tag{2.102}$$

So the energy increment is given by

$$dW = dq_1 V_1 = c_{11}V_1 dV_1 + C_{12}V_1 dV_2 + \ldots \tag{2.103}$$

Integration yields

$$W = \frac{1}{2}C_{11}V_1^2 + C_{12}V_1 V_2 + \ldots \tag{2.104}$$

Energy of the electrostatic field

Let us begin by finding the total energy for a distribution of point charges. We can imagine building up the distribution one charge at a time. The energy to bring in the first charge is naturally zero since the field is zero. The second charge, though, is repelled (or attracted) by the first charge so it requires more (less) energy to bring it in. In general the work required is given by

$$W_2 = Q_2\phi_{12} \tag{2.105}$$

where ϕ_{12} is the potential due to charge 1.

Since the electrostatic field is *conservative*, it does not matter how we bring in the second charge. Only its final position relative to the first charge is important. In terms of

$$W_2 = Q_2\phi_{12} = \frac{Q_2 Q_1}{4\pi \epsilon R_{12}} \tag{2.106}$$

where R_{12} is the final distance between the point charges. Likewise, when we bring in the third charge, the extra work required is

$$W_3 = Q_3\phi_{13} + Q_3\phi_{23} = \frac{Q_3 Q_1}{4\pi \epsilon R_{13}} + \frac{Q_3 Q_2}{4\pi \epsilon R_{23}} \tag{2.107}$$

We can therefore write in general that the electrostatic energy takes on the following form

$$4\pi \epsilon W = \frac{Q_1 Q_2}{R_{12}} + \frac{Q_1 Q_3}{R_{13}} + \frac{Q_1 Q_4}{R_{14}} + \cdots + \frac{Q_2 Q_3}{R_{23}} + \frac{Q_2 Q_4}{R_{24}} + \ldots \tag{2.108}$$

The general term has the form $\frac{Q_i Q_j}{4\pi R_{ij}}$, where i and j sum over all the particles in the system

$$W = \frac{1}{2}\sum\sum_{i \neq j} \frac{Q_i Q_j}{4\pi R_{ij}} \tag{2.109}$$

The factor of $\frac{1}{2}$ takes care of the double counting and enforcing $i \neq j$ ensures that we do not try to include the "self" energy of the particles.

We can rewrite the double sum into a more general form by observing that the inner sum is simply the potential due to all the particles evaluated at position of particle i

$$W = \frac{1}{2} \sum_{i \neq j} Q_i \sum_j \frac{Q_j}{4\pi R_{ij}} = \frac{1}{2} \sum Q_i \phi_i \tag{2.110}$$

If we now consider a charge distribution $\rho(r)$, it is easy to see how the above sum turns into an integral

$$W = \frac{1}{2} \int_V \rho \phi dV \tag{2.111}$$

The derived expression can be modified if we substitute $\rho = \nabla \cdot \mathbf{D}$ for charge

$$W = \frac{1}{2} \int_V \phi \nabla \cdot \mathbf{D} dV \tag{2.112}$$

and employ the chain rule

$$\nabla \cdot \phi \mathbf{D} = \phi \nabla \cdot \mathbf{D} + \mathbf{D} \cdot \nabla \phi \tag{2.113}$$

Since $\nabla \phi = -\mathbf{E}$, we have two volume integrals

$$W = \frac{1}{2} \int_V \nabla \cdot (\phi \mathbf{D}) dV + \frac{1}{2} \int_V \mathbf{D} \cdot \mathbf{E} dV \tag{2.114}$$

We can show that the first integral vanishes as follows. First apply the Divergence Theorem

$$\int_V \nabla \cdot (\phi \mathbf{D}) dV = \oint_S \phi \mathbf{D} \cdot \hat{n} dS \tag{2.115}$$

Now take a surface S that is very large, say a large sphere. If the sphere is very large and the charge distribution is of finite extent, then at some great distance from the source the actual charge distribution is immaterial as only the net charge matters. We know that the radial potential and fields for a charge density take on the following limiting forms $\phi \sim \frac{1}{r}$ and $D \sim \frac{1}{r^2}$. Since the surface area $S \sim r^2$, the integrand vanishes.

Therefore the electrostatic energy is reduced to

$$W = \frac{1}{2} \int_V \mathbf{D} \cdot \mathbf{E} dV \tag{2.116}$$

If we define the energy density w, we have

$$W = \frac{1}{2} \int_V w dV \tag{2.117}$$

$$w = \frac{1}{2} \mathbf{D} \cdot \mathbf{E} \tag{2.118}$$

We found this to be true for an ideal capacitor but now we see this is true in general.

2.3 Non-linear capacitance

When capacitors are built from ideal conductors, the charge is always a linear function of the terminal voltages. But there are many situations when the charge on a body is a non-linear function of the terminal voltage, $Q = q(V)$. In this case we shall find that the derivative of charge with respect to voltage, dq/dV, plays an important role.

Small-signal capacitance

Here we consider a non-linear capacitor of the form $Q = q(v)$. Consider the differential charge dQ flowing onto a non-linear capacitor in response to a small-signal excitation. Let the terminal voltage consist of a DC large signal plus a small-signal AC voltage $V_s + v_s(t)$. A Taylor series expansion yields the incremental charge flow

$$Q(V) = q(V_s + v_s(t)) \approx q(V_s) + \left.\frac{dq(V)}{dV}\right|_{V=V_s} v_s(t) \tag{2.119}$$

The first term $q(V_s)$ is a constant, so if we break the charge into a static and dynamic component, we have

$$Q = Q_0 + \delta q \approx q(V_s) + \left.\frac{dq(V)}{dV}\right|_{V=V_s} v_s(t) \tag{2.120}$$

Since $Q_0 = q(V_s)$, we have

$$\delta q = \left.\frac{dq(V)}{dV}\right|_{V=V_s} v_s(t) = C v_s(t) \tag{2.121}$$

where we have defined the small-signal capacitance by

$$C \triangleq \left.\frac{dq(V)}{dV}\right|_{V=V_s} \tag{2.122}$$

Large-signal capacitance

For an arbitrary signal, we must continue to use the general expression $Q = q(v)$. In particular, to find the current we note that

$$i = \frac{dQ}{dt} = \frac{dq(v)}{dt} = \frac{dq}{dV}\frac{dV}{dt} = C(V)\frac{dV}{dt} \tag{2.123}$$

We see that the important term is again dq/dV. Assuming that C(V) is analytic, we can perform a Taylor series expansion

$$i = (C_0 + C_1 V + C_2 V^2 + C_3 V^3 + \cdots)\frac{dV}{dt} \tag{2.124}$$

Again, for small signals, the C_0 term will dominate, which leads again to the same definition of small-signal capacitance

$$C_0 = C(V_s) = \left.\frac{dq(V)}{dV}\right|_{V=V_s} \tag{2.125}$$

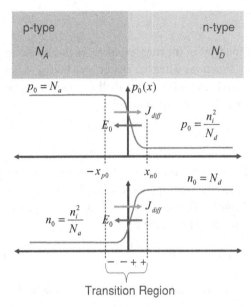

Transition Region

Figure 2.17 A pn-junction diode consisting of a n-type region with doping N_D and a p-type region with doping N_A. The concentration of electrons changes continuously from one majority carriers in the n-region to minority carriers in the p-region.

Junction diode

The capacitance of a pn-junction shown in Fig. 2.17 is a ubiquitous example of a non-linear capacitor. To calculate the capacitance of a reverse-biased junction diode, we need to find the charge density in a semiconducting material. As usual, assume that the p-type material is doped with N_A acceptor atoms and the n-type material is doped with N_D donor atoms. To start assume that no bias is applied across the terminals of the pn-junction.

From device physics, we know that when contact is made between these materials a built-in potential ϕ_{bi} builds up at the junction to prevent the diffusion of carriers across the junction. This potential develops in a small region around the junction, called the transition or depletion region. On the p-side of the depletion region, there are very few holes (hence the term *depletion*) and

$$\rho(x) \approx q(p_o - N_A) < 0 \tag{2.126}$$

Note that the net charge is negative since the acceptors are negatively ionized. Likewise, on the n-side, the opposite is true

$$\rho(x) \approx q(-n_0 + N_d) > 0 \tag{2.127}$$

Since the distribution of carriers is approximately governed by Boltzmann statistics, we have

$$\rho(x) = \begin{cases} q\left(n_i e^{-q\phi_0(x)/kT} - N_A\right) & -x_{p0} < x < 0 \\ q\left(N_D - n_i e^{q\phi_0(x)/kt}\right) & 0 < x < x_{n0} \end{cases} \tag{2.128}$$

Poisson's equation arms us with the tools necessary to solve this problem since

$$\frac{d^2\phi}{dx^2} = -\frac{\rho(x)}{\epsilon_s}$$ (2.129)

This is a difficult problem to solve since the potential appears on both sides of the equation. There is a much simpler solution which yields excellent results. Let us simply assume that the transition region is fully depleted of free carriers of charge. Then we simply have

$$\rho(x) = \begin{cases} -qN_A & -x_{p0} < x < 0 \\ qN_D & 0 < x < x_{n0} \end{cases}$$ (2.130)

Now the problem is pretty easy. We can simply integrate twice, apply some boundary conditions, and we're done. Let's do the math

$$E_0(x) = -\frac{d\phi}{dx} = \int_{-x_{p0}}^{x} \frac{\rho(x')}{\epsilon_s} dx' + E_0(-x_{p0})$$ (2.131)

The constant of integration is zero since the electric field outside of the transition region is zero. Thus

$$E_0(x) = -\frac{qN_a}{\epsilon_s}(x + x_{p0})$$ (2.132)

The electric field drops linearly as we traverse the junction. This is in contrast to a linear capacitor with constant electric field. This is totally anticipated by Gauss' Law without any recourse to mathematics. Gauss' Law tells us that electric fields diverge on charge. Therefore a uniform charge density will eat away at the field in a linear manner!

We can also do the integral from the n-side of the junction

$$E_0(x_{n0}) = 0 = \int_{x}^{x_{n0}} \frac{\rho(x')}{\epsilon_s} dx' + E_0(x) = \frac{qN_D}{\epsilon_s}(x_{n0} - x) + E_0(x)$$ (2.133)

or

$$E_0(x) = \frac{-qN_D}{\epsilon_s}(x_{n0} - x)$$ (2.134)

Since the field is continuous, at $x = 0$ we have

$$qN_A x_{p0} = qN_D x_{n0}$$ (2.135)

which simply states that the total charge on the n-side equals the total charge on the p-side, which is somewhat obvious. Gauss' Law guarantees this fact, since the electric field is zero outside the transition region. A plot of the electric field in the pn-junction is shown in Fig. 2.18.

If we now integrate again to arrive at potential, we find that on the p-side

$$\phi_0^p(x) = \phi_p + \frac{qN_A}{2\epsilon_s}(x + x_{p0})^2$$ (2.136)

and on the n-side

$$\phi_0^n(x) = \phi_n - \frac{qN_D}{2\epsilon_s}(x - x_{n0})^2$$ (2.137)

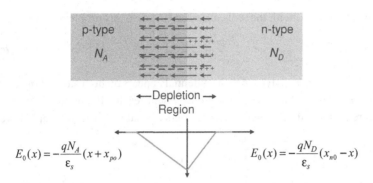

$$E_0(x) = -\frac{qN_A}{\epsilon_s}(x+x_{po})$$

$$E_0(x) = -\frac{qN_D}{\epsilon_s}(x_{n0}-x)$$

Figure 2.18 The built-in electric fields in a pn-junction diode.

At the interface, the potential must be a continuous function. This leads to

$$\phi_n - \frac{qN_D}{2\epsilon_s}x_{n0}^2 = \phi_p + \frac{qN_A}{2\epsilon_s}x_{p0}^2 \qquad (2.138)$$

The above equation, along with Eq. (2.135), allows us to solve for the depletion region depths

$$x_{n0} = \sqrt{\frac{2\epsilon_s\phi_{bi}}{qN_D}\left(\frac{N_A}{N_A+N_D}\right)} \qquad (2.139)$$

$$x_{p0} = \sqrt{\frac{2\epsilon_s\phi_{bi}}{qN_A}\left(\frac{N_D}{N_A+N_D}\right)} \qquad (2.140)$$

where $\phi_{bi} = \phi_n - \phi_p > 0$.

Now let's apply a reverse bias, or a potential that tends to increase the barrier height between the n- and p-regions. Certainly the current flow is essentially zero. In fact, for zero reverse bias, the built-in potential is precisely large enough to prevent majority carriers in one region from diffusing into the other region. But the barrier cannot prevent minority carriers from crossing the junction. So under a reverse bias, minority carriers that are thermally generated within a few diffusion constants from the junction will in fact give rise to a reverse current. This reverse current is essentially independent of the applied bias (until breakdown occurs). To see this intuitively, imagine blind mole rats that live near a cliff. Every once in a while a mole rat will come out of his hole and run along in a random direction. If he is so unlucky to run towards the cliff, then surely he will plunge to his death. Now, if we make the cliff twice as tall, the rate at which mole rats die will not change!

So back to the reverse biased pn-junction. When we apply an additional reverse bias, this voltage must be dropped across the transition region. This can only happen if the transition region *grows* in extent. In fact, we can now see that the equations derived under zero bias can be reused if we simply replace ϕ_{bi} with $\phi_{bi} - V_D$, where V_D is the additional reverse bias applied to the pn-junction. The depletion region can be written as a function of the reverse bias

$$x_n(V_D) = \sqrt{\frac{2\epsilon_s(\phi_{bi} - V_D)}{q N_D}\left(\frac{N_A}{N_A + N_D}\right)} = x_{n0}\sqrt{1 - \frac{V_D}{\phi_{bi}}} \tag{2.141}$$

$$x_p(V_D) = \sqrt{\frac{2\epsilon_s(\phi_{bi} - V_D)}{q N_A}\left(\frac{N_D}{N_A + N_D}\right)} = x_{p0}\sqrt{1 - \frac{V_D}{\phi_{bi}}} \tag{2.142}$$

The total depletion region depth is given by

$$X_d(V_D) = x_p(V_D) + x_n(V_D) = \sqrt{\frac{2\epsilon_s(\phi_{bi} - V_D)}{q}\left(\frac{1}{N_D} + \frac{1}{N_A}\right)} = X_{d0}\sqrt{1 - \frac{V_D}{\phi_{bi}}} \tag{2.143}$$

As we increase the reverse bias voltage, the depletion region grows to accommodate more charge. The charge on either side of the junction is given by

$$Q_j(V_D) = -q N_A x_p(V_D) = -q N_A X_{d0}\sqrt{1 - \frac{V_D}{\phi_{bi}}} \tag{2.144}$$

Charge is not a linear function of voltage, and so we have a non-linear capacitor! To see that this is indeed a capacitor, note that the pn-junction is initially neutral. When we apply a reverse bias voltage, the depletion region grows, exposing ionized donor and acceptor ions. Since the depletion region is free of mobile charge, a corresponding charge must be transferred from the battery in order to precisely balance the exposed charge. Hence, from the terminals of the battery, it appears that we have connected a non-linear capacitor.

What is the small-signal capacitance? Differentiating the expression for charge we arrive at

$$C_j = \left.\frac{dQ_j}{dV}\right|_{V=V_D} = \left.\frac{d}{dV}\left(-q N_a x_{p0}\sqrt{1 - \frac{V}{\phi_{bi}}}\right)\right|_{V=V_d} \tag{2.145}$$

or

$$C_j = \frac{C_{j0}}{\sqrt{1 - \frac{V_D}{\phi_{bi}}}} \tag{2.146}$$

where we have defined

$$C_{j0} = \frac{q N_A x_{p0}}{2\phi_{bi}} = \frac{q N_A}{2\phi_{bi}}\sqrt{\left(\frac{2\epsilon_s \phi_{bi}}{q N_A}\right)\left(\frac{N_D}{N_A + N_D}\right)} = \sqrt{\frac{2\epsilon_s}{2\phi_{bi}}\frac{N_A N_D}{N_A + N_D}} \tag{2.147}$$

Don't bother to memorize the expression for C_{j0}. You can derive it very easily if we notice that

$$C_{j0} = \epsilon_s\sqrt{\frac{q}{2\epsilon_s \phi_{bi}}\left(\frac{1}{N_D} + \frac{1}{N_A}\right)^{-1}} = \frac{\epsilon_s}{X_{d0}} \tag{2.148}$$

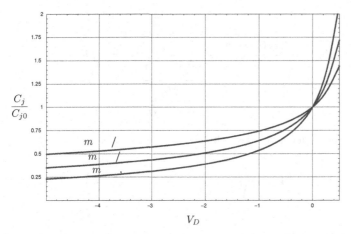

Figure 2.19 The normalized diode junction capacitance as a function of the terminal voltage for three different doping profiles. The $m = 1/2$ corresponds to the abrupt junction.

which is surprisingly the formula for a parallel plate capacitor! So the junction capacitance is simply

$$C_j(V_D) = \frac{\epsilon_s}{X_d(V_D)} \tag{2.149}$$

Again we can explain the simplicity of this result. When we apply an incremental voltage, the increment of charge comes from the outer edges of the depletion region, since this is the direction of the growth of the depletion region. Thus it appears that we have a parallel plate capacitor with plate separation of $X_d(V)$.

   This variation of this capacitance with voltage is plotted in Fig. 2.19. The shape of the curve can be altered by proper adjustment of the junction doping profile. In the above analysis we made the assumption that the doping profile is a step function across the junction. In practice, we assume that $X_d = X_{d0}\left(1 - \frac{V_D}{\phi_{bi}}\right)^m$ and experimentally determine the "grading coefficient" m.

MOS capacitor

The MOS capacitor, or MOS-C structure, shown in Fig. 2.20, is a capacitor constructed with one metal gate, an oxide insulator, and another semiconducting "plate." In practice, we often employ a poly-silicon gate, which acts like a metal. Let us assume that we short the terminals of the capacitor. Here the body is p-type, whereas the gate is a metal or heavily doped n-type material. Upon contact, electrons leave the gate and flow into the p-type material and the charge is negative. Likewise, holes from the body come into the gate and the charge is positive. By allowing the materials to exchange charge, we can see that a built-in potential difference will arise between the materials until an equilibrium potential difference has been established.

   Since all systems strive to reside in a minimum potential state, it is not surprising to find the positive charges on the gate residing above a depletion formed in the body, as shown in Fig. 2.21a. Remember that this is the equilibrium state. Initially free charge in the body near the contact will flow into the gate, forming a transient depletion region near the wire

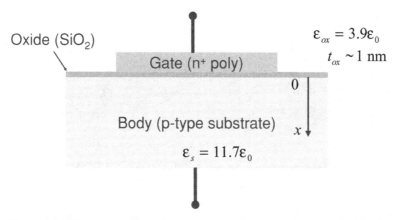

Figure 2.20 The cross section of a MOS-C capacitor structure. Permeability numbers are for Si and SiO$_2$.

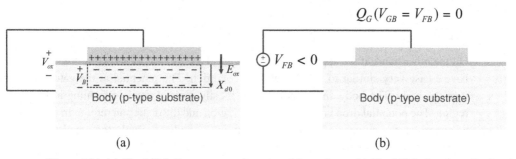

Figure 2.21 (a) The MOS-C structure under a zero-bias voltage. (b) The MOS-C under a flat-band voltage bias.

contact. But in the steady state, other holes will occupy these states, leaving the higher energy region under the gate depleted of carriers.

Flat band

Now if we apply a negative voltage V_{FB} to cancel the internal built-in potential, as shown in Fig. 2.21b, we see that the charge on the gate and the body will go to zero. This is the so-called "flat-band" condition, since the electric field (and hence the band bending) along the MOS interface is zero. The flat-band voltage necessary to do this is simply

$$V_{FB} = -(\phi_{n^+} - \phi_p) \tag{2.150}$$

for the p-type body and n^+ gate material.

Accumulation

If we further decrease the potential beyond the "flat-band" condition, we essentially have a parallel plate capacitor. As shown in Fig. 2.22a, since a negative bias attracts holes from the body to flow under the gate, and since holes are majority carriers, plenty of holes are

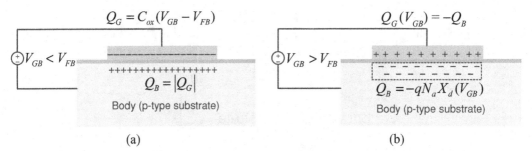

Figure 2.22 (a) The MOS-C structure in the accumulation region. (b) The MOS-C in the depletion region.

available to do this. Likewise plenty of electrons can flow into the gate through the n^+ material. Operated in a region with bias $V_{GB} < V_{FB}$ is called "accumulation," since holes accumulate at the surface of the MOS-C structure.

Depletion

On the other hand, if we apply a positive voltage $V_{GB} > V_{FB}$, as shown in Fig. 2.22b, we have a case very similar to the equilibrium condition. The potential at the gate is higher than the body. The body charge consists of mostly ionized carriers forming a depletion region. The potential drop is across the oxide region and the depletion region. In this region of operation, the number of minority carriers in the body is simply too small to have an appreciable impact on the behavior of the device.

Inversion

As we continue to apply a larger and larger gate voltage, we can see that the *surface potential* will continue to increase. Since the electron concentration is proportional to the potential, at the surface we have

$$n_s = n_i e^{\frac{q\phi_s}{kT}} \tag{2.151}$$

If the surface potential becomes large enough, the minority carrier concentration can become large enough to "invert" the material doping, as shown in Fig. 2.23a. When this occurs, the depletion region essentially stops growing and the extra charge is now readily provided by the inversion charge at the surface. For this reason, the potential at the surface is also "pinned" since electrons flow to the surface of the MOS-C structure and any extra voltage drop occurs in the oxide region. The gate voltage required to produce this surface inversion is called the threshold voltage V_T.

Threshold voltage

The threshold voltage for the p-body MOS-C structure is defined as the gate–body voltage necessary to cause the surface of the body to change from p-type to n-type. For this condition

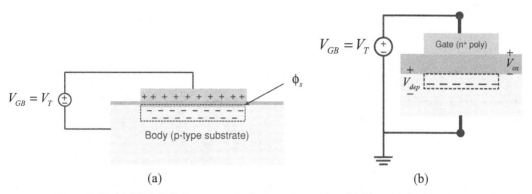

Figure 2.23 (a) The MOS-C structure in the inversion region. (b) The gate–body voltage consists of the built-in voltage, the oxide voltage, and the voltage drop across the depletion region.

to occur, the surface potential has to equal the negative of the p-type potential. Let's apply KVL around the loop shown in Fig. 2.23b

$$V_{GS} = V_{FB} + V_{ox} + V_{dep} \tag{2.152}$$

Note that $\phi_s = V_{BS} = -2\phi_p$ to cause inversion. The factor of two is somewhat confusing since $n_s = N_A$ implies

$$n_s = n_i e^{\frac{q\phi_s}{kT}} = N_A \tag{2.153}$$

or $\phi_s = -\phi_p$. So the potential difference between the inverted surface and the body is $-\phi_p - \phi_p = -2\phi_p$. Since the body is grounded, this is precisely the potential we need at the surface to invert it.

The term V_{ox} can be replaced by applying the normal field boundary condition

$$V_{ox} = E_{ox} t_{ox} = \frac{\epsilon_s}{\epsilon_{ox}} t_{ox} E_s \tag{2.154}$$

The electric field at the surface of the silicon can be found by application of Gauss' Law to the depletion region. Since we assume that the electric field is zero inside the body, the electric flux at the surface must equal the charge stored inside the depletion region

$$E_s = \frac{q N_A x_{dep}}{\epsilon_s} = \frac{q N_A}{\epsilon_s} \sqrt{\frac{2\epsilon_s}{q N_A} \phi_s} = \sqrt{\frac{2q N_A(-2\phi_p)}{\epsilon_s}} \tag{2.155}$$

These substitutions yield the following value for the threshold voltage

$$V_{TH} = V_{FB} - 2\phi_p + \frac{1}{C_{ox}} \sqrt{2q\epsilon_s N_A(-2\phi_p)} \tag{2.156}$$

MOS Q-V and C-V curves

Given the above calculations, we can derive the MOS Q-V curve. A simple approximation is to assume that once inversion occurs, the depletion region stops growing. This is a good approximation since charge is an exponential function of the surface potential. Under this

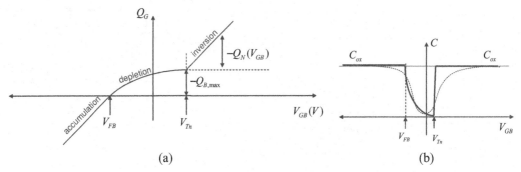

Figure 2.24 (a) The MOS-C charge voltage or Q-V curve. (b) The MOS-C capacitance versus voltage curve. Note this is simply a derivative of the Q-V curve.

assumption

$$Q_G(V_{TH}) \approx -Q_{B,max} \tag{2.157}$$

or

$$Q_G(V_{GB}) = C_{ox}(V_{GB} - V_{TH}) - Q_{B,max} \tag{2.158}$$

The MOS Q-V curve is shown in Fig. 2.24a. In accumulation $V_{GB} < V_{FB}$, we have neglected the minority carriers and thus the charge is proportional to the applied bias, much like a regular linear capacitor. But note carefully that the charge is not zero for $V_{GB} = 0$ but rather for V_{FB}. In the region between flat band and threshold, $V_{FB} < V_{GB} < V_{TH}$, we are in the depletion region. Again, if we ignore the effects of the minority carriers, the charge is due to the depletion region only. Then at $V > V_{TH}$ we assume that the depletion region grows imperceptibly and all of the charge appears at the surface of the body through the inversion charge.

The MOS-C small-signal C-V curve is derived by simply taking the derivative of the Q-V curve. This is shown in Fig. 2.24b. In the accumulation and inversion regions, the capacitance is essentially given by C_{ox}, or the parallel plate capacitance of the MOS-C. In the depletion region, the capacitance drops from C_{ox} as the depletion region grows larger and larger. But then suddenly, and quite artificially, inversion occurs and the capacitance jumps back to the value of C_{ox}. Of course, in reality, the minority carriers do not play a negligible role in the MOS-C structure. Below but near the threshold voltage, their presence is very important for the small-signal capacitance, even if their numbers are small, since the rate at which these charges appear with increasing voltage is exponential. A more realistic C-V curve, as sketched with the dashed line in the figure, shows a smooth transition across inversion and flat band.

In accumulation mode the capacitance is just due to the voltage drop across t_{ox}. In inversion, the incremental charge comes from the inversion layer (the depletion region stops growing). In the depletion region, the voltage drop is across the oxide and the depletion region, so we have two series connected capacitors. To calculate the capacitance in the depletion region directly, note that an equivalent circuit for the MOS-C structure in this region is

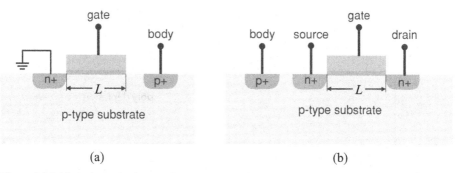

(a)   (b)

Figure 2.25 (a) An integrated MOS-C structure is constructed with a surface body contact and an n$^+$ "source" layer to provide electrons for inversion. (b) A MOSFET is a natural inversion mode MOS-C structure. If the source and drain are tied together, both regions can become a source for electrons.

two series capacitors, an oxide capacitance C_{ox} in series with the depletion capacitance

$$C_{dep} = \frac{\epsilon_s}{x_{dep}} \tag{2.159}$$

Taking the capacitors in series, we have

$$C_{tot} = \frac{C_{dep}C_{ox}}{C_{dep} + C_{ox}} = \frac{C_{ox}}{1 + \frac{\epsilon_s}{\epsilon_{ox}}\frac{t_{ox}}{x_{dep}}} \tag{2.160}$$

In the above discussion you may have wondered, "Where do the electrons come from to form the inversion layer?" In the body of the MOS-C structure, electrons are minority carriers and few and far between. So when inversion occurs, where do we find all the electrons necessary to invert the surface? Well, there was a subtle assumption that if we apply a change in gate voltage, we wait long enough for thermal generation to create a sufficient number of electrons to form the surface layer. We may have to wait a very long time! In other words, if we apply a fast enough signal to the gate, there isn't enough time for the minority carriers to be generated and thus the capacitance remains at the low value given by depletion.

While the depletion region can respond very quickly to our gate voltage since it is formed by majority carriers, the minority carrier generation is slow. There is a simple way to solve this problem, as shown in Fig. 2.25, where a n$^+$ grounded contact is placed adjacent to the gate. Normally electrons are prevented from entering the body, like any good pn-junction. But as we raise the surface potential, electrons can easily diffuse into the surface of the structure. Since the energy distribution of electrons in thermal equilibrium is exponential,[2] changing the potential barrier linearly results in an exponential increase in the number of electrons that can cross the n$^+$-surface junction and likewise an exponential increase in surface conductors.

There are a few physical effects that we have neglected to mention. The most important physical effects are the poly-silicon depletion region and the quantum mechanical effects. The poly-depletion region is easy enough to understand. Since the gate material is usually

---

[2]   Or, if you like, the tail of the distribution is approximately exponential.

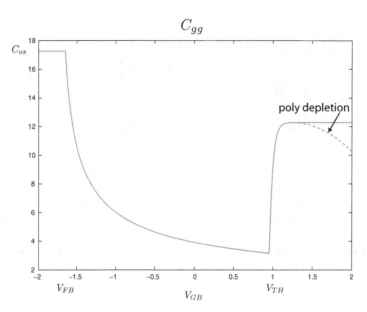

Figure 2.26 A more realistic MOS C-V curve takes into account the role of minority carriers below threshold, the quantum mechanical confinement, and the poly-silicon gate depletion.

constructed with a heavily doped poly-silicon, rather than a metal, it will also form a small depletion region. As the device t_{ox} is reduced in short-channel devices, this depletion region width is an appreciable fraction of t_{ox}, resulting in a lower effective gate bias. As shown in Fig. 2.26, as we increase the V_{GB} beyond V_T, the depletion region width tends to increase, lowering the capacitance.

The quantum mechanical confinement is also easily understood if we solve Poisson's equation along with Schrodinger's equation in a consistent fashion. The result is that the peak charge does not occur at the surface of the silicon body, but slightly displaced from the surface. In effect, the oxide thickness increases, also resulting in a smaller capacitance.

2.4 References

Some of the material was put together while teaching an Electromagnetics course at UC Berkeley. Useful references include Feynman's Lectures [18], Cheng [6], Reitz [53], and Inan and Inan's book [62]. Some of the device physics material was drawn from lecture notes put together by Howe and Sodini [49].

3 Resistance

You may be tempted to skip this chapter. After all, resistors are pretty elementary, right? Actually, below the surface resistors are complicated beasts. In fact, resistors apparently defy one of the Newton's fundamental law of physics, as charges in resistors move with constant velocity when subjected to an external force. We will probe into the inner workings of resistors in order to answer this riddle.

3.1 Ohm's Law

Let's begin with the venerable Ohm's Law. Is it trivial that $V = I \times R$? Perhaps all that is going on is that

$$V = f(I) = f(0) + f'(0)I + \frac{1}{2}f''(0)I^2 + \cdots \approx f'(0)I \tag{3.1}$$

where we have kept the linear term in the Taylor series expansion of the voltage–current relationship.[1] If this is the case, what's the range of validity of the above approximation. What's surprising is that Ohm's Law is valid over an enormous range.

Note that Ohm's Law is also equivalently expressed as

$$\mathbf{J} = \sigma \mathbf{E} \tag{3.2}$$

We assert that the ratio between the voltage and current in a conductor is always constant. We see this stems from $J = \sigma E$

$$I = \int_A \mathbf{J} \cdot d\mathbf{S} = \int_A \sigma \mathbf{E} \cdot d\mathbf{S} \tag{3.3}$$

$$V = \int_C \mathbf{E} \cdot d\mathbf{l} \tag{3.4}$$

The ratio is called the resistance

$$R = \frac{V}{I} = \frac{\int_C \mathbf{E} \cdot d\mathbf{l}}{\int_A \sigma \mathbf{E} \cdot d\mathbf{S}} \tag{3.5}$$

---

[1] The expansion point is about zero current since we experimentally observe that the average current in a conductor is zero for zero applied voltage.

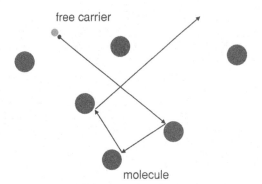
free carrier

molecule

Figure 3.1 Simple billiard-ball model of collision between a free carrier and a molecule.

What is disturbing about Ohm's Law is that current, which is proportional to the velocity, is proportional to the electric field, which is imparting a force on the charge. Is Newton rolling in his grave?

Conductivity of a gas

Electrical conduction is due to the motion of positive and negative charges. For water with pH $= 7$, the concentration of hydrogen H^+ ions (and OH^-) is

$$10^{-7}\text{mole/L} = 10^{-10}\text{mole/cm}^3 = 10^{-10} \times 6.02 \times 10^{23}\text{cm}^{-3} = 6 \times 10^{13}\text{cm}^{-3} \quad (3.6)$$

Typically, the concentration of charged carriers is much smaller than the concentration of neutral molecules. The motion of the charged carriers (electrons, ions, molecules) gives rise to electrical conduction.

At a temperature T, each charged carrier will move in a random direction and velocity until it encounters a neutral molecule or another charged carrier. Since the concentration of charged carriers is much less than molecules, it will most likely encounter a molecule. For a gas, the molecules are widely separated (~ 10 molecular diameters). After colliding with the molecule, there is some energy exchange and the charge carrier will come out with a new velocity and new direction. For simplicity, assume elastic collisions and ignore atomic level force interactions.

This is shown schematically in Fig. 3.1. The key observation is that the initial velocity and direction is lost (randomized) after a few collisions. When we apply an electric field, during each "free flight," the carriers will gain a momentum of $q\mathbf{E} \times t$. Therefore, after t seconds, the momentum is given by

$$M\mathbf{u} + q\mathbf{E}t \quad (3.7)$$

If we take the average momentum of all particles at any given time, we have

$$M\overline{\mathbf{u}} = \frac{1}{N}\sum_j \left(M\mathbf{u}_j + \mathbf{E}qt_j\right) \quad (3.8)$$

where we assume N carriers, an initial momentum $M\mathbf{u}_j$ before the collision, and a time t_j between collisions.

When we sum over all the initial random velocities of the particles, we are averaging over a large number of random variables with zero mean, and so the average is zero

$$M\bar{\mathbf{u}} = \frac{1}{N} \sum_j \left(M\bar{\mathbf{u}}_j + Eqt_j \right) \tag{3.9}$$

The momentum gain is therefore given by the second term

$$M\bar{\mathbf{u}} = \frac{1}{N} \sum_j Eqt_j = Eq\tau \tag{3.10}$$

where τ is the average time between collisions.

Current is defined as the amount of charge crossing a particular cross-sectional area per unit time. Consider a density of N carriers carrying a charge q and moving with velocity v_d. How many cross a perpendicular surface A? In a time Δt, carriers move a distance of $v_d\Delta t$. So any carriers in the volume $Av_d\Delta t$ will cross the surface if they are all moving towards it. The current is therefore

$$I = \frac{Av_d\Delta t N q}{\Delta t} = v_d N q A \tag{3.11}$$

The vector current density J is therefore

$$\mathbf{J} = Nq\mathbf{v}_d \tag{3.12}$$

Using Eq. (3.10) we arrive at

$$\mathbf{J} = Nq\bar{\mathbf{u}} = Nq \left(\frac{Eq\tau}{M} \right) = Nq^2 \frac{\tau}{M} \mathbf{E} = \sigma \mathbf{E} \tag{3.13}$$

Since current is contributed by positive and negative charge carriers

$$\mathbf{J} = \mathbf{J}^+ - \mathbf{J}^- = e \left(\frac{N^+ e\tau^+}{M^+} - \frac{-N^- e\tau^-}{M^-} \right) \mathbf{E} \tag{3.14}$$

or

$$\mathbf{J} = \sigma \mathbf{E} \tag{3.15}$$

The conductivity of a gas is therefore given by

$$\sigma = e^2 \left(\frac{N^+\tau^+}{M^+} + \frac{N^-\tau^-}{M^-} \right) \tag{3.16}$$

Conduction in metals

The high conductivity of metals is due to a large concentration of free electrons. These electrons are not attached to the solid but are free to move about the solid. This sea of electrons, or an electron gas, in effect keeps the solid together. In metal sodium, for instance, each atom contributes a free electron. The ionized molecules are evenly spaced and form a

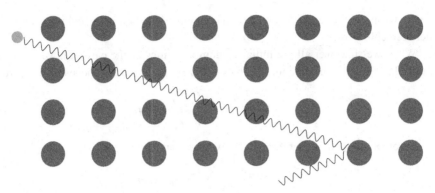

Figure 3.2 An electron can penetrate through potential barriers of ions due to the periodic nature of the crystal.

crystal lattice. From the measured value of conductivity (easy to do), we can back calculate the mean free time

$$\tau = \frac{\sigma m}{Ne^2} = \frac{(1.9 \times 10^7)(9 \times 10^{-28})}{(2.5 \times 10^{22})(23 \times 10^{-20})} = 3 \times 10^{-12} \text{sec} \qquad (3.17)$$

This value of mean free time is surprisingly long. The mean velocity for an electron at room temperature is roughly given by

$$\frac{mv^2}{2} = \frac{3}{2}kT \qquad (3.18)$$

or $v = 3 \times 10^7$cm/sec. At this speed, the electron travels a distance $v\tau = 3 \times 10^{-7}$cm. The molecular spacing between adjacent ions is only about 3.8×10^{-8}cm. Why is it that the electron is on average zooming by ten positively charged ions?

The answer to this deep puzzle requires us to admit the wave nature of electrons. As shown schematically, in Fig. 3.2, the free carrier can penetrate right through positively charged host atoms. Quantum mechanics explains all of this. For a periodic arrangement of potential fluctuations, the electron does not scatter. It is as if our electron is free to travel around the crystal, but its mass is not the same as the electron mass, but an "effective" mass determined by the crystal. So why does the electron scatter at all?

First, at a temperature T, the atoms are in random motion and thus upset the periodicity, as shown in Fig. 3.3a. Even at low temperatures, though, the presence of impurities upsets the periodicity, as seen in Fig. 3.3b.

Free carrier mobility

Drift

As we have seen, when a carrier in a conductor is accelerated by an external field, it initially travels in the direction of the field. But after scattering with an impurity or phonon, the electron's new trajectory is random and independent of the field. We have shown that the average velocity of carriers increases proportional to the field E

$$v_d = \frac{qt_c}{m}E = \mu E \qquad (3.19)$$

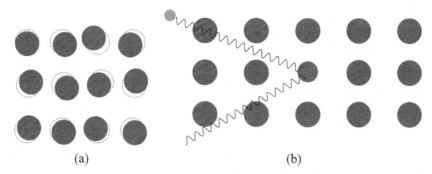

Figure 3.3 Electron scattering occurs due to (a) the random vibration of the molecules which upset the periodicity of the crystal lattice and (b) due to the presence of impurities in the lattice.

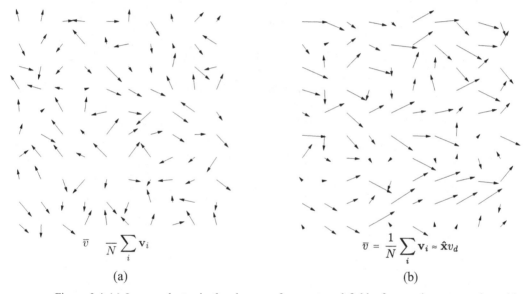

Figure 3.4 (a) In a conductor in the absence of any external fields, free carriers are moving with random velocity in random directions. (b) In response to an external electric field, there is a net motion, or drift, in the direction of the field.

The constant μ is the mobility, and the "drift" velocity v_d is so named because most of the motion is still random as $v_d \ll v_{th}$.

In Fig. 3.4a, in the absence of an external field, electrons are moving with random velocity (net motion is zero). In Fig. 3.4b, electrons are moving with random velocity plus a drift component to the right (drift has been exaggerated).

Charge conservation

One of the fundamental facts of nature is charge conservation. This has been experimentally observed and verified in countless experiments. Therefore, if we look at a volume bounded by surface S, as shown in Fig. 3.5, if there is net current leaving this volume, it must be

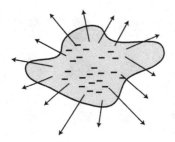

Figure 3.5 By charge conservation, in a region with net positive current flow *out* of the region, a net negative charge builds up inside over time.

accompanied by a charge pile-up in the region of opposite polarity

$$\oint_S \mathbf{J} \cdot \mathbf{dS} = -\frac{\partial}{\partial t} \int_V \rho dV \tag{3.20}$$

We can transform this simple observation by employing the Divergence Theorem

$$\oint_S \mathbf{J} \cdot \mathbf{dS} = \int_V \nabla \cdot \mathbf{J} dV \tag{3.21}$$

For any volume we have found that

$$\int_V \nabla \cdot \mathbf{J} dV - \frac{\partial}{\partial t} \int_V \rho dV = \int_V \left(\nabla \cdot \mathbf{J} + \frac{\partial \rho}{\partial t} \right) dV = 0 \tag{3.22}$$

If this is true for any volume V, it must be true that

$$\nabla \cdot \mathbf{J} + \frac{\partial \rho}{\partial t} = 0 \tag{3.23}$$

Since $\mathbf{J} = \sigma \mathbf{E}$, we have

$$\nabla \cdot \sigma \mathbf{E} + \frac{\partial \rho}{\partial t} = 0 \tag{3.24}$$

If the conductivity is constant, we have

$$\sigma \nabla \cdot \mathbf{E} + \frac{\partial \rho}{\partial t} = 0 \tag{3.25}$$

It is experimentally observed that Gauss' Law is satisfied in all reference frames of constant velocity so $\nabla \cdot \mathbf{E} = \frac{\rho}{\epsilon}$

$$\frac{\sigma}{\epsilon} \rho + \frac{\partial \rho}{\partial t} = 0 \tag{3.26}$$

If we assume that ρ is a function of time only, we have the following differential equation

$$\frac{\sigma}{\epsilon} \rho + \frac{d\rho}{dt} = 0 \tag{3.27}$$

Letting $\tau = \sigma/\epsilon$, we solve the above equation

$$\rho = \rho_0 e^{-\frac{t}{\tau}} \tag{3.28}$$

Thus the charge density in a conductor decays exponentially to zero with a rate determined by the time constant $\tau = \sigma/\epsilon$, or the *relaxation time* of the material. Earlier we deduced that conductors have zero volume charge density under electrostatic conditions. Now we can estimate the time it takes to reach this equilibrium.

Relaxation time for good conductors

In fact, a good conductor like copper has $\sigma \sim 10^8 \text{S/m}$. It is hard to measure the dielectric constant of a good conductor, since the effect is masked by the conductive response. For $\epsilon \sim \epsilon_0$, we can estimate the relaxation time

$$\tau \sim \frac{10 \times 10^{12}}{10^8} \text{s} \sim 10^{-19} \text{s} \tag{3.29}$$

This is an *extremely* short period of time! How do electrons move so fast? Note this result is independent of the size of the conductor. The answer is that they don't. Only a slight displacement in the charge distribution can balance the net charge and thus transfer it to the border region.

3.2 Conduction in semiconductors

Semiconductors have formed the foundation for modern society. Historians may call our age the "sand age" since virtually all of our machines are made of Si semiconductors. The conductivity of semiconductors, as their name suggests, lies somewhere between conductors and insulators. We will explain this with a very simple band model.

From quantum mechanics we know that the allowed energy levels for an atom are discrete (two electrons can occupy a state with opposite spin). When atoms are brought into close proximity, these energy levels split. If there are a large number of atoms, such as in a solid crystal, the discrete energy levels form a seemingly continuous band. In our simple model, each material solid is characterized by a valence band and a conduction band, with a forbidden region called the band gap.

The valence band corresponds to bound electronic states. In other words, electrons with this energy level are bound to their host atoms. Usually these electrons are part of a covalent bond. Conduction band electrons, on the other hand, are free to move around in the solid. Of course they are not free to leave the material as there is an additional amount of energy required for an electron to leave the solid. This is characterized by the work function of the material. The energy distribution of the electrons is determined by Fermi-Dirac statistics and the Fermi level E_F of the material. At low temperature, most electrons fill energy levels below the Fermi level.

One of the most important properties of a solid is the energy difference between the valence band and the conduction band, or the band gap. This is important because a valence band electron can become a conduction band electron if it can attain this energy level, either through incoming radiation (a photon), or through thermal energy. If the band gap is very small, then even at low temperatures, many electrons will occupy the conduction band.

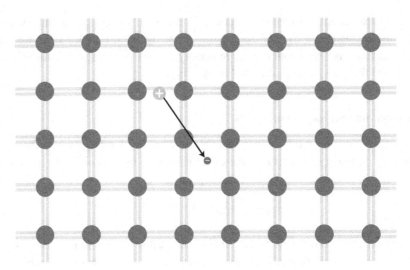

Figure 3.6 A cartoon two-dimensional Si lattice. Circles represent Si atoms and lines represent covalent bonds formed by valence band electrons. Broken bonds give rise to electrons and holes.

Insulators have a very large band gap, about 3–7 eV, so that the probability that an electron has sufficient energy to occupy the conduction band is extremely small. And semiconductors have an intermediate band gap. Si, for instance, has a band gap of about 1.12 eV. This is actually quite large compared to the thermal energy of kT, which is about 26meV at room temperature.

Electrons and holes

When early researchers measured the conductivity of Si and other semiconductors, they were in for a surprise. The conductivity numbers were often a factor of 2 from their predictions. It was as if there were twice as many free carriers in the material. Well, in fact, that is precisely what happens in a semiconductor. In addition to electrons, "holes" also conduct current.

In Fig. 3.6, we have a cartoon diagram of a two-dimensional crystal such as Si that forms four covalent bonds with its neighbors, represented by lines in the figure. In this figure valence band electrons are the covalent bonds in the crystal. If we break one of these bonds to free an electron, there is in essence a hole of positive charge left behind. It is important to realize that this hole is in fact free to move. This is because it represents a state of low energy for another electron. So in fact it is very easy for another nearby valence band electron to hop into this hole. In doing so it creates another hole and thus the hole travels! So electrons moving in the valence band from bond to bond corresponds to the equivalent motion of a positive charge. A very crude analogy is a bubble in a sealed liquid. This bubble corresponds to our hole, and it moves around the liquid as if it were a real particle.

When a conduction band electron encounters a hole, the process is called recombination. The electron and hole annihilate one another thus depleting the supply of carriers. In thermal equilibrium, a generation process counterbalances to produce a steady stream of carriers. We can see that the thermal generation of free carriers produces an equal amount of electrons

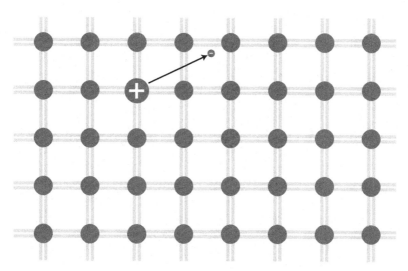

Figure 3.7 A donor dopant atom in the Si lattice contributes a free electron.

and holes, or $n_0 = p_0$. If the thermal generation rate G_{th} is equal to the recombination rate R, we have

$$G_{th} = R = kn \times p \tag{3.30}$$

where we assert that the regeneration rate is proportional to the product of holes and electrons. This is logical because for every electron the probability of recombination is proportional to the number of holes, and vice versa. Since at a given temperature the thermal generation rate is constant, the product of $n \times p$ is also a constant

$$np = n_i^2 \tag{3.31}$$

which is the intrinsic concentration of electrons (and holes) in a material at a given temperature. For Si at 300K, this is about 10^{10}cm^{-3}.

Doping

Thus far our discussion of semiconductors, in particular Si, has been limited to pure or intrinsic Si. In practice, pure Si is very hard to produce as even trace quantities of impurities can dramatically alter the behavior of the resulting material. This is in fact one of the great benefits of Si in that its behavior is very well controlled through selective doping.

Imagine adding a trace amount of a group V element such as phosphorous or arsenic to a Si crystal, as shown in Fig. 3.7. Since Si is a group IV element, it needs four electrons to fill its shell. A P atom can pose as a Si atom and form four covalent bonds if it can somehow rid itself of one electron. So in fact with very little provocation, it will do so and "donate" an electron to the crystal. The group V element is thus an ionized donor atom. It is important to note that each dopant atom creates an electron, but not a hole. The ionized atom is an *immobile* positive charge. The material is said to become n-type.

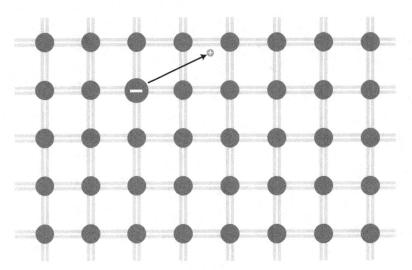

Figure 3.8 An acceptor dopant atom in the Si lattice contributes a free hole.

In contrast, a group III element, such as boron, is an "acceptor" dopant, as shown in Fig. 3.8. Since it has only three electrons in a four electron covalent bond world, it has a hole, so to speak. This hole is an energetically favorable state so other nearby valence electrons are likely to occupy this state, thus creating an ionized negative immobile charge site. In stealing an electron from the crystal it creates a hole that can travel through the crystal as discussed before. The material is said to become p-type.

In practice we actually dope a material with both acceptors and donors as we grow several layers of n-type and p-type material. The difference between the concentration of donors and acceptors will determine the n- or p-typeness of the material. It is also important to note that our arguments so far assume that the temperature T is large enough so that thermal energy is sufficient to ionize the dopants. If the temperature becomes too low, a process of freeze-out occurs, dramatically altering the conductivity of the material.

The motion of both electrons and holes in a crystal is governed by the band structure of the material. Fortunately, we can ignore these complexities and simply assign an effective mass to holes and electrons and treat them as free particles.

Acceptor and donor accounting

Let us begin by counting electrons in the donor case. The total charge density in the material is given by

$$\rho = \underbrace{-qn_0}_{\text{free electrons}} + \underbrace{qp_0}_{\text{free holes}} + \underbrace{qN_d}_{\text{immobile ions}} = 0 \tag{3.32}$$

By the law of mass action, in equilibrium, the product of electrons and holes is a constant, or $np = n_i^2(T)$. Substituting for the holes we have

$$-qn_0 + q\frac{n_i^2}{n_0} + qN_d = 0 \tag{3.33}$$

or

$$-n_0^2 + N_d n_0 + n_i^2 = 0 \tag{3.34}$$

Solving the quadratic equation, we arrive at the physical solution

$$n_0 = \frac{N_d + \sqrt{N_d + 4n_i^2}}{2} \approx N_d \tag{3.35}$$

where we have assumed that $N_d \gg n_i$, since the doping $N_d \sim 10^{11} - 10^{18}$, usually several orders of magnitude larger than n_i. The same conclusion holds for acceptors, so that for a p-type material we have

$$p_0 \approx N_a \tag{3.36}$$

In the case of a compensated material,[2] it is the net doping that produces the free carriers

$$p_0 \approx N_a - N_d \tag{3.37}$$

Free carrier mobility

Since the conductivity of a material depends on the number of free carriers (Eq. (3.16))

$$\sigma = q\mu N \tag{3.38}$$

we have in effect a technique to produce a material of desired resistivity. If the material is n-type, this is simply

$$\sigma = q\mu_n N_d \tag{3.39}$$

where μ_n is the electron mobility. Unfortunately, the mobility is not a constant as we change the material doping. As we have remarked earlier, any disturbance to the periodic structure of a crystal acts as a scattering site and thus, not surprisingly, as we increase the doping, we decrease the mobility of free carriers. This is confirmed by the experimental plot of the bulk mobility of electrons and holes in Si, as shown in Fig. 3.9. As expected, the total doping concentration, $N_d + N_a$, affects the mobility, not the net doping, $|N_d - N_a|$.

Free carrier mobility model

The mobility of free carriers plays an important role in the design of devices and ultimately in the performance of circuits. It is therefore very important to produce an accurate model for the mobility. The equation

$$J = \sigma E = qn\mu_n E \tag{3.40}$$

is too good to be true. It implies that we can create current without bound by simply making E sufficiently large. In modern nanoscale semiconductor devices, we can in fact

---

[2] A compensated material is doped with both donors and acceptors impurities so that the overall characteristics are a function of the net doping.

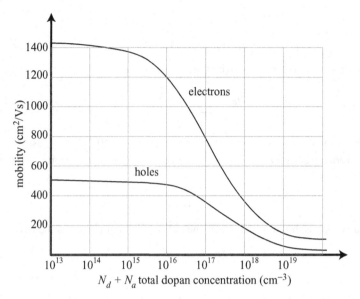

Figure 3.9 The mobility of electrons and holes as a function of total doping concentration.

create enormously large electric fields by simply applying a few volts on to our structures. Our various assumptions about the mobility of a free carrier begin to breakdown, though, under large field conditions. Most importantly, when creating the drift model of carrier current, we assumed that the drift velocity is a small fraction of the thermal velocity. As the fields are made very large, though, a linear mobility model predicts that we can in fact impart enough momentum to produce ultra high-speed carriers. Not only is the speed of light a limitation, but in practice we find that there is a limit due to increased phonon scattering at high fields, as shown in Fig. 3.10. We say that carriers become velocity saturated at high fields. This effect can be modeled by introducing a field dependent mobility model

$$\mu = \frac{\mu_0}{\left(1 + \left(\frac{E}{E_{sat}}\right)^n\right)^{1/n}} \tag{3.41}$$

where the coefficient n is fit to experimental data, with $n = 2$ for electrons and $n = 1$ for holes. The saturation velocity for electrons is about 8×10^6 cm/s and 6×10^6 cm/s for holes. From this we can solve for E_{sat}

$$v_{sat} = \mu E_{sat} = \mu_0 \frac{E_{sat}}{2^{1/n}} \tag{3.42}$$

There are other field dependencies in the mobility model. In particular, if current flow occurs near the surface of a semiconductor–oxide interface, such as in a MOSFET, then the mobility is found to be much lower at the surface. This is due to enhanced scattering sites due to the non-perfect interface between Si and SiO_2. A vertical field that tends to pull carriers towards the surface, therefore, will tend to lower the effective mobility.

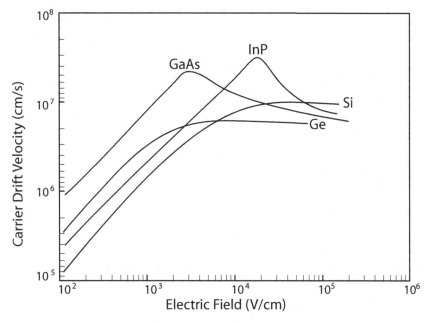

Figure 3.10 Velocity saturation for various materials. *From [57].*

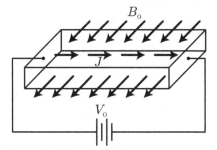

Figure 3.11 A rectangular conductor is placed inside a magnetic field.

On the other hand, in a MOSFET structure, we also find that the mobility drops at low vertical fields due to enhanced Coulomb scattering.[3] This occurs since the field effect produces inversion, which masks the background dopant sites at the surface. At low fields, though, the dopants are exposed, causing enhanced scattering.

Hall effect

In this section we will review an important technique for measurement of free carrier density based on the Hall effect. Consider a conductor placed inside a magnetic field, as shown in Fig. 3.11. When current is traveling through a conductor, at any instant it experiences a

---

[3] See Chapter 2 for a review of inversion in a MOSFET structure.

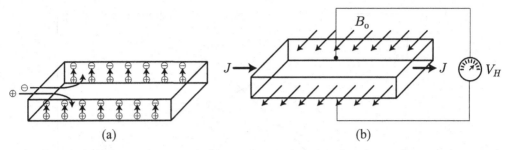

(a)   (b)

Figure 3.12 (a) Due to the magnetic force on free carriers, there is a charge pile-up on the top and bottom of the conductor. (b) This charge creates an internal electric field that opposes the movement of charges vertically.

force given by the Lorentz equation

$$F = q\mathbf{E} + q\mathbf{v} \times \mathbf{B} \tag{3.43}$$

The force $q\mathbf{E}$ leads to conduction along the length of the bar (due to momentum relaxation) with average speed v_d but the magnetic field causes a downward deflection

$$F = q\hat{x}E_0 - q\hat{y}v_d B_0 \tag{3.44}$$

In the steady state, the movement of charge down (or electrons up) creates an internal electric field which must balance the downward pull, as shown in Fig. 3.12. We expect a "Hall" voltage to develop across the top and bottom faces of the conducting bar

$$V_H = E_y d = v_d B_0 d \tag{3.45}$$

Since $v_d = \mu E_x$, and $J_x = \sigma E_x$, we can write $v_d = \mu J_x / \sigma$

$$V_H = \frac{\mu J_x}{\sigma} B_0 d \tag{3.46}$$

Since the conductivity of a material is given by $\sigma = q N \mu$, where q is the unit charge, N is the density of mobile charge carriers, and μ is the mobility of the carriers. Then

$$V_H = \frac{J_x B_0 d}{q N} \tag{3.47}$$

or

$$N = \frac{J_x B_0 d}{q V_H} = \frac{I B_0 d}{Aq V_H} \tag{3.48}$$

Notice that all the quantities on the right-hand side are either known or easily measured. Thus the density of carriers can be measured indirectly through measuring the Hall voltage.

3.3   Diffusion

The process of diffusion is quite intuitive and well known. In Fig. 3.13, imagine that we fill the left chamber with a gas at temperate T. If we suddenly remove the divider, what happens? The gas will fill the entire volume of the new chamber.

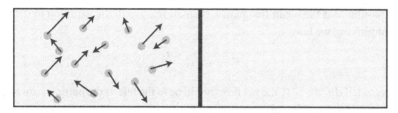

Figure 3.13 A chamber is separated into two sections by a partition. In the left section there is a gas at temperature T. If the partition is removed, we know by experience that the gas on the left will fill the entire chamber.

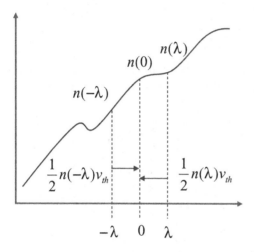

Figure 3.14 A concentration variation of free particles results in a flux proportional to the gradient of the concentration.

Whenever there exists a concentration gradient, we know from experience that over time the region of high concentration and low concentration balance as particles from the high concentration region move to the region of low concentration. How does this actually happen so naturally?

If each particle moves on average left or right, then eventually half will be in the right chamber. If the molecules were charged (or electrons), then there would be a net current flow. The diffusion current flows from high concentration to low concentration.

Let us quantify the diffusion current. Assume that the mean free path is λ. Find the flux of carriers crossing the $x = 0$ plane in Fig. 3.14

$$F = \frac{1}{2} v_{th}(n(-\lambda) - n(\lambda)) \tag{3.49}$$

where we assume that particles are moving with velocity v_{th}. The factor of $1/2$ accounts for the fact that half of the particles will be moving to the right into the $x = 0$ plane. If there are more particles at $x = -\lambda$ than $x = +\lambda$, then a net flux of particles will cross the $x = 0$ plane to the right. Expanding $n(x)$ about $x = 0$, we have

$$F = \frac{1}{2} v_{th} \left(\left[n(0) - \lambda \frac{dn}{dx} \right] - \left[n(0) + \lambda \frac{dn}{dx} \right] \right) \tag{3.50}$$

where we assume that the mean free path λ is small relative to the variation in the concentration. Simplifying we have

$$F = -v_{th}\lambda \frac{dn}{dx} \tag{3.51}$$

As expected, if $dn/dx > 0$, the net flux should be to the left. If the particles are electrons, then the current produced by this flux is given by

$$J = -qF = qv_{th}\lambda \frac{dn}{dx} \tag{3.52}$$

Since the thermal energy per degree of freedom is given by $kT/2$, we have

$$\frac{1}{2}m_n^* v_{th}^2 = \frac{1}{2}kT \tag{3.53}$$

The mean free path is related to the mean free time, or the average time between collisions, by $\lambda = v_{th}\tau_c$. We are also careful to use the effective mass m_n^* for electrons in a crystal. Manipulating our equations a bit

$$v_{th}\lambda = v_{th}^2\tau_c = kT\frac{\tau_c}{m_n^*} = \frac{kT}{q}\left(\frac{q\tau_c}{m_n^*}\right) \tag{3.54}$$

The parenthetical term is just the mobility for free carriers

$$J = qv_{th}\lambda \frac{dn}{dx} = q\left(\frac{kT}{q}\mu_n\right)\frac{dn}{dx} = qD_n\frac{dn}{dx} \tag{3.55}$$

The constant D_n is known as the diffusion constant and we see that it is related to the mobility. This relationship is known as the Einstein relation

$$D_n = \left(\frac{kT}{q}\right)\mu_n \tag{3.56}$$

which you will never forget, because it rhymes like a simple poem, *kay-tee over q equals dee over mu.*

3.4 Thermal noise

With all the random motion in a typical conductor, you would not be surprised to hear that every resistor is a source of noise. This noise, often referred to as Johnson noise, Nyquist noise, or simply thermal noise, is present in every conductor and linearly proportional to the temperature T. The key point about the noise is that it has zero average value but non-zero RMS, with a noise voltage as shown in Fig. 3.15a. At any given instant, if we sample the noise voltage, it has a Gaussian amplitude distribution. The mathematical description of such a function is not an easy task. We will avoid many of the complicated issues and just work with RMS values.

Johnson was the first to observe this noise experimentally. Nyquist was probably the first to explain it. We will also demonstrate that this form of noise is consistent with the theory of thermodynamics.

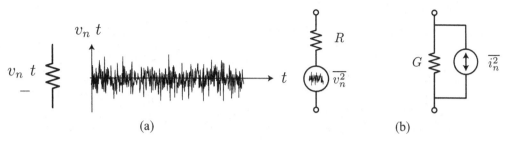

Figure 3.15 (a) The noise voltage of a resistor as a function of time. (b) Equivalent circuit representation as a noiseless resistor in series or shunt with a noise voltage or current.

The average value of the noise waveform is zero

$$\overline{v_n}(t) = <v_n(t)> = \frac{1}{T}\int_T v_n(t)dt = 0 \tag{3.57}$$

The mean is also zero if we freeze time and take an infinite number of samples from identical resistors. The variance, though, is non-zero. Equivalently, we may say that the signal power is non-zero

$$\overline{v_n(t)^2} = \frac{1}{T}\int_T v_n^2(t)dt \neq 0 \tag{3.58}$$

The RMS (root-mean-square) voltage is given by

$$v_{n,rms} = \sqrt{\overline{v_n(t)^2}} \tag{3.59}$$

The power spectrum of the noise shows the concentration of noise power at any given frequency. Many noise sources are "white" in that the spectrum is flat (up to extremely high frequencies). In such cases the noise waveform is totally unpredictable as a function of time. In other words, there is absolutely no correlation between the noise waveform at time t_1 and some later time $t_1 + \delta t$, no matter how small we make δt.

As shown in Fig. 3.15b, we can represent the noise as a voltage source in series with the resistor of the RMS value given by

$$\overline{v_R^2} = 4kTRB \tag{3.60}$$

where B is the bandwidth of observation. Equivalently, we can represent this with a shunt current source

$$\overline{i_n^2} = 4kTGB \tag{3.61}$$

Here B is the bandwidth of observation and kT is Boltzmann's constant times the temperature of observation. This result comes from thermodynamic considerations, thus explaining the appearance of kT. Often we speak of the "spot noise," or the noise in a specific narrow-band δf

$$\overline{v_n^2} = 4kTR\delta f \tag{3.62}$$

Since the noise is white, the shape of the noise spectrum is determined by the external elements (Ls and Cs).

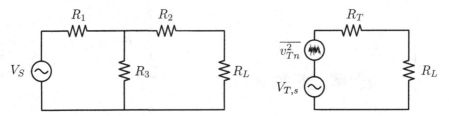

Figure 3.16 Any arbitrary resistive circuit can be reduced to a single equivalent noise source.

Example 4

Suppose that $R = 10k\Omega$ and $T = 20°C = 293K$.

$$4kT = 1.62 \times 10^{-20} \tag{3.63}$$

$$\overline{v_n^2} = 1.62 \times 10^{-16} \times B \tag{3.64}$$

$$v_{n,rms} = \sqrt{\overline{v_n(t)^2}} = 1.27 \times 10^{-8}\sqrt{B} \tag{3.65}$$

If we limit the bandwidth of observation to $B = 10^6 \text{MHz}$, then we have

$$v_{n,rms} \approx 13\mu V \tag{3.66}$$

This represents the limit for the smallest voltage we can resolve across this resistor in this bandwidth.

If we put two resistors in series, then the mean square noise voltage is given by

$$\overline{v_n^2} = 4kT(R_1 + R_2)B = \overline{v_{n1}^2} + \overline{v_{n2}^2} \tag{3.67}$$

We add noise powers, *not* the noise voltages. Likewise, for two resistors in parallel, we can add the mean square currents

$$\overline{i_n^2} = 4kT(G_1 + G_2)B = \overline{i_{n1}^2} + \overline{i_{n2}^2} \tag{3.68}$$

This holds for any pair of independent noise sources (zero correlation).

For an arbitrary resistive circuit, such as Fig. 3.16, we can find the equivalent noise by using a Thevenin (Norton) equivalent circuit or by transforming all noise sources to the output by the appropriate *power* gain (e.g. voltage squared or current squared)

$$V_{T,s} = V_S \frac{R_3}{R_1 + R_3} \tag{3.69}$$

$$\overline{v_{Tn}^2} = 4kT R_T B = 4kT(R_2 + R_1 \| R_3)B \tag{3.70}$$

For a general linear circuit, such as the black box shown in Fig. 3.17, the mean square noise voltage (current) at any port is given by the equivalent input resistance (conductance)

$$\overline{v_{eq}^2} = 4kT\Re(Z(f))\delta f \tag{3.71}$$

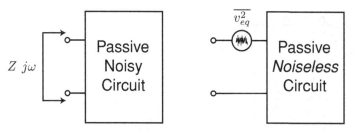

Figure 3.17 The equivalent noise of an arbitrary linear one-port circuit.

This is the "spot" noise. If the network has a filtering property, then we integrate over the band of interest

$$\overline{v_{T,eq}^2} = 4kT \int_B \Re(Z(f))df \tag{3.72}$$

Unlike resistors, ideal (lossless) Ls and Cs *do not* generate noise. They do shape the noise due to their frequency dependence.

Example 5

Noise of an RC circuit

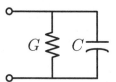

Figure 3.18 An example RC circuit for calculation of the total thermal noise.

To find the equivalent mean square noise voltage of an RC circuit, begin by calculating the impedance

$$Z = \frac{1}{Y} = \frac{1}{G + j\omega C} = \frac{G - j\omega C}{G^2 + \omega^2 C^2} \tag{3.73}$$

Integrating the noise over all frequencies, we have

$$\overline{v_n^2} = \frac{4kT}{2\pi} \int_0^\infty \frac{G}{G^2 + \omega^2 C^2} d\omega = \frac{kT}{C} \tag{3.74}$$

Notice the result is *independent* of R. Since the noise/bandwidth is proportional/inversely proportional to R, its influence cancels out.

Thermodynamic origin of noise

As shown in Fig. 3.19, imagine a black body held at temperature T. Inside this body we insert an antenna with characteristic impedance $Z_0 = R$ and through a small opening, we

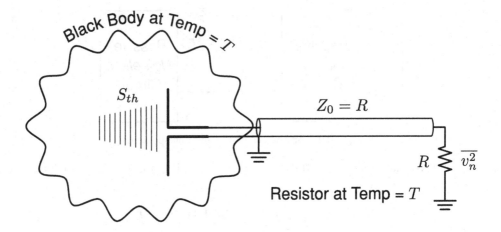

Figure 3.19 A black body held at temperature T exchanges thermal energy with a termination resistor R.

route a transmission line to a termination resistor R, also held at a temperature T, with a small opening where we insert a transmission line.

From thermodynamic considerations, we have the black body radiation density

$$S = \frac{2kT}{\lambda^2} \tag{3.75}$$

Our antenna will capture a fraction of this energy $A_e S$, where $A_e = \frac{\lambda^2}{4\pi}$ is the antenna effective cross. If we integrate over the solid angle, we have

$$E_L = \frac{1}{2} S A_e 4\pi \tag{3.76}$$

where only half of the energy is captured by the antenna due to the random polarization of the incoming radiation. Substituting in the above equation, we have the important results

$$E_L = \frac{1}{2} \frac{2kT}{\lambda^2} \frac{\lambda^2}{4\pi} 4\pi = kT \tag{3.77}$$

If we observe a bandwidth B, we capture an amount of power kTB. So we see that the resistor captures kTB Watts of power from the black body. But since the black body is in thermodynamic equilibrium with the resistor, the resistor must likewise deliver the same power back to the black body!

Now let us imagine that the resistor is a noiseless resistor in series with a voltage source. Under matched conditions, the RMS power delivered to a load is given by

$$P_L = \frac{\left(\frac{v_n}{2}\right)^2}{R} = \frac{\overline{v^2}}{4R} \tag{3.78}$$

If we equate this with the power delivered by a noise resistor

$$kT\delta f = \frac{\overline{v^2}}{4R} \tag{3.79}$$

or

$$\overline{v^2} = 4kTR\delta f \tag{3.80}$$

which is in fact the physically observed amount of noise generated by a resistor.

3.5   References

A lot of this material was put together while teaching a device physics undergraduate course at UC Berkeley and thus some of the material is drawn from lecture notes put together by Howe and Sodini [27]. The discussion on the conductivity of a gas and metal borrows heavily from Purcell [49]. Purcell's book on electromagnetics is wonderfully good and very accessible. The presentation from thermodynamic origin of noise comes from Hagen's excellent book [26]. This book is a very good source for learning the fundamentals of RF circuits, presented for a general audience but with a lot of great insight for even the expert.

4 Ampère, Faraday, and Maxwell

4.1 Ampère: static magnetic fields

Experimental observations

Consider a pair of parallel wires carrying steady currents I_1 and I_2. Since steady currents imply zero net charge distribution, there should be no electrostatic force between these current-carrying wires. But experimentally we do observe a force which tends to be attractive if the currents are in the same direction and repulsive if the currents are in the opposite direction. This new force is in fact an electrostatic force if we consider the problem from a relativistic point of view! Even though the net charge on each current-carrying conductor is zero in a static reference frame, in a moving reference frame there is net charge density and hence a force exists.

Magnetic force

Through careful observations, Ampère demonstrated that this force can be computed using the following equation

$$dF_m = \frac{\mu_0}{4\pi} \frac{I_2 d\ell_2 \times I_1 d\ell_1 \times \hat{\mathbf{R}}}{R^2} \tag{4.1}$$

The resemblance to the Coulomb force equation is notable. Both forces fall like $1/R^2$. For steady currents, $\nabla \cdot \mathbf{J} = 0$ implies that the currents must flow in loops. Thus we can calculate the force between two loops as follows

$$F_m = \oint_{C_1} \oint_{C_2} \frac{\mu_0}{4\pi} \frac{I_2 d\ell_2 \times I_1 d\ell_1 \times \hat{\mathbf{R}}}{R^2} \tag{4.2}$$

Magnetic field

Just as in the case of electric forces, the concept of "action at a distance" is disturbing and counterintuitive. Thus we prefer to think of the current in loop C_1 generating a "field" and then we say that this field interacts with the current in loop C_2 to generate a force. Simply reordering the magnetic force equation gives

$$F_m = \oint_{C_2} I_2 d\ell_2 \times \underbrace{\frac{\mu_0}{4\pi} \oint_{C_1} \frac{I_1 d\ell_1 \times \hat{\mathbf{R}}}{R^2}}_{\mathbf{B}} \tag{4.3}$$

where we have interpreted the second term as a magnetic field $\mathbf{B}$. Loop C_2 is the source of the field and loop C_1 is the field point under consideration. The unit of $\mathbf{B}$ is the Tesla (T), where $1T = 10^4 G$, in terms of the CGS units of Gauss (G).

Units of magnetic field

The Tesla (T) and Gauss (G) are derived units. Since $F \propto I^2 \mu$, the units of μ are simply $N \cdot A^{-2}$. This is more commonly known as $H \cdot m^{-1}$. The units of the magnetic field are therefore

$$[B] = [\mu]A \cdot m \cdot m^{-2} = H \cdot A \cdot m^{-2} \tag{4.4}$$

Note that the units of D are $C \cdot m^{-2}$, which can be written as $F \cdot V \cdot m^{-2}$. From circuit theory we know that voltage is proportional to $\omega L I$, so $L I$ has units of $\frac{V}{\omega}$. So the unit of $[B]$ is $V \cdot s \cdot m^{-2}$. For reference, the magnetic field of the earth is only .5G, so 1T is a very large field.

Direction of magnetic force

Due to the vector cross product, the direction of the force of the magnetic field is perpendicular to the direction of motion and the magnetic field. Use the right-hand rule to figure out the direction of $\mathbf{F}_m$ in any given situation.

E and B duality

For a point charge dq, the electric force is given by

$$\mathbf{F}_e = q\mathbf{E} \tag{4.5}$$

For the magnetic force for a point charge in a current loop, we have

$$\mathbf{F}_m = Id\ell \times \mathbf{B} = qNd\ell\mathbf{v} \times \mathbf{B} \tag{4.6}$$

The equations for $\mathbf{E}$ and $\mathbf{B}$ are also similar when we consider an arbitrary current density J and charge density ρ

$$\mathbf{E} = \frac{1}{4\pi\epsilon} \int_V \frac{\rho(r')\mathbf{R}}{R^2} dV' \tag{4.7}$$

$$\mathbf{B} = \frac{1}{4\pi\mu^{-1}} \int_V \frac{\mathbf{J}(\mathbf{r}') \times \hat{\mathbf{R}}}{R^2} dV' \tag{4.8}$$

Magnetic charge

We may now compare the magnetic field to the electric field and look for similarity and differences. In this book we shall not discuss the relativistic viewpoint that explains the link between electric and the magnetic field. Instead, we shall assume that the magnetic field is an entity of its own. Apparently, the source of magnetic field is moving charge (currents), whereas the source of electric fields is charge. But what about magnetic charges? Is there any reason to believe that nature should be asymmetric and give us electrical charge and not magnetic charge? If magnetic charge existed, then the argument for Gauss' Law would apply

$$\oint_S \mathbf{B} \cdot \mathbf{dS} = Q_m \tag{4.9}$$

where Q_m is the amount of magnetic charge inside the volume V bounded by surface S. But no one has ever observed any magnetic charge! So, for all practical purposes, we can assume that $Q_m \equiv 0$ and so Gauss' Law applied to magnetic fields yields

$$\oint_S \mathbf{B} \cdot \mathbf{dS} = 0 \tag{4.10}$$

Divergence of B

By the Divergence Theorem, locally this relation translates into

$$\oint_S \mathbf{B} \cdot \mathbf{dS} = \int_V \nabla \cdot \mathbf{B} dV = 0 \tag{4.11}$$

Since this is true for any surface S, the integrand must be identically zero

$$\nabla \cdot \mathbf{B} = 0 \tag{4.12}$$

A vector field with zero divergence is known as a solenoidal field. We have already encountered such a field since $\nabla \cdot \mathbf{J} = 0$. Such a field does not have any sources and thus always curls back onto itself. **B** fields are thus always loops.

Divergence of curl

Let's calculate the divergence of the curl of an arbitrary vector field **A**, $\nabla \cdot \nabla \times \mathbf{A}$. Let's compute the volume of the above quantity and apply the Divergence Theorem

$$\int_V \nabla \cdot \nabla \times \mathbf{A} dV = \oint_S \nabla \times \mathbf{A} \cdot \mathbf{dS} \tag{4.13}$$

To compute the surface integral, consider a new surface S' with a hole in it. The surface integral of $\nabla \times \mathbf{A}$ can be written as the line integral using Stoke's Theorem

$$\int_{S'} \nabla \times \mathbf{A} \cdot \mathbf{dS} = \oint_C \mathbf{A} \cdot \mathbf{dl} \tag{4.14}$$

where C is the perimeter of the hole. As we shrink this hole to a point, the right-hand side goes to zero and the surface integral turns into the closed-surface integral. Thus

$$\int_V \nabla \cdot \nabla \times \mathbf{A} dV = \oint_S \nabla \times \mathbf{A} \cdot \mathbf{dS} = 0 \tag{4.15}$$

Since this is true for any volume V, it must be that

$$\nabla \cdot \nabla \times \mathbf{A} = 0 \tag{4.16}$$

Thus a solenoidal vector can always be written as the curl of another vector. Thus the magnetic field **B** can be written as

$$\mathbf{B} = \nabla \times \mathbf{A} \tag{4.17}$$

Ampère's Law

One of the fundamental relations for the magnetic field is Ampère's Law. It is analogous to Gauss' Law. We can derive it by taking the curl of the magnetic field

$$\nabla \times \mathbf{B} = \nabla \times \frac{1}{4\pi \mu^{-1}} \int_V \frac{\mathbf{J}(\mathbf{r}') \times \hat{\mathbf{R}}}{R^2} dV' \tag{4.18}$$

After some *painful* manipulations, this can be simplified to the famous Ampère's Law

$$\nabla \times \mathbf{B} = \mu \mathbf{J} \tag{4.19}$$

Now apply Stoke's Theorem

$$\int_S \nabla \times \mathbf{B} \cdot d\mathbf{S} = \oint_C \mathbf{B} \cdot d\mathbf{l} = \int_S \mu \mathbf{J} \cdot d\mathbf{S} = \mu I \tag{4.20}$$

Application of Ampère's Law

Ampère's Law is very handy in situations involving cylindrical symmetry. This is analogous to applying Gauss' Law to problems with spherical symmetry. For example, consider the magnetic field due to a long wire. The field should have no r or z dependence (by symmetry) so the integral of B over a circle enclosing the wire is simply a constant times the perimeter

$$\oint_C \mathbf{B} \cdot d\mathbf{l} = 2\pi r B_r = \mu I \tag{4.21}$$

So the magnetic field drops as $1/r$

$$B(r) = \frac{\mu I}{2\pi r} \tag{4.22}$$

Example 6

Consider the magnetic field of a two-dimensional round wire of radius a carrying a uniform current I shown in Fig. 4.1a. From symmetry we deduce that the magnetic field is circumferential. Invoking Ampère's Law, the closed line integral of the field around the wire is equal to the current enclosed by the loop. Let's take a loop inside the wire so that $r < a$

$$\oint_C \mathbf{B} \cdot d\mathbf{l} = \int_0^{2\pi} B_\theta d\theta = 2\pi r B_\theta(r) = \mu I \frac{\pi r^2}{\pi a^2} \tag{4.23}$$

since $B(r) = B_\theta(r)$ is a constant as a function of θ, and a circle of radius $r < a$ encloses a fraction r^2/a^2 of the surface area of the wire. Hence the magnetic field

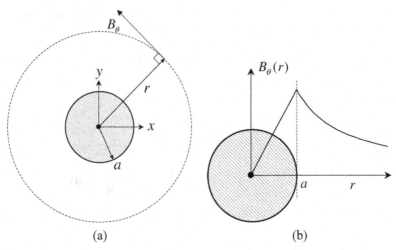

Figure 4.1 (a) A two-dimensional round wire carrying a current I out of the surface of the page. (b) The radial dependence of the magnetic field inside and outside of the wire.

inside the wire increases linearly as a function of r

$$B_\theta(r) = \frac{\mu I r}{2\pi a^2} \tag{4.24}$$

Outside the wire, $r > a$, we enclose the entire current and so

$$B_\theta(r) = \frac{\mu I}{2\pi r} \tag{4.25}$$

The functional dependence of $B_\theta(r)$ is plotted in Fig. 4.1b.

Example 7

Consider the magnetic field between two infinitely long wires carrying oppositely directed currents. The geometry is shown in Fig. 4.2a where a positive current coming out of the plane of the page is shown at $x = -d$ and a negative current into the plane of the page is at $x = +d$. We will solve this problem by invoking symmetry and superposition. From symmetry it is clear that the magnetic field of a single filament is circularly directed. From Ampère's Law, the closed line integral of the field around the wire is equal to the current enclosed by the loop

$$\oint_C \mathbf{B} \cdot \mathbf{dl} = \int_0^{2\pi} B_\theta d\theta = 2\pi r B_\theta(r) = \mu I \tag{4.26}$$

since $B(r) = B_\theta(r)$ is a constant as a function of θ. Thus we have the result that

$$B_\theta = \frac{\mu I}{2\pi r} \tag{4.27}$$

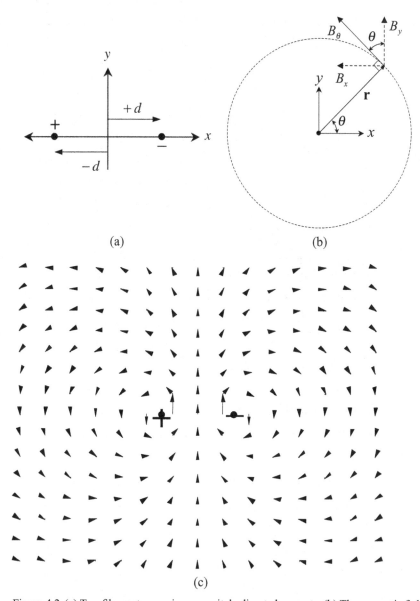

(a)   (b)

(c)

Figure 4.2 (a) Two filaments carrying oppositely directed currents. (b) The magnetic field of a single filament. (c) The magnetic field of the two filament system.

Writing the above result in Cartesian coordinates (see Fig. 4.2b for the geometry) we have

$$B_x = -B_\theta \sin\theta = \frac{-\mu I y}{2\pi(x^2 + y^2)}$$

$$B_y = B_\theta \cos\theta = \frac{\mu I x}{2\pi(x^2 + y^2)} \tag{4.28}$$

By superposition the field for two filaments is simply the vectorial summation of the field of the individual filaments displaced to $x = d$ and $x = -d$

$$B_x = \frac{-\mu I y}{2\pi \left((x+d)^2 + y^2\right)} + \frac{\mu I y}{2\pi \left((x-d)^2 + y^2\right)}$$

$$B_y = \frac{-\mu I (x+d)}{2\pi \left((x+d)^2 + y^2\right)} - \frac{\mu I (x-d)}{2\pi \left((x-d)^2 + y^2\right)} \tag{4.29}$$

This field is plotted in Fig. 4.2c using the following commands in *Mathematica*

```
In[1]:= Bx = -y/((x+1)^2+y^2) + y/((x-1)^2+y^2)
In[2]:= By = (x+1)/((x+1)^2+y^2) - (x-1)/((x-1)^2+y^2)
In[3]:= B = {Bx,By};
In[4]:= <<Graphics'PlotField'
In[5]:= bfield = PlotVectorField[B,{x,-5,5},{y,-4,4}];
In[6]:= cur1 = Point[{-1,0}];
In[7]:= cur2 = Point[{1,0}];
In[8]:= cur = {cur1,cur2};
In[9]:= Show[bfield,Graphics[{PointSize[.02],cur}]]   }
```

Magnetic vector potential

Earlier we showed that we can also define a vector $\mathbf{A}$ such that $\mathbf{B} = \nabla \times \mathbf{A}$. Since $\nabla \times \mathbf{B} = \mu \mathbf{J}$

$$\nabla \times \nabla \times \mathbf{A} = \mu \mathbf{J} \tag{4.30}$$

We can apply the vector identity for $\nabla \times \nabla \times \mathbf{A}$

$$\nabla \times \nabla \times \mathbf{A} = \nabla (\nabla \cdot \mathbf{A}) - \nabla^2 \mathbf{A} \tag{4.31}$$

Given that the divergence of $\mathbf{A}$ is arbitrary, let us choose the most convenient value. In magnetostatics that is $\nabla \cdot \mathbf{A} = 0$. Then we have

$$\nabla^2 \mathbf{A} = -\mu \mathbf{J} \tag{4.32}$$

We have met this equation before!

Equations for potential

The vector Laplacian can be written as three scalar Laplacian equations (using rectangular coordinates). For instance, the x-component is given by

$$\nabla^2 A_x = -\mu J_x \tag{4.33}$$

By analogy with the scalar potential, therefore, the solution is given by

$$A_x = \frac{\mu}{4\pi} \int_V \frac{J_x(\mathbf{r}')dV'}{|\mathbf{r} - \mathbf{r}'|} \tag{4.34}$$

and, in general, the total vector potential is given by

$$\mathbf{A} = \frac{\mu}{4\pi} \int_V \frac{\mathbf{J}(\mathbf{r}')dV'}{|\mathbf{r} - \mathbf{r}'|} \tag{4.35}$$

Why use vector potential?

The vector potential $\mathbf{A}$ is easier to calculate than $\mathbf{B}$ since each component is a simple scalar calculation. Also, the direction of $\mathbf{A}$ is easy to determine since it follows $\mathbf{J}$. At this point the vector potential seems like a mathematical creation and it does not seem to have any physical relevance. This is compounded by the fact that $\nabla \cdot \mathbf{A}$ is arbitrary. Later on, we will see that $\mathbf{A}$ has a lot of physical relevance and in some ways it is more fundamental than the vector $\mathbf{B}$.

From vector A to B

Now that we have an equation for $\mathbf{A}$, we can verify that it is indeed consistent with the experimentally observed equation for $\mathbf{B}$

$$\nabla \times \mathbf{A} = \nabla \times \frac{\mu}{4\pi} \int_V \frac{\mathbf{J}(\mathbf{r}')dV'}{|\mathbf{r} - \mathbf{r}'|} \tag{4.36}$$

Interchanging the order of integration and differentiation

$$\nabla \times \mathbf{A} = \frac{\mu}{4\pi} \int_V \nabla \times \frac{\mathbf{J}(\mathbf{r}')}{|\mathbf{r} - \mathbf{r}'|} dV' \tag{4.37}$$

We now need some fancy footwork to go further.

Yet another vector identify

It is relatively easy to show that for a scalar function ϕ times a vector field $\mathbf{F}$

$$\nabla \times \phi\mathbf{F} = \nabla\phi \times \mathbf{F} + \phi\nabla \times \mathbf{F} \tag{4.38}$$

Applying this to our case, note that ∇ operates on the coordinates $\mathbf{r}$ whereas $\mathbf{J}(\mathbf{r}')$ is a function of the primed coordinates, and hence a constant

$$\nabla \times \frac{\mathbf{J}(\mathbf{r}')}{|\mathbf{r} - \mathbf{r}'|} = \nabla\frac{1}{|\mathbf{r} - \mathbf{r}'|} \times \mathbf{J}(\mathbf{r}') = -\frac{\hat{\mathbf{R}}}{|\mathbf{r} - \mathbf{r}'|^2} \times \mathbf{J}(\mathbf{r}') \tag{4.39}$$

where we have used

$$\nabla\frac{1}{|\mathbf{r} - \mathbf{r}'|} = \frac{\hat{\mathbf{R}}}{|\mathbf{r} - \mathbf{r}'|^2} \tag{4.40}$$

The expression for B matches up with the experimentally observed equation

$$\mathbf{B} = \nabla \times \mathbf{A} = \frac{\mu}{4\pi} \int_V \frac{\mathbf{J}(\mathbf{r}') \times \hat{\mathbf{R}}}{R^2} dV' \tag{4.41}$$

4.2 Magnetic materials

Magnetization vector

We would like to study magnetic fields in magnetic materials. Let's define the magnetization vector as

$$\mathbf{M} = \lim_{\Delta V \to 0} \frac{\sum_k \mathbf{m}_k}{\Delta V} \tag{4.42}$$

where $\mathbf{m}_k$ is the magnetic dipole of an atom or molecule. The vector potential due to these magnetic dipoles in a differential volume dv' is given by

$$d\mathbf{A} = \mu_0 \frac{\mathbf{M} \times \hat{\mathbf{r}}}{4\pi R^2} dv' \tag{4.43}$$

so

$$\mathbf{A} = \frac{\mu_0}{4\pi} \int_{V'} \frac{\mathbf{M} \times \hat{\mathbf{r}}}{R^2} dv' \tag{4.44}$$

Using

$$\nabla' \left(\frac{1}{R} \right) = \frac{\hat{\mathbf{r}}}{R^2} \tag{4.45}$$

$$\mathbf{A} = \frac{\mu_0}{4\pi} \int_{V'} \mathbf{M} \times \nabla' \left(\frac{1}{R} \right) dv' \tag{4.46}$$

Consider the vector identity

$$\nabla' \times \left(\frac{\mathbf{M}}{R} \right) = \frac{1}{R} \nabla' \times \mathbf{M} + \nabla' \left(\frac{1}{R} \right) \times \mathbf{M} \tag{4.47}$$

We can thus break the vector potential into two terms

$$\mathbf{A} = \frac{\mu_0}{4\pi} \int_{V'} \frac{\nabla' \times \mathbf{M}}{R} dv' - \frac{\mu_0}{4\pi} \int_{V'} \nabla' \times \left(\frac{\mathbf{M}}{R} \right) dv' \tag{4.48}$$

Another Divergence Theorem

Consider the vector $\mathbf{u} = \mathbf{a} \times \mathbf{v}$, where $\mathbf{a}$ is an arbitrary constant. Then

$$\nabla \cdot \mathbf{u} = \nabla \cdot (\mathbf{a} \times \mathbf{v}) = (\nabla \times \mathbf{a}) \cdot \mathbf{v} - (\nabla \times \mathbf{v}) \cdot \mathbf{a} = -(\nabla \times \mathbf{v}) \cdot \mathbf{a} \tag{4.49}$$

Now apply the Divergence Theorem to $\nabla \cdot \mathbf{u}$

$$\int_V -(\nabla \times \mathbf{v}) \cdot \mathbf{a} \, dV = \oint_S (\mathbf{a} \times \mathbf{v}) \cdot d\mathbf{S} = \oint_S ((\mathbf{a} \times \mathbf{v}) \cdot \hat{\mathbf{n}}) \, dS \tag{4.50}$$

where $\hat{\mathbf{n}}$ is a unit vector perpendicular to the surface. Reordering the vector triple product

$$\oint_S (\mathbf{a} \cdot \mathbf{v} \times \hat{\mathbf{n}}) \, dS \tag{4.51}$$

Since the vector $\mathbf{a}$ is constant, we can pull it out of the integrals

$$\mathbf{a} \cdot \int_V (-\nabla \times \mathbf{v}) \, dV = \mathbf{a} \cdot \oint_S \mathbf{v} \times \hat{\mathbf{n}} \, dS \tag{4.52}$$

The vector $\mathbf{a}$ is arbitrary, so we have

$$\int_V (\nabla \times \mathbf{v}) \, dV = -\oint_S \mathbf{v} \times \hat{\mathbf{n}} \, dS \tag{4.53}$$

Applying this to the second term of the vector potential

$$\int_{V'} \nabla' \times \left(\frac{\mathbf{M}}{R}\right) dv' = -\oint_S \frac{\mathbf{M} \times \hat{\mathbf{n}}}{R} \, dS \tag{4.54}$$

Vector potential due to magnetization

The vector potential due to magnetization has a volume component and a surface component

$$\mathbf{A} = \frac{\mu_0}{4\pi} \int_{V'} \frac{\nabla' \times \mathbf{M}}{R} \, dv' + \frac{\mu_0}{4\pi} \oint_S \frac{\mathbf{M} \times \hat{\mathbf{n}}}{R} \cdot \mathbf{dS} \, dv' \tag{4.55}$$

We can thus define an equivalent magnetic volume current density

$$\mathbf{J}_m = \nabla \times \mathbf{M} \tag{4.56}$$

and an equivalent magnetic surface current density

$$\mathbf{J}_s = \mathbf{M} \times \hat{\mathbf{n}} \tag{4.57}$$

Volume and surface currents

In Fig. 4.3, we can see that for uniform magnetization, all the internal currents cancel and only the magnetization vector on the boundary (surface) contributes to the integral.

Relative permeability

We can include the effects of materials on the macroscopic magnetic field by including a volume current $\nabla \times \mathbf{M}$ in Ampère's equation

$$\nabla \times \mathbf{B} = \mu_0 (\mathbf{J} + \nabla \times \mathbf{M}) \tag{4.58}$$

or

$$\nabla \times \left(\frac{\mathbf{B}}{\mu_0} - \mathbf{M}\right) = \mathbf{J} \tag{4.59}$$

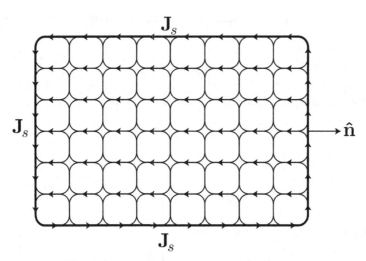

Figure 4.3 The magnetization current flowing in a uniform conductor.

We thus have defined a new quantity **H**

$$\mathbf{H} = \frac{\mathbf{B}}{\mu_0} - \mathbf{M} \tag{4.60}$$

The units of **H**, the *magnetic field*, are A/m.

Ampère's equation for media

We can thus state that for any medium under static conditions

$$\nabla \times \mathbf{H} = \mathbf{J} \tag{4.61}$$

equivalently

$$\oint_C \mathbf{H} \cdot \mathbf{dl} = I \tag{4.62}$$

Since linear materials respond to the external field in a linear fashion

$$\mathbf{M} = \chi_m \mathbf{H} \tag{4.63}$$

so

$$\mathbf{B} = \mu(1 + \chi_m)\mathbf{H} = \mu\mathbf{H} \tag{4.64}$$

or

$$\mathbf{H} = \frac{1}{\mu}\mathbf{B} \tag{4.65}$$

The table in Fig. 4.4 can serve as a memory aid as it shows the duality between the electric and magnetic field.[1]

---

[1] I personally don't like this choice since to me **E** and **B** are "real" and so the equations should be arranged to magnify this analogy. Unfortunately, the equations are not organized this way (partly due to choice of units) so we'll stick with convention.

E	H	ρ	J
D	B	V	A
ϵ	μ^{-1}	$\cdot$	$\times$
P	M	$\times$	$\cdot$

Figure 4.4 The duality between electric and magnetic fields.

Magnetic materials

Magnetic materials are classified as follows:

- Diamagnetic: $\mu_r \leq 1$, usually χ_m is a small negative number
- Paramagnetic: $\mu_r \geq 1$, usually χ_m is a small positive number
- Ferromagnetic: $\mu_r \gg 1$, thus χ_m is a large positive number

Most materials in nature are diamagnetic. To fully understand the magnetic behavior of materials requires a detailed study (and quantum mechanics). In this book we mostly assume $\mu \approx \mu_0$.

The magnetic flux density **B** is related to the magnetic field **H** by

$$\mathbf{B} = \mu\mathbf{H} = \mu_r\mu_0\mathbf{H}$$
$$\mu_0 = 4\pi \times 10^{-7}\,\text{H/m}$$

The relative permeability μ_r measures the effect of constituent atomic and/or molecular magnetic dipole moments. Most materials in nature are diamagnetic. The induced magnetic fields *oppose* the applied field. But the response is usually very weak and so $\mu \sim 1$. This is due primarily to the response of the electron "orbit" in an atom.

Paramagnetic materials

Some materials have a natural net magnetic dipole moment. Such materials give a paramagnetic response, which can arise from dipole moments in an atom, molecules, crystal defects, and conduction electrons. The dipole moments tend to align with the magnetic field but are deflected from complete alignment by their thermal activity. The fields resulting from the partial alignment *adds* to the applied field so $\mu \geq 1$. It can be shown that this effect is also relatively weak so that at room temperature $\mu \approx 1 + 10^{-5}$ [27].

Ferromagnetics and ferrimagnetics

Materials with residual magnetization exist where it is energetically favorable for internal magnetic dipoles to align spontaneously below a certain (Curie) temperature. These *ferromagnetic* and *ferrimagnetic* materials exhibit non-linear and large μ factors $\sim 10^3 - 10^6$. As shown in Fig. 4.5, these magnetic moments align in response to the external field. Note that it is energetically favorable to align in the direction of the field, thus enhancing the magnetic field. In diamagnetic materials, by contrast, the effect of the material is to reduce the field.

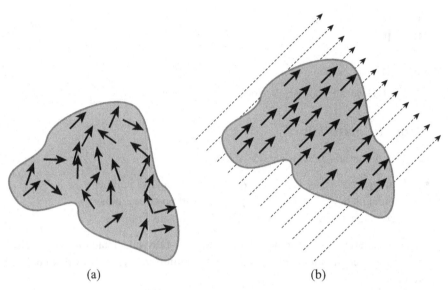

(a)   (b)

Figure 4.5 For ferromagnetic materials, (a) in the absence of an external field, the magnetic domains point in random directions. (b) The application of the field aligns the magnetic dipoles.

Boundary conditions for a magnetic field

We have now established the following equations for a static magnetic field

$$\nabla \times \mathbf{H} = \mathbf{J} \tag{4.66}$$

$$\nabla \cdot \mathbf{B} = 0 \tag{4.67}$$

$$\oint_C \mathbf{H} \cdot \mathbf{dl} = \int_S \mathbf{J} \cdot \mathbf{dS} = I \tag{4.68}$$

$$\oint_S \mathbf{B} \cdot \mathbf{dS} = 0 \tag{4.69}$$

And for linear materials, we find that $\mathbf{H} = \mu^{-1}\mathbf{B}$

Tangential H

The appropriate boundary conditions follow immediately from our previously established techniques. As shown in Fig. 4.6a, take a small loop intersecting with the boundary and take the limit as the loop becomes tiny

$$\int_C \mathbf{H} \cdot \mathbf{dl} = (H_{t1} - H_{t2})d\ell = 0 \tag{4.70}$$

So the tangential component of H is continuous

$$H_{t1} = H_{t2} \tag{4.71}$$

$$\mu_1^{-1} B_{t1} = \mu_2^{-1} B_{t2} \tag{4.72}$$

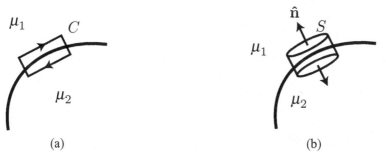

Figure 4.6 The setup for the calculation of the boundary condition for the B field.

Note that B is discontinuous because there is an effective surface current due to the change in permeability. Since B is "real," it reflects this change. If, in addition, a surface current is flowing in between the regions, then we need to include it in the above calculation.

Normal B

Consider a pillbox cylinder enclosing the boundary between the layers, as shown in Fig. 4.6b. In the limit that the pillbox becomes small, we have

$$\oint \mathbf{B} \cdot \mathbf{dS} = (B_{1n} - B_{2n})dS = 0 \tag{4.73}$$

and thus the normal component of B is continuous

$$B_{1n} = B_{2n} \tag{4.74}$$

Boundary conditions for a conductor

If a material is a very good conductor, then we will show that it can only support current at the surface of the conductor. In fact, for an ideal conductor, the current lies entirely on the surface and it is a true surface current. In such a case the current enclosed by an infinitesimal loop is finite

$$\oint_C \mathbf{H} \cdot \mathbf{dl} = (H_{t1} - H_{t2})d\ell = J_s d\ell \tag{4.75}$$
$$H_{t1} - H_{t2} = J_s \tag{4.76}$$

This can be expressed compactly as

$$\hat{\mathbf{n}} \times (\mathbf{H}_1 - \mathbf{H}_2) = \mathbf{J}_s \tag{4.77}$$

But for a perfect conductor, we will see that $H_2 = 0$, so $H_{1t} = J_s$

$$\hat{\mathbf{n}} \times \mathbf{H}_1 = \mathbf{J}_s \tag{4.78}$$

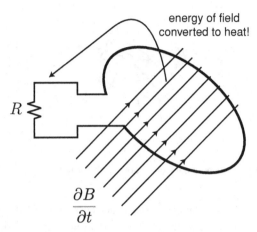

energy of field
converted to heat!

R

$\dfrac{\partial B}{\partial t}$

Figure 4.7 A changing magnetic field cuts through a loop and generates a current. The magnetic energy is converted to heat in the load resistor.

4.3 Faraday's big discovery

In electrostatics, we learned that $\oint \mathbf{E} \cdot \mathbf{dl} = 0$. Let's use the analogy between B and D (and E and H). Since $q = Cv$ and $\psi = Li$, and $i = \dot{q} = C\dot{v}$, should we not expect that $\dot{\psi} = L\dot{i} = v$? In fact, this is true! Faraday was able to show this experimentally

$$\oint_C \mathbf{E} \cdot \mathbf{dl} = -\frac{d\psi}{dt} = -\frac{d}{dt}\int_S \mathbf{B} \cdot \mathbf{dS} \tag{4.79}$$

The electric force is no longer strictly conservative, $\mathbf{E} \neq -\nabla\phi$.

Faraday's Law in differential form

Using Stoke's Theorem

$$\int_C \mathbf{E} \cdot \mathbf{dl} = \int_S \nabla \times \mathbf{E} \cdot \mathbf{dS} = -\frac{d}{dt}\int_S \mathbf{B} \cdot \mathbf{dS} = -\int_S \frac{\partial \mathbf{B}}{\partial t} \cdot \mathbf{dS} \tag{4.80}$$

Since this is true for any arbitrary curve C, this implies that

$$\nabla \times \mathbf{E} = -\frac{\partial \mathbf{B}}{\partial t} \tag{4.81}$$

Faraday's Law is true for any region of space, including free space. In particular, if C is bounded by an actual loop of wire, then the flux cutting this loop will induce a voltage around the loop. A simple experimental setup is shown in Fig. 4.7 where the magnetic energy is converted to heat.

Example 8

Transformers
A transformer is constructed as shown in Fig. 4.8. In a transformer, by definition the flux in the "primary" side is given by $\psi_1 = L_1 I_1$. Likewise, the flux crossing the

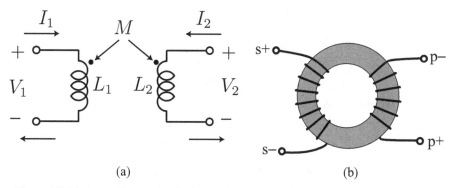

(a)   (b)

Figure 4.8 (a) A transformer circuit element is constructed using (b) two windings around a common magnetic core.

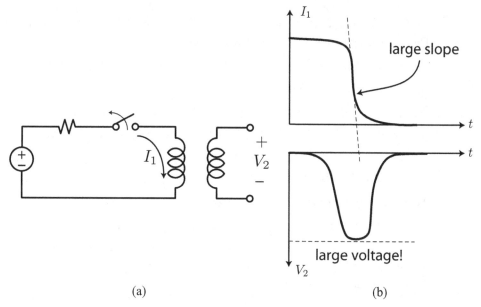

(a)   (b)

Figure 4.9 (a) A simple spark–gap circuit works by interrupting the current in the primary, thus generating a large magnetic flux change and (b) a large voltage is induced in the secondary terminals.

"secondary" side is given by $\psi_2 = M_{21} I_1 = M_{12} I_1 = M I_1$ (assuming $I_2 = 0$). Thus if the current in the primary changes, a voltage is induced in the secondary

$$V_2 = \dot{\psi}_2 = M \dot{I}_1 \qquad (4.82)$$

Generating sparks!

Since the voltage at the secondary side is proportional to the rate of change of current in loop 1, we can generate very large voltages at the secondary side by interrupting the current with a switch. This is shown schematically in Fig. 4.9. This is in fact a simplified version of how spark plugs are fired in an automobile engine.

The return of the vector potential

Since $\nabla \cdot \mathbf{B} \equiv 0$, we can write $\mathbf{B} = \nabla \times \mathbf{A}$. Thus

$$\nabla \times \mathbf{E} = -\frac{\partial \mathbf{B}}{\partial t} = -\frac{\partial \nabla \times \mathbf{A}}{\partial t} = -\frac{\partial \mathbf{A}}{\partial t} \tag{4.83}$$

If we group terms, we have

$$\left(\mathbf{E} + \frac{\partial \mathbf{A}}{\partial t} \right) = 0 \tag{4.84}$$

So, as we saw in electrostatics, we can likewise write

$$\mathbf{E} + \frac{\partial \mathbf{A}}{\partial t} = -\nabla \phi \tag{4.85}$$

We choose a negative sign for ϕ to be consistent with electrostatics. Since if $\frac{\partial}{\partial t} = 0$, this equation breaks down to the electrostatic case and then we identify ϕ as the scalar potential. This gives us some insight into the electromagnetic response as

$$\mathbf{E} = -\nabla \phi - \frac{\partial \mathbf{A}}{\partial t} \tag{4.86}$$

$$\mathbf{E} = \underbrace{-\nabla \phi}_{\text{electric response}} \underbrace{-\frac{\partial \mathbf{A}}{\partial t}}_{\text{magnetic response}} \tag{4.87}$$

Is the vector potential real?

We can now re-derive Faraday's Law as follows

$$V = \oint_C \mathbf{E} \cdot d\mathbf{l} = -\oint_C \nabla \phi \cdot d\mathbf{l} - \frac{\partial}{\partial t} \oint_C \mathbf{A} \cdot d\mathbf{l} \tag{4.88}$$

The line integral involving $\nabla \phi$ is zero by definition, so we have the induced emf equal to the line integral of $\mathbf{A}$ around the loop in question

$$V = -\frac{\partial}{\partial t} \oint_C \mathbf{A} \cdot d\mathbf{l} \tag{4.89}$$

We also found that equivalently

$$V = -\frac{\partial}{\partial t} \int_S \mathbf{B} \cdot d\mathbf{S} \tag{4.90}$$

Eq. (4.89) is somewhat more satisfying that Faraday's Law in terms of the flux. Although it is mathematically equivalent, it explicitly shows us the shape of the loop's role in determining the induced flux. The flux equation, though, depends on a surface bounding the loop, and *any* surface will do (this is discussed in more detail in the next chapter). Sometimes it is even difficult to imagine the shape of such a surface (e.g. a coil).

Consider the magnetic coupling between a solenoid and a large loop surrounding the solenoid, as shown in Fig. 4.10. For an ideal solenoid, $\mathbf{B} = 0$ outside of the cylinder. Certainly we can assume that $\mathbf{B} \approx 0$ outside of this region.

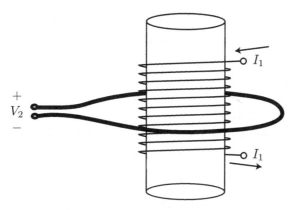

Figure 4.10 An ideal solenoid has zero magnetic field outside of the loop.

Then the voltage induced into the outer loop only depends on the constant flux generated within the center section coincident with the solenoid. What is disturbing is that, even though $\mathbf{B} = 0$ along the loop, there is a force pushing electrons inside the metal. The force would therefore appear to arise *not* from a magnetic field since $\mathbf{B} = 0$. The viewpoint with vector potential, though, does not pose any problems since $\mathbf{A} \neq 0$ outside of the loop. Therefore, when we integrate $\mathbf{A}$ outside of the loop, there is a non-zero result.

4.4 Maxwell's displacement current

Let's summarize what we have learned thus far. There are no magnetic charges, so

$$\nabla \cdot \mathbf{B} = 0 \tag{4.91}$$

and electric fields diverge on physical charge

$$\nabla \cdot \mathbf{D} = \rho \tag{4.92}$$

Faraday's Law tells us that

$$\nabla \times \mathbf{E} = -\frac{\partial B}{\partial t} \tag{4.93}$$

and Ampère's Law relates magnetic fields to currents by

$$\nabla \times \mathbf{H} = J \tag{4.94}$$

Are these equations complete and self consistent? In other words, do they over specify the problem or are some equations still missing? Furthermore, are they self consistent? Mathematics tells us that $\nabla \cdot (\nabla \times \mathbf{H}) = 0$, which implies that

$$\nabla \cdot \mathbf{J} = 0 \tag{4.95}$$

But this can only hold for steady fields. In general, by conservation of charge we know that

$$\nabla \cdot \mathbf{J} = -\frac{\partial \rho}{\partial t} \tag{4.96}$$

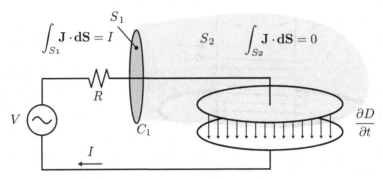

Figure 4.11 The magnetic field can be computed by surface S_1 or S_2. The surface S_1 cuts through a wire, and thus by Ampére's Law generates a magnetic field. But the surface S_2 does not cut through any DC current.

In other words, we have to add something to the right-hand side of Ampére's equation to make it self consistent! Maxwell was the first to make this observation. Since $\nabla \cdot \mathbf{D} = \rho$, it is natural to add a *displacement current* to the Ampére's equation

$$\nabla \times \mathbf{H} = J + \frac{\partial \mathbf{D}}{\partial t} \tag{4.97}$$

This now makes our equation self consistent since

$$\nabla \cdot \nabla \times \mathbf{H} = 0 = \nabla \cdot \mathbf{J} + \frac{\partial \mathbf{D}}{\partial t} = \nabla \cdot \mathbf{J} + \frac{\partial \rho}{\partial t} \tag{4.98}$$

Magnetic field of a capacitor

Now we can resolve a contradiction in Ampére's equation. If we consider the magnetic field of the circuit shown in Fig. 4.11, we know that there is a magnetic field around loop C_1 since current cuts through surface S_1. But Ampére's Law says that *any* surface bounded by C_1 can be used to calculate the magnetic field. If we use surface S_2, then the current cutting through this surface is zero, which would yield a zero magnetic field!

Displacement current of a capacitor

The answer to this contradiction is displacement current. If current is flowing in this circuit, then the electric field between the capacitor plates must be changing. Thus $\frac{\partial \mathbf{D}}{\partial t} \neq 0$. So the displacement current cutting surface S_2 must be the same as the conductive current cutting through surface S_1

$$\int_{S_1} \mathbf{J}_c \cdot d\mathbf{S} = \int_{S_2} \frac{\partial \mathbf{D}}{\partial t} \cdot d\mathbf{S} \tag{4.99}$$

Maxwell's equations

We have now studied the complete set of Maxwell's equations. In Integral form

$$\oint_C \mathbf{E} \cdot d\mathbf{l} = -\frac{d}{dt} \int_S \mathbf{B} \cdot d\mathbf{S} \tag{4.100}$$

$$\oint_C \mathbf{H} \cdot d\mathbf{l} = \frac{d}{dt} \int_S \mathbf{D} \cdot d\mathbf{S} + \int_S \mathbf{J} \cdot d\mathbf{S} \tag{4.101}$$

$$\oint_S \mathbf{D} \cdot d\mathbf{S} = \int_V \rho \, dV \tag{4.102}$$

$$\oint_S \mathbf{B} \cdot d\mathbf{S} = 0 \tag{4.103}$$

The fields are related by the following material parameters

$$\mathbf{B} = \mu \mathbf{H} \tag{4.104}$$

$$\mathbf{D} = \epsilon \mathbf{E} \tag{4.105}$$

$$\mathbf{J} = \sigma \mathbf{E} \tag{4.106}$$

For most materials we assume that these are scalar relations. In differential form

$$\nabla \times \mathbf{E} = -\frac{\partial \mathbf{B}}{\partial t} \tag{4.107}$$

$$\nabla \times \mathbf{B} = \epsilon_0 \mu_0 \frac{\partial \mathbf{E}}{\partial t} + \mu_0 \left(\mathbf{J} + \frac{\partial P}{\partial t} + \nabla \times \mathbf{M} \right) \tag{4.108}$$

$$\nabla \cdot \mathbf{E} = \frac{1}{\epsilon_0}(\rho - \nabla \cdot \mathbf{P}) \tag{4.109}$$

$$\nabla \cdot \mathbf{B} = 0 \tag{4.110}$$

Source-free regions

In source-free regions $\rho = 0$ and $J = 0$. Assume the material is uniform (no bound charges or currents)

$$\nabla \times \mathbf{E} = -\frac{\partial B}{\partial t} \tag{4.111}$$

$$\nabla \times \mathbf{B} = \epsilon_0 \mu_0 \frac{\partial E}{\partial t} \tag{4.112}$$

$$\nabla \cdot \mathbf{E} = 0 \tag{4.113}$$

$$\nabla \cdot \mathbf{B} = 0 \tag{4.114}$$

As shown in Fig. 4.12, we can intuitively see that $\frac{\partial E}{\partial t} \rightarrow \frac{\partial B}{\partial t} \rightarrow \frac{\partial E}{\partial t} \rightarrow \ldots$, in other words wave motion is possible.

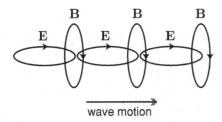

wave motion

Figure 4.12 The propagation of an electromagnetic wave in a source-free region occurs as the solenoidal electric field generates a magnetic field, and a magnetic field in turn generates an electric field.

Time-harmonic Maxwell's equation

Under time-harmonic conditions (many important practical cases are time harmonic, or nearly so, or else Fourier analysis can handle non-harmonic cases)

$$\nabla \times \mathbf{E} = -j\omega\mathbf{B} \tag{4.115}$$

$$\nabla \times \mathbf{H} = \mathbf{J} + j\omega\mathbf{D} \tag{4.116}$$

$$\nabla \cdot \mathbf{E} = \rho/\epsilon \tag{4.117}$$

$$\nabla \cdot \mathbf{B} = 0 \tag{4.118}$$

These equations are not all independent. Take the divergence of the curl, for instance

$$\nabla \cdot (\nabla \times \mathbf{E}) \equiv 0 = -j\omega\nabla \cdot \mathbf{B} \tag{4.119}$$

In other words, the non-existence of magnetic charge is built in to our curl equation. If magnetic charge is ever observed, we would have to modify our equations. This is analogous to the displacement current that Maxwell introduced to make the curl of H equation self consistent. Working with (4.116)

$$\nabla \cdot (\nabla \times \mathbf{H}) \equiv 0 = \nabla \cdot \mathbf{J} + j\omega\nabla \cdot \mathbf{D} \tag{4.120}$$

but

$$\nabla \cdot \mathbf{J} = -\frac{\partial \rho}{\partial t} = -j\omega\rho \tag{4.121}$$

This implies that $\nabla \cdot \mathbf{D} = \rho$, so Gauss' Law is built-in to our curl equations as well.

Tangential boundary conditions

The boundary conditions on the E-field at the interface of two media is

$$\hat{n} \times (\mathbf{E}_1 - \mathbf{E}_2) = 0 \tag{4.122}$$

Or equivalently, $E_{1t} = E_{2t}$. If magnetic charges are ever found, then this condition will have to include the possibility of a surface magnetic current. The boundary conditions on **H** are similar

$$\hat{n} \times (\mathbf{H}_1 - \mathbf{H}_2) = \mathbf{J}_s \tag{4.123}$$

For the interface of a perfect conductor, for example, a surface current flows so that ($H_2 = 0$)

$$\hat{n} \times \mathbf{H}_1 = \mathbf{J}_s \tag{4.124}$$

Boundary conditions for current

Applying the "pillbox" argument to the divergence of current

$$\int_V (\nabla \cdot \mathbf{J}) dV = \oint_S \mathbf{J} \cdot d\mathbf{S} = -\int_V \frac{\partial \rho}{\partial t} dV \tag{4.125}$$

in the limit

$$J_{1n} - J_{2n} = -\frac{\partial \rho_s}{\partial t} \tag{4.126}$$

where ρ_s is the surface current. In the static case

$$J_{1n} = J_{2n} \tag{4.127}$$

implies that $\sigma_1 E_1 = \sigma_2 E_2$. This implies that $\rho_s \neq 0$ since $\epsilon_1 E_1 \neq \epsilon_2 E_2$ (unless the ratios of σ match the ratio of ϵ perfectly!)

4.5 References

This chapter draws from many sources, particularly Feynman [18], Cheng [6], and Inan and Inan [62].

5 Inductance

5.1 Introduction

Inductance is an intrinsic part of every physical circuit in which currents flow. Even the simplest circuit shown in Fig. 5.1a cannot be fully understood unless the inductance is taken into account. Even though no inductor is drawn in the figure explicitly, there is a net electromagnetic force

$$V_L = L \frac{dI}{dt} \tag{5.1}$$

induced around the loop that tends to oppose any change in the current magnitude. The inductance L is a property of the loop related to the physical geometry of the circuit. We can represent this induced voltage by an element that has the voltage–current relationship embodied in Eq. (5.1), as shown schematically in Fig. 5.1b. In this case, the inductance is a property of the loop and it does not stem from a separate coil (as suggested by the inductor's symbol) introduced into the circuit. While we can calculate the resistance of the loop by breaking the loop into three separate conductors, as suggested by $R_1 - R_3$, this cannot be done easily with the inductance because the currents in the conductors interact magnetically.

But is this intrinsic inductance really important? Using techniques of this chapter we can determine the magnitude of L for a given circuit. From the application at hand, we can estimate dI/dt or ωL in a steady-state periodic circuit and from Eq. (5.1) we can estimate the magnitude of the induced voltage. If L is sufficiently small, then we can ignore the magnetic effects of the circuit. Clearly then, for circuits operating at higher frequencies or at high switching rates dI/dt, we cannot ignore the inductive behavior. But even at lower frequencies, if the circuit is physically large then the L term is significant and the induced voltages are important.

It is interesting to note that in this circuit there is no one point where this voltage is induced but rather it is induced around the entire loop. In fact, voltages are induced in the entire space and not just along the circuit path. This gives rise to *eddy currents*, or magnetically induced currents that flow in other conducting bodies. Furthermore, the voltage is not induced uniformly along the path of the circuit, but in a distributed fashion where the induction is stronger at points where more magnetic energy is stored.

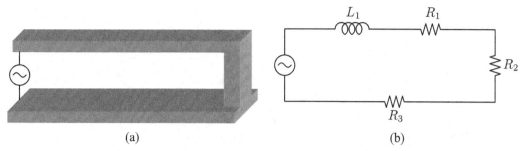

(a)   (b)

Figure 5.1 (a) A simple circuit with inductance. (b) Schematic representation of circuit.

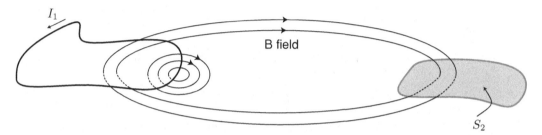

Figure 5.2 The flux linkage between two current loops.

Is the distributed nature of inductance important? Consider the input impedance of the above circuit seen from the terminals of the voltage source. Isn't it simply $j\omega L + R$? Surprisingly, no! In later chapters we will compute the input impedance of this distributed circuit and find the conditions under which the value deviates from the lumped circuit approximation.

5.2   Inductance

Magnetic flux

Magnetic flux plays an important role in many EM problems (in analogy with electric charge)

$$\Psi = \int_S \mathbf{B} \cdot \mathbf{dS} \tag{5.2}$$

Due to the absence of magnetic charge

$$\Psi = \oint_S \mathbf{B} \cdot \mathbf{dS} \equiv 0 \tag{5.3}$$

but net flux can certainly cross an open surface. The magnetic flux crossing a loop is shown in Fig. 5.2 by the integral of B crossing the shaded surface. Magnetic flux is independent of the surface but only depends on the curve bounding the surface. This is easy to show

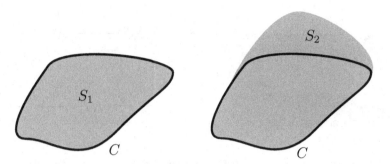

Figure 5.3 Any open surface bounded by the circuit can be used to calculate the magnetic flux.

since

$$\Psi = \int_S \mathbf{B} \cdot \mathbf{dS} = \int_S \nabla \times \mathbf{A} \cdot \mathbf{dS} = \oint_C \mathbf{A} \cdot \mathbf{dl} \qquad (5.4)$$

$$\Psi = \int_{S_1} \mathbf{B} \cdot \mathbf{dS} = \int_{S_2} \mathbf{B} \cdot \mathbf{dS} \qquad (5.5)$$

So in Fig. 5.2, we can select either S_1 or S_2 to calculate the flux.

Flux linkage

In Fig. 5.3, consider the flux-crossing surface S_2 due to a current I_1 flowing in loop S_1

$$\Psi_{21} = \int_{S_2} \mathbf{B}_1 \cdot \mathbf{dS} \qquad (5.6)$$

Likewise, the "self-flux" of a loop is defined by the flux crossing the surface of a path when a current is flowing in the path

$$\Psi_{11} = \int_{S_1} \mathbf{B}_1 \cdot \mathbf{dS} \qquad (5.7)$$

The flux is linearly proportional to the current and otherwise only a function of the geometry of the path. To see this, let us calculate Ψ_{21} for filamental loops

$$\Psi_{21} = \oint_{C_2} \mathbf{A}_1 \cdot \mathbf{dl}_2 \qquad (5.8)$$

but

$$\mathbf{A}_1 = \frac{1}{4\pi \mu_0^{-1}} \oint_{C_1} \frac{I_1 \mathbf{dl}_1}{R - R_1} \qquad (5.9)$$

Substituting, we have a double integral

$$\Psi_{21} = \frac{I_1}{4\pi \mu_0^{-1}} \oint_{C_2} \left(\oint_{C_1} \frac{\mathbf{dl}_1}{R - R_1} \right) \cdot \mathbf{dl}_2 \qquad (5.10)$$

The geometry of this double integral is shown in Fig. 5.4. We thus have a simple formula that only involves the magnitude of the current and the average distance between every two

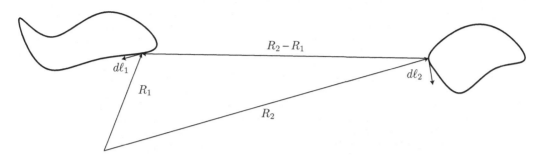

Figure 5.4 The distance between two arbitrary points in the loop.

points on the loops

$$\Psi_{21} = I_1 \times \frac{1}{4\pi\mu_0^{-1}} \oint_{C_2} \oint_{C_1} \frac{\mathbf{dl}_1 \cdot \mathbf{dl}_2}{R_2 - R_1} \tag{5.11}$$

Mutual and self inductance

Since the flux is proportional to the current by a geometric factor, we may write

$$\Psi_{21} = M_{21} I_1 \tag{5.12}$$

We call the factor M_{21} the mutual inductance

$$M_{21} = \frac{\Psi_{21}}{I_1} = \frac{1}{4\pi\mu_0^{-1}} \oint_{C_2} \oint_{C_1} \frac{\mathbf{dl}_1 \cdot \mathbf{dl}_2}{R_2 - R_1} \tag{5.13}$$

The above expression is known as Neumann's equation. It illustrates compactly the geometric nature of mutual inductance. In the above computation R takes on the values of the distance between every pair of points on the loops. In other words, mutual inductance is the weighted average value of $1/R$ between every pair of points on the loops weighted by $\cos(\theta)$, where θ is the local angle between the points in the loop. Thus, points lying on segments mutually perpendicular do not contribute to the mutual inductance.

An interesting observation is that mutual coupling is reciprocal. In other words

$$M_{12} = \frac{\mu}{4\pi} \oint_{C_1} \oint_{C_2} \frac{1}{R} \mathbf{dl}_2 \cdot \mathbf{dl}_1 = M_{21} \tag{5.14}$$

since the weighted "averaging" may be done in any order. This is a profound result that has many implications in circuit theory. Consider the example shown in Fig. 5.5 where a small loop couples to a larger loop. Even though the magnetic flux of loop B is small, loop A "captures" a large portion of the flux. Similarly, even though loop A has a large magnetic flux, only a small portion of it is captured by loop B and thus $M_{AB} = M_{BA}$.

The units of M are H since $[\mu] = $ H/m. The "self-flux" mutual inductance is simply called the self inductance and donated by $L_1 = M_{11}$.

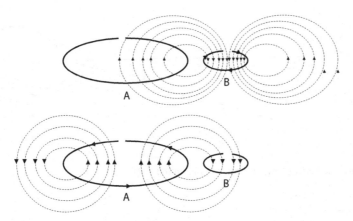

Figure 5.5 Illustration of reciprocity through the *forward* and *reverse* coupling between two vastly different loops.

System of mutual inductance equations

If we generalize to a system of current loops we have a system of equations

$$\Psi_1 = L_1 I_1 + M_{12} I_2 + \ldots M_{1N} I_N$$

$$\vdots$$

$$\Psi_N = M_{N1} I_1 + M_{N2} I_2 + \ldots L_N I_N$$

Or in matrix form $\psi = M\mathbf{i}$, where M is the inductance matrix. This equation resembles $\mathbf{q} = C\mathbf{v}$, where C is the capacitance matrix.

Example 9

Coaxial conductor

In transmission line problems, we need to compute inductance/unit length. Consider the shaded area in Fig. 5.6, from $r = a$ to $r = b$. The magnetic field in the region between conductors is easily computed

$$\oint \mathbf{B} \cdot \mathbf{dl} = B_\phi 2\pi r = \mu_0 I \tag{5.15}$$

The external flux (excluding the volume of the ideal conductors) is given by

$$\psi' = \int_a^b B_\phi dr = \frac{\mu_0 I}{2\pi} \int_a^b \frac{dr}{r} = \frac{\mu_0 I}{2\pi} \ln\left(\frac{b}{a}\right) \tag{5.16}$$

The inductance per unit length is therefore

$$L' = \frac{\mu_0}{2\pi} \ln\left(\frac{b}{a}\right) \quad [\text{H/m}] \tag{5.17}$$

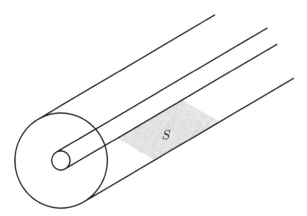

Figure 5.6 The partial inductance of a coaxial conductor is computed by calculating the flux in the shaded area.

Recall that the product of inductance and capacitance per unit length for a transmission line is a constant (see Ch. 9)

$$L'C' = \frac{1}{c^2} \tag{5.18}$$

where c is the speed of light in the medium. Thus we have calculated the capacitance per unit length without any extra work.

5.3   Magnetic energy and inductance

Energy for a system of current loops

We would like to calculate the magnetic energy associated with a system of current loops, shown in Fig. 5.7. In the electrostatic case, we assembled our charge distribution one point charge at a time and used electric potential to calculate the energy. This can be done for the magnetostatic case but there are some complications.

Energy for two loops

As we move in our second loop with current I_2, we would be cutting across flux from loop 1, as shown in Fig. 5.8a, and therefore an induced voltage around loop 2 would change the current. When we bring the loop to rest, the induced voltage would drop to zero. To maintain a constant current, therefore, we would have to supply a voltage source in series to cancel the induced voltage. The work done by this voltage source represents the magnetostatic energy in the system.

A simpler approach is to bring in the two loops with zero current and then increase the current in each loop one at a time, shown in Fig. 5.8b. First, let us increase the current in

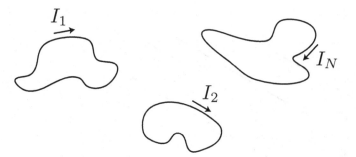

Figure 5.7 A system of current loops store magnetic energy.

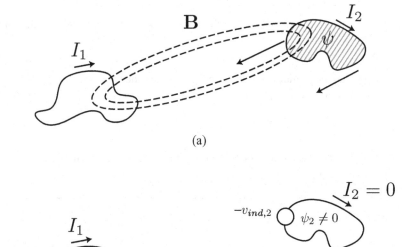

Figure 5.8 (a) When moving loop 2 into our system, the rate of change of flux in loop 2 generates a current in loop 2, which changes the field configuration. (b) Instead, we bring in loops 1 and 2 without any current, and turn on the current in each loop one at a time.

loop 1 from zero to I_1 in some time t_1. Note that at any instant of time, a voltage is induced around loop number 1 due to its changing flux

$$v_{ind,1} = -\frac{d\psi}{dt} = -L_1\frac{di_1}{dt} \tag{5.19}$$

where i_1 represents the instantaneous current.

Note that this induced voltage will tend to decrease the current in loop 1. This is a statement of Lenz's Law. In other words, the induced voltage in loop 1 tends to create a magnetic field to oppose the field of the original current! To keep the current constant in loop 1, we must connect a voltage source to cancel the induced voltage. The work done by this voltage source is given by

$$w_1 = \int_0^{t_1} p(\tau)d\tau = -\int_0^{t_1} v_{ind,1} i_1(\tau)d\tau \tag{5.20}$$

where $p(t) = -v_{ind,1} i_1(t) = L_1 i_1 \frac{di_1}{dt}$. The net work done by the source is simply

$$w_1 = L_1 \int_0^{t_1} i_1 \frac{di_1}{d\tau}d\tau = L_1 \int_0^{I_1} i_1 di_1 = \frac{1}{2}L_1 I_1^2 \tag{5.21}$$

Note that to keep the current in loop 2 equal to zero, we must also provide a voltage source to counter the induced voltage

$$v_{ind,2}(t) = -M_{21}\frac{di_1}{dt} \tag{5.22}$$

This voltage source does not do any work since $i_2(t) = 0$ during this time.

By the same argument, if we increase the current in loop 2 from 0 to I_2 in time t_2, we need to do work equal to $\frac{1}{2}L_2 I_2^2$. But is that all? No, since to keep the current in loop 1 constant at I_1 we must connect a voltage source to cancel the induced voltage

$$-v_{ind,1} = \frac{d\psi_1}{dt} = M_{12}\frac{di_2}{dt} \tag{5.23}$$

The additional work done is therefore

$$w_1' = \int_0^{t_2} M_{12}\frac{di_2}{d\tau} I_1 d\tau = M_{12} I_1 I_2 \tag{5.24}$$

The total work to bring the current in loop 1 and loop 2 to I_1 and I_2 is therefore

$$W = \frac{1}{2}L_1 I_1^2 + \frac{1}{2}L_2 I_2^2 + M_{12} I_1 I_2 \tag{5.25}$$

But the energy should not depend on the order we turn on the currents. Thus we can immediately conclude that $M_{12} = M_{21}$. We already saw this when we derived an expression for M_{12} using the Neumann equation.

Generalize to N loops

We can now pretty easily generalize our argument for two loops to N loops

$$W = \frac{1}{2}\sum_i L_i I_i^2 + \sum \sum_{i>j} M_{ij} I_i I_j \tag{5.26}$$

The first term represents the "self" energy for each loop and the second term represents the interaction terms.

$$W = \frac{1}{2}\sum \sum_{i=j} M_{ij} I_i I_j + \frac{1}{2}\sum \sum_{i \neq j} M_{ij} I_i I_j \tag{5.27}$$

$$W = \frac{1}{2}\sum_i \sum_j M_{ij} I_i I_j \tag{5.28}$$

The factor $1/2$ accounts for the double counting in the double summation. Let's rewrite this equation and combine terms

Energy in terms of vector potential

We derived the mutual inductance between two filamentary loops using Neumann's equation

$$M_{ij} = \frac{\mu_0}{4\pi} \oint_{C_i} \oint_{C_j} \frac{d\mathbf{l}_i \cdot d\mathbf{l}_j}{R_{ij}} \tag{5.29}$$

Let's substitute the above relation into the expression for energy

$$W = \frac{1}{2} \sum_i I_i \left[\sum_j M_{ij} I_j \right] \tag{5.30}$$

$$W = \frac{1}{2} \sum_i I_i \left[\sum_j I_j \frac{\mu_0}{4\pi} \oint_{C_i} \oint_{C_j} \frac{d\mathbf{l}_i \cdot d\mathbf{l}_j}{R_{ij}} \right] \tag{5.31}$$

Let's change the order of integration and summation

$$W = \frac{1}{2} \sum_i I_i \oint_{C_i} \left[\sum_j I_j \frac{\mu_0}{4\pi} \oint_{C_j} \frac{d\mathbf{l}_j}{R_{ij}} \right] \cdot d\mathbf{l}_i \tag{5.32}$$

Each term of the bracketed expression represents the vector potential due to loop j evaluated at a position on loop i. By superposition, the sum represents the total voltage potential due to all loops

$$W = \frac{1}{2} \oint_{C_i} I_i \mathbf{A} \cdot d\mathbf{l}_i \tag{5.33}$$

We derived this for filamental loops. Generalize to an arbitrary current distribution and we have

$$W_m = \frac{1}{2} \int_V \mathbf{J} \cdot \mathbf{A} dV \tag{5.34}$$

Compare this to the expression for electrostatic energy

$$W_e = \frac{1}{2} \int_V \rho \phi dV \tag{5.35}$$

Thus the vector potential $\mathbf{A}$ really does represent the magnetic potential due to a current distribution in an analogous fashion as ϕ represents the electric potential.

Energy in terms of fields

Let's replace $\mathbf{J}$ by Ampère's Law $\nabla \times \mathbf{H} = \mathbf{J}$

$$W_m = \frac{1}{2} \int_V (\nabla \times \mathbf{H}) \cdot \mathbf{A} dV \tag{5.36}$$

Using the identity

$$\nabla \cdot (\mathbf{H} \times \mathbf{A}) = (\nabla \times \mathbf{H}) \cdot \mathbf{A} + \mathbf{H} \cdot (\nabla \times \mathbf{A}) \tag{5.37}$$

$$W_m = \frac{1}{2} \int_V \nabla \cdot (\mathbf{H} \times \mathbf{A}) dV + \frac{1}{2} \int_V (\nabla \times \mathbf{A}) \cdot \mathbf{H} dV \tag{5.38}$$

Apply the Divergence Theorem to the first term to give

$$W_m = \frac{1}{2} \int_S \mathbf{H} \times \mathbf{A} \cdot d\mathbf{S} + \frac{1}{2} \int_V (\nabla \times \mathbf{A}) \cdot \mathbf{H} dV \tag{5.39}$$

We would like to show that the first term is zero. To do this, consider the energy in all of space $V \to \infty$. Use a large sphere of radius r and take the radius to infinity. We know that if we are sufficiently far from the current loops, the potential and field behave like $A \sim r^{-1}$ and $H \sim r^{-2}$. The surface area of the sphere goes like r^2. The surface integral, therefore, gets smaller and smaller as the sphere approaches infinity.

The remaining volume integral represents the total magnetic energy of a system of currents

$$W_m = \frac{1}{2} \int_V (\nabla \times \mathbf{A}) \cdot \mathbf{H} dV \tag{5.40}$$

But $\nabla \times \mathbf{A} = \mathbf{B}$

$$W_m = \frac{1}{2} \int_V \mathbf{B} \cdot \mathbf{H} dV \tag{5.41}$$

And the energy density of the field is seen to be

$$w_m = \mathbf{B} \cdot \mathbf{H} \tag{5.42}$$

Recall that $w_e = \mathbf{D} \cdot \mathbf{E}$

Another formula for inductance

The self inductance of a loop is given by

$$L = \frac{1}{I} \int_S \mathbf{B} \cdot d\mathbf{S} \tag{5.43}$$

Since the total magnetic energy for a loop is $\frac{1}{2}LI^2$, we have an alternate expression for the inductance

$$\frac{1}{2}LI^2 = \frac{1}{2} \int_V \mathbf{B} \cdot \mathbf{H} dV \tag{5.44}$$

$$L = \frac{1}{I^2} \int_V \mathbf{B} \cdot \mathbf{H} dV \tag{5.45}$$

This alternative expression is sometimes easier to calculate. We will see this when we compute the *internal* inductance of a round wire.

Self inductance of filamentary loops

We have tacitly assumed that the inductance of a loop is a well-defined quantity. But for a filamentary loop, we can expect trouble. By definition

$$L = \frac{1}{I} \int_S \mathbf{B} \cdot d\mathbf{S} = \frac{1}{I} \oint_C \mathbf{A} \cdot d\mathbf{l} \tag{5.46}$$

$$L = \frac{\mu}{4\pi} \oint_C \oint_C \frac{d\ell' \cdot d\ell}{R} \tag{5.47}$$

This is just Neumann's equation with $C_1 = C_2$. But for a filamental loop, $R = 0$ when both loops traverse the same point. The integral is thus not defined for a filamental loop!

Magnetic energy of a circuit

In the previous section we derived that the total magnetic energy in a circuit is given by

$$E_m = \frac{1}{2}L_1 I_1^2 + \frac{1}{2}L_2 I_2^2 + M I_1 I_2 \tag{5.48}$$

We would like to show that this implies that $M \leq \sqrt{L_1 L_2}$. Let's re-write the above into the following positive definite form

$$E_m = \frac{1}{2}L_1 \left(I_1 + \frac{M}{L_1} I_2\right)^2 + \frac{1}{2}\left(L_2 - \frac{M^2}{L_1}\right) I_2^2 \tag{5.49}$$

An important observation is that regardless of the current I_1 or I_2, the magnetic energy is non-negative, so $E_m \geq 0$.

Consider the current $I_2 = \frac{-L_1}{M} I_1$, which cancels the first term in E_m

$$E_m = \frac{1}{2}\left(L_2 - \frac{M^2}{L_1}\right) I_2^2 \geq 0 \tag{5.50}$$

Since $I_2^2 \geq 0$, we have

$$L_2 - \frac{M^2}{L_1} \geq 0 \tag{5.51}$$

Therefore it is now clear that this implies

$$L_1 L_2 \geq M^2 \tag{5.52}$$

Coupling coefficient

Usually we express this inequality as

$$M = k\sqrt{L_1 L_2} \tag{5.53}$$

where k is the *coupling coefficient* between two circuits, with $|k| \leq 1$. If two circuits are perfectly coupled (all flux from circuit 1 crosses circuit 2), $k = 1$ (ideal transformer). Note that $M < 0$ implies that $k < 0$, which is totally reasonable as long as k lies on the unit interval $-1 \leq k \leq 1$. Negative coupling just means that the flux gets inverted before crossing the second circuit. This is easily achieved by winding the circuits with opposite orientation.

5.4   Discussion of inductance

As we have seen in the previous section, there are many equivalent ways to define inductance. All of the following forms are equally important as they shed new insights into the problem. In this section we discuss these definitions in further detail.

Magnetostatics and quasistatics

Magnetostatics is the study of magnetic fields under time-invariant conditions. Clearly, with no time variation, Maxwell's equations take on very simple forms. While these conditions seem oversimplified for practical applications, especially when disallowing any time variation, we shall see later that many seemingly difficult non-static problems admit solutions when seen in this light. For instance, when studying transmission lines, structures where electric and magnetic energy is in constant dynamic exchange, we shall find that the cross-sectional waveforms take on static field configurations. In other problems, the static behavior will dominate up to very high frequencies and thus we can solve such problems using the *quasistatic* assumption. At a given frequency, or equivalently for a given wavelength of operation, we shall show that this is especially true for structures that are small relative to the wavelength. For instance, even for signals varying tens of billions of times per second, the wavelength is of the order of centimeters, an order of magnitude larger than most integrated circuit structures.

Under the magnetostatic conditions, currents flow uniformly in the volume of the problem space. Thus the net current flow into any volume element V is zero. If it were non-zero, then net charge would continuously accumulate in this volume element, violating the steady-state assumption.

The net current flowing into the volume element V is given by integrating the component of current density $\mathbf{J}$ perpendicular to the bounding surface of the volume S. By the Divergence Theorem we have

$$0 = \oint_S \mathbf{J} \cdot d\mathbf{S} = \int_V \nabla \cdot \mathbf{J} \, dV \tag{5.54}$$

Assuming Ohm's Law to be valid in the conductive regions and further assuming that the material is isotropic implies the current density is proportional and parallel to the local electric field $\mathbf{J} = \sigma \mathbf{E}$. Substituting in the above equation we have

$$0 = \int_V \nabla \cdot \mathbf{J} \, dV = \int_V \sigma \nabla \cdot \mathbf{E} \, dV \tag{5.55}$$

where for simplicity we assume that the material conductivity σ is independent of position. But from Maxwell's equations, in particular Gauss' Law, the divergence of the electric field can only arise from charge. We therefore have

$$0 = \int_V \frac{\sigma}{\epsilon} \rho \, dV \tag{5.56}$$

Again, let us assume that the permitivity does not vary with position. Take an arbitary small conductive volume ΔV. Since $\sigma \neq 0$, this forces the net charge to equal zero, or $\rho = 0$. Since the volume element we chose was arbitrary, this implies that all conductors in the problem must have a net charge of zero.

But current flow is a result of charge in motion and from the outset we assume that $\mathbf{J} \neq 0$. Is $\rho = 0$ a contradiction? No, because at any given moment we assume that all moving charges are neutralized by charges of opposite sign. These charges can also be in motion, such as holes in a semiconductor, or they can be immobile ionized host atoms. In

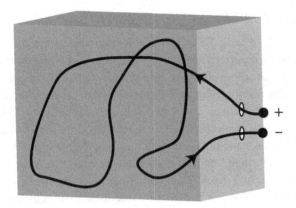

Figure 5.9 A conductive box containing an inductor. The inductor wires are insulated from the box. The inductor may or may not self-intersect.

reality, this is an approximation for finite values of σ since the atoms are in constant thermal agitation causing a net time-varying multi-pole charge moment.

But doesn't $\rho = 0$ imply also that $\mathbf{E} = 0$? No, only that $\nabla \cdot \mathbf{E} = 0$. Since we have assumed steady-state conditions, the electric field has only a curl-free component

$$\mathbf{E} = -\nabla\phi \qquad (5.57)$$

The magnetic field does not interact with the electric field since this can only occur under time-varying situations. Maxwell's equations lead to a source-free Poisson's equation

$$\nabla^2\phi = -\frac{\rho}{\epsilon} = 0 \qquad (5.58)$$

Since there are no sources in the above equation, the scalar potential can be non-zero only if the boundary conditions are non-zero. If we expand the volume space of the problem to infinity where $\phi = 0$, then we do indeed have a contraction. Since in this case we have $\phi \equiv 0$ and thus $\mathbf{E} \equiv 0$, hence no currents can flow. This would also be the case for a grounded closed cavity with walls constructed from perfect conductors. Without any time variation, current cannot flow in the interior of the cavity. Thus, we have to assume that on the boundary of our problem space there are points where $\phi \neq 0$.

As shown in Fig. 5.9, we must break the circuit loop under consideration and apply a field between the terminals of the loop.

Magnetic flux

Inductance defined in terms of magnetic flux is the most common approach, but it requires some subtle observations. As we have seen, the voltage induced in a loop can be obtained by integrating the electric field along the path of the circuit. By Stoke's Theorem we have

$$V_L = -\oint_C \mathbf{E} \cdot \mathbf{dl} = -\int_S (\nabla \times \mathbf{E}) \cdot \mathbf{dS} \qquad (5.59)$$

Figure 5.10 Several candidate surfaces bounded by the same curve C.

Note that this is true for any surface S bounded by the circuit path C. From Maxwell's equations we have

$$\nabla \times \mathbf{E} = -\frac{d\mathbf{B}}{dt} \tag{5.60}$$

so that

$$V_L = \frac{d}{dt} \int_S \mathbf{B} \cdot d\mathbf{S} \tag{5.61}$$

The right-hand side is the magnetic flux through the surface bounded by the current. We defined the magnetic flux ψ by

$$\psi \equiv \int_S \mathbf{B} \cdot d\mathbf{S} \tag{5.62}$$

and it follows that

$$V_L = \frac{d\psi}{dt} \tag{5.63}$$

comparing with Eq. (5.1) we have

$$L \equiv \frac{\psi}{I} \tag{5.64}$$

In computing the flux, which surface do we use in the above integral? In Fig. 5.10 we show several candidate surfaces. Clearly, there are an infinite number of surfaces bounded by the closed curve. Is there a unique surface we can attribute to the path? We might try by limiting the surface to the minimal surface (the so-called "soap film" surface generated by dipping the circuit in a soap bath). But the first step of the above manipulation is justified by Stoke's Theorem which admits *any* surface as long as it is enclosed by the path of the circuit.

The magnetic field $\mathbf{B}$ is computed under magnetostatic conditions described above. For a closed circuit path under magnetostatic conditions, this implies a zero electric field in the volume of perfect conductors. Since the voltage drop around a loop of perfectly conducting material is zero, no voltages can be induced in the loop and so the rate of change flux ψ must equal zero. A realistic example of this occurs in a superconductor where the static state is literally frozen into space.

Most practical inductors are constructed from materials of finite non-zero resistivity. We break the loop and introduce an electric field at the terminals. Since the electric field is

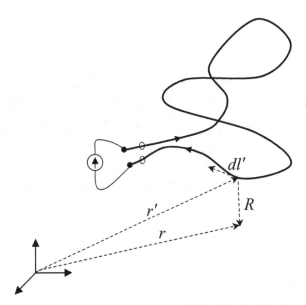

Figure 5.11 Computing the magnetic field using the Biot-Savart equations.

zero or nearly zero along the path of the conductive portion of the loop, the entire induced voltage calculated above appears at the terminals of the circuit.

Under magnetostatic conditions, we can calculate $\mathbf{B}$ from the so-called Biot-Savart Law [52]. This result is easily derived from Ampère's Law in Maxwell's equations

$$\mathbf{B}(r) = \mu \oint_C \frac{I(\mathbf{r}')\mathbf{dl}' \times \mathbf{R}}{4\pi R^3} = \frac{1}{4\pi \mu^{-1}} \oint_C \frac{I(\mathbf{r}')\mathbf{dl}' \times \hat{\mathbf{R}}}{R^2} \tag{5.65}$$

where, as shown in Fig. 5.11, the vector $\mathbf{R}$ is drawn from the "source" point r' along the path of integration on the circuit to the "field" point r. The convention of designating the source point with primed coordinates is an arbitrary but common practice.

The above equation is insightful as it states that the influence of current elements on magnetic field follows a inverse-square law dependence similar to the influence of charge on the electric field.

Now the second subtlety enters in the definition of inductance. In computing the surface integral of Eq. (5.62), where does the boundary of the surface lie? Eq. (5.65) clearly shows that the magnetic field is unbounded as we approach a filamental wire so that the integral is undefined for this case. Therefore, we conclude that the inductance of an infinitesimally thin wire is infinite. Should the boundary be taken as only the inside edge of the wire or should the boundary also include the wire itself?

Internal and external inductance

Let us partition the flux as follows

$$\psi = \psi_i + \psi_o \tag{5.66}$$

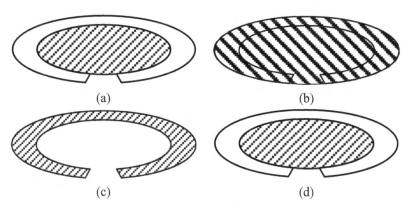

Figure 5.12 (a) The *inner* flux of a circuit. (b) The *outer* flux of a circuit. (c) The *internal* flux of a circuit. (d) The *external* flux of a circuit (same as (a)).

where ψ_i is the inner flux shown in Fig. 5.12a and is computed from the surface integral when the boundary is taken at the internal edge of the loop. The outer flux ψ_o is shown in Fig. 5.12b, where the area of the flux is bounded by the outer edge of the wire forming the loop.

The voltage induced due to $d\psi_i/dt$ is smaller since we are integrating over a smaller area and the flux is non-zero in the volume of the conductors. Comparatively the voltage induced due to $d\psi_o/dt$ is larger. The difference between the voltage induced on the inner edge and outer edge is thus greater the larger the width of the wire. For finite σ we can see then that the current distribution could be non-uniform along the width of the conductor. For good conductors under quasistatic conditions, the current flow is uniform through the width of conductors and the flux is therefore non-zero in the volume of conductors. The voltage induced, therefore, includes both the inner flux ψ_i and the outer flux ψ_o.

The external flux ψ_{ext} is identical with the "inner" flux ψ_i we defined above. The "internal" flux is given by

$$\psi_{int} = \psi_o - \psi_i \qquad (5.67)$$

and contains the flux bounded by the inner and outer edges of the conductor. The total inductance is derived by including both terms. This partitioning is convenient once we consider the frequency variation of inductance. As we shall see, at high frequencies the current flow in the volume of conductors decreases (due to induced eddy currents) and current density peaks at the surface of conductors. Then the internal flux ψ_{int} vanishes and the inductance is given by the external flux ψ_{ext}. The external flux is approximately independent of frequency, whereas the internal flux tends to diminish at increasing frequency.

Internal inductance of a round wire

Usually if the wire radius is small relative to the loop area, $L_{int} \ll L_{ext}$. We shall see that at high frequencies, the magnetic field decays rapidly in the volume of conductors and thus the $\psi_{int} \to 0$ and $L(f \to \infty) = L_{ext}$. Consider a round wire carrying uniform current. We

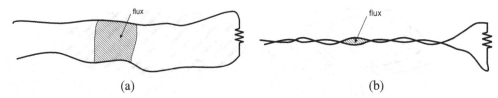

Figure 5.13 (a) Flux of a loop inductor. (b) The magnetic flux of a twisted pair circuit.

can easily derive the magnetic field through Ampère's Law

$$B_{inside} = \frac{\mu_0 I r}{2\pi a^2} \tag{5.68}$$

Using this expression, we can find the internal inductance

$$W_m = \frac{1}{2} \int_V \mathbf{B} \cdot \mathbf{H} dV = \frac{1}{2} \int_{V_{inside}} \mathbf{B} \cdot \mathbf{H} dV + \frac{1}{2} \int_{V_{outside}} \mathbf{B} \cdot \mathbf{H} dV \tag{5.69}$$

$$W_m = \frac{1}{2} L_{int} I^2 + \frac{1}{2} L_{ext} I^2 \tag{5.70}$$

The "inside" term is easily evaluated

$$W_{m,int} = \frac{1}{2} \frac{\mu_0 I^2}{(2\pi a^2)^2} \int_0^a r^2 2\pi r dr = \frac{1}{2} \frac{\mu_0 I^2}{8\pi} \tag{5.71}$$

The internal inductance per unit length is thus

$$L_{int} = \frac{\mu_0}{8\pi} \tag{5.72}$$

Numerically, this is 50pH/mm, a pretty small inductance. Recall that this is only the inductance due to energy stored inside of the wires. The external inductance is likely to be much larger.

Inductance for a general structure

We are now in a position to comment on the inductance of an arbitrary circuit. For instance, to increase the inductance of a circuit we should increase the flux of the circuit. Since the current is divided out of flux, it does not come into play for linear materials where the magnetic permeability μ is independent of current. Hence, the inductance is a purely geometric property of the circuit and we may speak of the inductance in the same way that we speak of other geometric properties, such as the perimeter or surface area of the circuit.

Since the inductance is an integral over the surface of the circuit, it seems as if a circuit that spans a larger area should have higher inductance. For instance, compare the flux area of the two loops shown in Fig. 5.13. In part (a), we show a circuit that spans a large area since the "return current" flows in a wire physically far from the incoming current. In Fig. 5.13b, we keep the return path in close proximity to the incoming current by twisting the wires together. In the limit that the wires are just touching (with some insulation in between to prevent a short), the inductance is driven to arbitrarily low values.

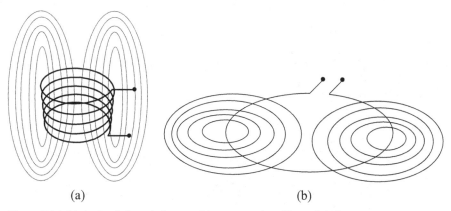

Figure 5.14 (a) A simple loop inductor with area equal to (b) a coil inductor.

Twisting the wires together has another benefit, besides low inductance. The magnetic pickup from a twisted pair is much smaller than a two-wire line with the equivalent inductance. This is because any flux linking with the two-wire line generates an emf, whereas in a twisted pair, for a uniform incoming B field, the flux tends to cancel out since the wire is twisted. In one section of the wire, the flux pickup is positive, but in the next it is negative. In this way, incoming magnetic fields generate a net flux of nearly zero on the twisted pair.

Let us compare two circuits with equal surface area but very different values of inductance. In Fig. 5.14a we see a traditional coil inductor. To help visualize the surface of integration, imagine the portion in the coil as a spiraling ramp resembling a multi-level parking lot structure (Fig. 5.15). For comparison we have also shown a large single loop inductor designed so that the area of the loop is equal to the surface area of the coil. The coil has larger inductance, though, since the magnetic field at any given point inside the coil has additive contributions from several adjacent "loops." Since the surface area of the single loop is equal to the surface area of the coil, the flux will be larger if the average magnetic field is larger in the coil. Intuitively, it seems like this is the case due to the additive nature of the field from one loop to adjacent loops.

Mutual inductance

The vector nature of the magnetic field leads to an interesting property of magnetic coupling. If the magnetic flux generated by one loop impinges perpendicularly on another loop, it does not generate an induced voltage and hence the mutual coupling $M_{ij} = 0$ for such a case. This principle can be used to isolate structures in a circuit where coupling is undesired.

The net results of mutual inductance can lead to many interesting effects. For instance, loops wound in the opposite orientation add negatively and thus tend to cancel the flux and lower the induced voltage. This is illustrated in Fig. 5.16, where we use the right-hand convention to orient the magnetic fields. Clearly, loop B tends to cancel the self flux of loop A, whereas loop C adds to the flux of loop A.

A nice application of this concept is the ground-fault interrupt circuit (GFI) shown in Fig. 5.17. This circuit is designed to detect a *ground fault*, a condition where an electric

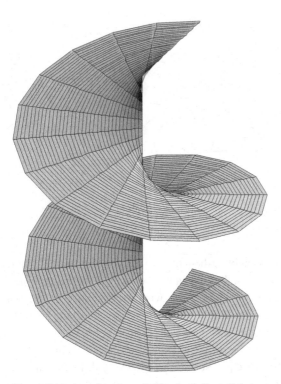

Figure 5.15 A surface bounded by the helix coil inductor.

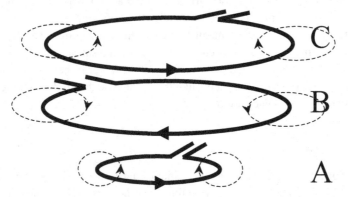

Figure 5.16 Three mutually coupled loops of current. The flux of loop C adds to the flux of loop A, whereas the flux of loop B abstracts from A.

appliance faults by shorting a signal line to ground. Under normal operation all the current that enters the appliance should return to ground via the ground line. If there is any imbalance between the signal line current and ground current, this signifies that a parasitic path to ground has been established, and possibly via conduction current through an unfortunate victim (say a person grounded accidentally touches the high voltage signal line). Under normal operation the voltage on the GFI loop is zero since the flux of the incoming

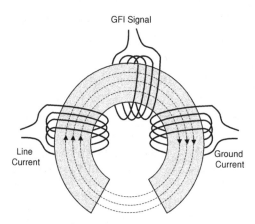

Figure 5.17 Schematic representation of a ground-fault interrupt (GFI) circuit.

current is canceled by the ground current. If any imbalance occurs, a net voltage will be induced at the GFI node and this condition can be detected to quickly shut off power to the appliance.

Magnetic energy perspective

From electromagnetic theory we know that the power stored due to an electromagnetic field is given by [52]

$$P = \frac{1}{2}\frac{d}{dt}\int_V \left(\frac{1}{\mu}|\mathbf{B}|^2 + \epsilon|\mathbf{E}|^2 + \mathbf{E}\cdot\mathbf{J}\right)dV \tag{5.73}$$

where the integral is over the entire volume space of the problem where the fields are non-zero. In a good "inductor," the magnetic energy dominates over the electrical energy, or $\mu^{-1}|B|^2 \gg \epsilon|E|^2$. Thus, the above parenthetical expression reduces to the total stored magnetic energy

$$E_L = \frac{1}{2}\int_V \frac{1}{\mu}|\mathbf{B}|^2 dV \tag{5.74}$$

Using equation E_L we derived an alternate expression for inductance

$$L = \frac{1}{I^2}\int_V \frac{1}{\mu}|\mathbf{B}|^2 dV \tag{5.75}$$

This viewpoint emphasizes the fact that the voltages induced across an inductor terminal or around a circuit loop are due to the magnetic stored energy. Thus an induced voltage can only sustain itself as long as $\mathbf{B}$ is non-zero. In a time periodic circuit the magnetic energy is continuously "pumped" as the induced voltage changes sign. When $dI/dt > 0$ energy is supplied to the inductor. Alternatively, when $dI/dt < 0$ energy returns to the rest of the circuit. Naturally, an inductor constructed from materials with zero resistance must conserve energy.

The inductance computed from the definition of flux must equal the value computed from the energy of the field. This implies that

$$\frac{1}{I} \int_V \frac{1}{\mu} |\mathbf{B}|^2 dV = \int_S \mathbf{B} \cdot d\mathbf{S} \tag{5.76}$$

Example 10

In this example we compute the partial inductance of a two wire transmission line. Assume each wire is round and of radius a and the origins of the wires are separated by a distance d.

From previous examples we already have the necessary tools to compute the external inductance. Since the magnetic field outside of a round wire is equivalent to the field of a filament placed at the origin of the wire, we can reuse the results from the previous example for the magnetic field of two filaments. For the external inductance, we integrate this expression between the wires (excluding the conductors)

$$L = \int_a^{d-a} B_y(y = 0)dx \tag{5.77}$$

Recall that $B_y(0)$ is given by

$$B_y = \frac{-\mu I(x+d)}{2\pi \left((x+d)^2 + y^2\right)} - \frac{\mu I(x-d)}{2\pi \left((x-d)^2 + y^2\right)}\bigg|_{y=0} \tag{5.78}$$

Substituting for B_y we have the following easy integral

$$L = \frac{\mu I}{2\pi} \int_a^{d-a} \left(\frac{1}{x} - \frac{1}{x-d}\right) dx \tag{5.79}$$

so that the external inductance per unit length is given by

$$L = \frac{\mu I}{\pi} \ln\left(\frac{d-a}{a}\right) \tag{5.80}$$

The above expression grows logarithmically as a function of separation d. To compute the internal inductance, let us first consider an isolated conductor, or equivalently take $d \gg a$. Then the volume integral of $|B|^2$ inside the wire is given by

$$\int_V B^2 dV = \int_0^{2\pi} \int_0^a B(r)^2 r \, dr \, d\theta \tag{5.81}$$

where

$$B(r)^2 = \frac{\mu^2 I^2 r^2}{4\pi^2 a^4} \tag{5.82}$$

Performing the trivial integral (there aren't too many in this business!), we have

$$\int_V B^2 dV = \frac{\mu^2 I^2}{8\pi} \tag{5.83}$$

and from the energy definition of the inductance, the internal inductance per unit length for the conductor is given by

$$L = \frac{\mu}{8\pi} \tag{5.84}$$

This expression is interestingly independent of the radius of the wire. So for two widely separated wires we double the above result for the internal inductance. The exact solution is more complicated but follows in a similar fashion.

Magnetic vector potential

As we have seen, the concept of the magnetic vector potential comes about directly as a consequence of Maxwell's equations. The magnetic vector **B** is divergenceless

$$\nabla \cdot \mathbf{B} \equiv 0 \tag{5.85}$$

due to the absence of magnetic monopoles. Since $\nabla \cdot \nabla \times \mathbf{F} \equiv 0$ is an identity for any vector function **F**, it follows that

$$\mathbf{B} = \nabla \times \mathbf{A} \tag{5.86}$$

We call the vector **A** the vector potential in analogy with the electric scalar potential ψ. At first, though, it seems like the vector **A** is superfluous since it too is a vector and not a scalar as in the case of ψ. Furthermore there is some physical ambiguity in **A** since its divergence is arbitrary.

The utility in defining **A**, though, is clear when we compare the defining equation for **A** compared with **B**. Under magnetostatic conditions the vector function **A**(r) is given by

$$\mathbf{A}(r) = \mu \int_{V'} \frac{\mathbf{J}(\mathbf{r}')}{4\pi R} dV' \tag{5.87}$$

As before R is the distance between the source and field point and the volume of integration is over all space where $\mathbf{J} \neq 0$.

Compare this equation with the Biot-Savart Law shown in Eq. (5.65). The vector **A** is in the direction of the current and no cross products are involved in the computation. In fact, if we partition the vector $\mathbf{J} = J_x \hat{\mathbf{x}} + J_y \hat{\mathbf{y}} + J_z \hat{\mathbf{z}}$ into Cartesian coordinates, then Eq. (5.87) reduces to three scalar integrals identical in form to the calculation of the electric scalar potential in terms of charge density.

The vector potential **A** is related to inductance as follows. Substitute $\mathbf{B} = \nabla \times \mathbf{A}$ into Maxwell's equation

$$\nabla \times \mathbf{E} = -\frac{\partial \mathbf{B}}{\partial t} = -\nabla \times \frac{\partial \mathbf{A}}{\partial t} \tag{5.88}$$

Since the curl operator is associative, we have

$$\nabla \times \left(\mathbf{E} + \frac{\partial \mathbf{A}}{\partial t} \right) = 0 \tag{5.89}$$

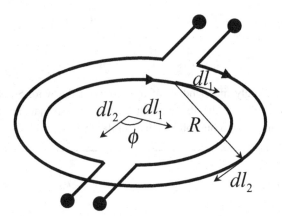

Figure 5.18 Two magnetically coupled filamentary loops.

and this implies that the curl-free expression in the parentheses, similar to the electrostatic field, can be written as the gradient of a scalar function

$$E + \frac{\partial A}{\partial t} = -\nabla \phi \qquad (5.90)$$

The sign of the scalar potential is arbitrarily chosen to correspond to the electrostatic case. In other words, the electric field at any point in space has two components

$$E = -\frac{\partial A}{\partial t} - \nabla \phi \qquad (5.91)$$

If we integrate the above expression around a closed loop, the second term is zero and we are only left with the integral of the first term

$$V_L = \oint_C E \cdot dl = -\frac{d}{dt} \oint_C A \cdot dl \qquad (5.92)$$

leading to an expression for the inductance in terms of the magnetic vector potential

$$L = \frac{1}{I} \oint_C A \cdot dl \qquad (5.93)$$

We could have derived the above result directly from Stoke's Theorem

$$\int_S B \cdot dS = \oint_C (\nabla \times B) \cdot dl \qquad (5.94)$$

but the above derivation gives us more physical insight into the magnetic vector potential. In fact, Eq. (5.91) justifies the name *potential*. Under general circumstances, the electric field is the gradient of the scalar potential and the time derivative of the vector potential. Both operations involve differentiation. The time derivative with the vector potential reminds us that magnetic voltages are only induced under dynamic conditions. From Eq. (5.91) the SI units of A must be volt seconds per meter (V · s/m, or since the units of B can be specified in Tesla (T) or weber per square meter (W/m²), we also have Tesla meter (T · m) or weber per meter (W/m).

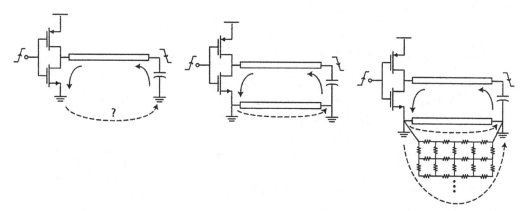

Figure 5.19 (a) An integrated circuit wire carries AC current to a capacitor load. (b) The return current can flow through an explicit path, or (c) through a distributed path formed by the substrate and ground network.

5.5 Partial inductance and return currents

Often people speak of the inductance of an "open" path, such as a piece of wire or a metal trace in an integrated circuit. Quite often, you hear about the inductance of a bond wire or a lead on a package. What are they talking about?

In fact, people are speaking of *partial* inductance. In other words, consider the inductance of the integrated circuit trace shown in Fig. 5.19a. Imagine that we drive this wire with a voltage source at one end through a capacitor load at the other end. Where does the "return current" come from? This depends on how the circuit is laid out. For instance, a separate trace in parallel to the first can carry the return current back to the source, and thus the inductance is defined by this loop (Fig. 5.19b). But often an explicit ground return is not included, and the return current flows through the ground network. Since the substrate is often grounded, the return current can also flow partially through the substrate and thus the inductance is now different since the "loop" is different (Fig. 5.19c). In fact, with the ground and substrate we have a dilemma since we don't know the exact current distribution. At high frequency, we imagine that most of the current will flow close to the surface in the vicinity of the trace. But at low frequencies, the current flow is more widespread.

Nevertheless, people still say that this trace has "so and so" inductance. They are talking about the inductance of the loop assuming the return current is at infinity. In other words, imagine calculating the inductance through the vector potential as follows. As suggested by Fig. 5.20, if we integrate Eq. (5.93) *partially* through several open paths, then the voltage induced at the terminals of the ith curve is given by

$$V_i = \int_{C_i} \mathbf{E} \cdot \mathbf{dl} \tag{5.95}$$

If we ignore the contribution from ϕ in Eq. (5.91) and only consider the contribution from the vector potential, we then have

$$\hat{V}_i = -\frac{d}{dt} \int_{C_i} \mathbf{A} \cdot \mathbf{dl} \tag{5.96}$$

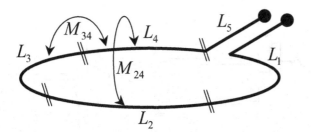

Figure 5.20 Breaking the inductance into partial self and mutual inductance terms.

Due to linearity, we can partition **A** into several terms

$$\mathbf{A} = \mathbf{A}_1 + \mathbf{A}_2 + \cdots \tag{5.97}$$

where each term represents the contribution of segment i to the magnetic vector potential

$$\mathbf{A_i} = \frac{\mu}{4\pi} \int_{V_i} \frac{\mathbf{J}_i}{R} dV_i \tag{5.98}$$

The total current I_i flowing into segment i can be taken out of the integral as long as the current distribution is retained in the integral, $\mathbf{J(r)}_i = \mathbf{k(r)}I_i$. Let us define the *partial mutual inductance* between segment i and j as follows

$$M_{ij} = \frac{1}{I_j} \int_{C_i} \mathbf{A}_j \cdot \mathbf{dl} \tag{5.99}$$

so that the voltage around the entire structure is simply given by

$$V = \sum_i \sum_j M_{ij} \frac{d I_i}{dt} \tag{5.100}$$

From the above definition we define the *partial inductance matrix* as the $N \times N$ matrix M with elements M_{ij}. Given vectors v and i representing the segment voltages and currents, we extend Eq. (5.1)

$$v = M \frac{di}{dt} \tag{5.101}$$

5.6 Impedance and quality factor

The circuit concept of impedance can be stated in terms of power

$$Z_{in} = \frac{V}{I} = \frac{V \cdot I^*}{|I|^2} = \frac{P}{\frac{1}{2}|I|^2} = R + jX \tag{5.102}$$

Applying Poynting's Theorem to a "black box," we can write this as

$$Z_{in} = \frac{P_0 + P_\ell + 2j\omega(W_m - W_e)}{\frac{1}{2}|I|^2} \tag{5.103}$$

Note that the resistive component has a radiation term, an ohmic loss term, and possibly a dielectric or permeability loss term

$$R = \frac{P_0 + P_\ell}{\frac{1}{2}|I|^2} \tag{5.104}$$

The reactance is positive if $W_m > W_e$, and negative otherwise

$$X = \frac{2\omega(W_m - W_e)}{\frac{1}{2}|I|^2} \tag{5.105}$$

The quality factor for a "black box" (usually resonator) is defined as follows

$$Q = 2\pi \frac{\text{Peak energy stored}}{\text{Energy loss per cycle}} \tag{5.106}$$

The denominator can be reformulated in terms of the average power loss to give

$$Q = \omega \frac{\text{Peak energy stored}}{P_\ell} \tag{5.107}$$

From Poynting's Theorem, the net stored energy is given by $W_m - W_e$, so we may be tempted to write

$$Q \overset{?}{=} \omega \frac{W_m - W_e}{P_\ell} = \frac{1}{2} \frac{\Im(Z_{in})}{\Re(Z_{in})} \tag{5.108}$$

But the peak energy is different from the *net* energy. In a LC resonator, the energy is shared between the inductor and capacitor in a time-varying fashion. The peak energy can be computed from the capacitor when $i = 0$ or from the inductor when $v_C = 0$.

$$Q = \omega \frac{|W_m| + |W_e|}{P_\ell} = \omega \frac{W_m^{peak}}{P_\ell} = \omega \frac{W_e^{peak}}{P_\ell} = \frac{\omega L}{R} = \frac{1}{\omega C R} = \frac{X}{R} \tag{5.109}$$

For a single one-port element *not* in resonance, we often define the Q factor as

$$Q = \frac{\Im(Z_{in})}{\Re(Z_{in})} = \frac{X}{R} \tag{5.110}$$

We see that this is correct under the assumption that the one-port forms a resonant circuit!

5.7 Frequency response of inductors

Skin effect

At low frequencies when the DC resistance of an inductor dominates over the reactance, for example when $R \gg \omega L$, then current flows evenly through the cross section of a uniform material of a given conductivity. Thus when computing the low-frequency inductance (strictly the DC inductance[1]) we have non-zero flux in the volume of conductors and to

---

[1] The concept of DC may seem meaningless since at DC there is no induction and thus inductance plays no part. However, we shall see that inductance is a weak function of frequency so that even at moderate frequencies, the DC inductance remains valid.

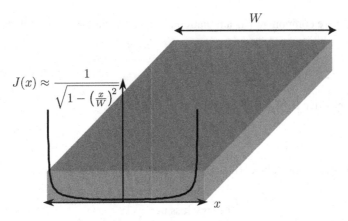

Figure 5.21 Non-uniform current flow through conductors at high frequency. The shown functional dependence is an approximation to the edge discontinuity for rectangular conductors.

calculate the inductance correctly we have to take this into account. At increasing frequency, the internal current distribution redistributes, changing the *internal impedance* of conductors. Ignoring displacement current, the internal impedance is given by

$$Z_{int} = \frac{V_{res}}{I} = \frac{1}{I} \int_C \frac{\mathbf{J}}{\sigma} \cdot \mathbf{dl} \tag{5.111}$$

The above equation is simply the line integral of the electric field $\mathbf{E}$ along the path of a resistive material of conductivity σ. At low frequency when $J = I/A$ is uniform we have the familiar Ohm's Law expression

$$Z_{int} = \frac{1}{I} \int_C \frac{Idl}{\sigma A} = R \tag{5.112}$$

At high frequency $\mathbf{J} = \sigma \mathbf{E}_{tot}$ where $\mathbf{E}_{tot}$ is the total vectorial sum field, including the applied and induced fields. As we alluded to earlier when discussing the internal and external inductance, the induced field is stronger along the outer edge of a loop (which encloses more flux) than the inner loop. Since the induced field opposes the applied field, as shown in Fig. 5.21, we expect the current density to peak near the inner edge of the conductor. In fact, this is exactly what is observed experimentally or by solving Maxwell's equations directly.

We shall show that at increasingly high frequencies the non-uniformity of the current distribution dominates completely and the internal flux is driven to zero. Thus, the infinite frequency asymptote for inductance should equal the value obtained by the external flux

$$L_\infty = \frac{\psi_{ext}}{I} \tag{5.113}$$

Note that the DC inductance

$$L_{DC} = \frac{\psi_{ext} + \psi_{int}}{I} \tag{5.114}$$

is larger only by ψ_{int}. Since for usual geometries $\psi_{ext} \gg \psi_{int}$ and $\psi_{int}(f)$ is a decreasing function of frequency, the inductance decreases only slightly with frequency. For instance,

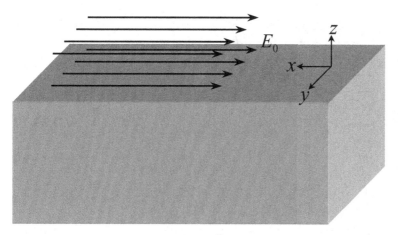

Figure 5.22 Current flow along the surface of a semi-infinite plane.

if the internal flux only accounts for 5% of the total flux, then the high-frequency inductance will only differ by 5% from the DC value.

Surface impedance

In this section we will justify some of the above arguments by solving Maxwell's equations directly. For a simple analytic solution we have to limit ourselves to an unrealistic one-dimensional geometry shown in Fig. 5.22 but the insight afforded by this simple example applies in a surprising number of more general situations.

For a semi-infinite plane supporting a conductive current, we have the following time-harmonic Maxwell equations

$$\nabla \times \mathbf{H} = (\sigma + j\omega\epsilon)\mathbf{E} \tag{5.115}$$

The displacement current $j\omega\epsilon\mathbf{E}$ is negligible in comparison to the conductive current $\sigma\mathbf{E}$. This is true for good conductors and in fact a useful definition for a good conductor. Taking the curl of Maxwell's curl equation for an electric field we have

$$\nabla \times \nabla \times \mathbf{E} = -j\omega\mu\nabla \times \mathbf{H} \tag{5.116}$$

and substituting for the curl of $\mathbf{H}$ from above while dropping the displacement current

$$\nabla \times \nabla \times \mathbf{E} = -j\omega\mu\sigma\mathbf{E} \tag{5.117}$$

The above differential equation describes the spacial variation of the components of the vector $\mathbf{E}$. For simplicity we assume that $\mathbf{E}$ is x-directed so that $\mathbf{E} = \hat{x}E_x$ and furthermore we assume that $\mathbf{E}$ is uniformly incident on the plane. Simplifying the above equation we are led to the following scalar differential equation

$$-\left(\frac{\partial^2 E_x}{\partial y} + \frac{\partial^2 E_x}{\partial z}\right) = -j\omega\mu\sigma E_x \tag{5.118}$$

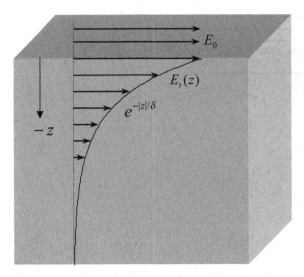

Figure 5.23 The decay of electric field into the depth of a semi-infinite plane conductor.

We assume a uniform wave incident on the plane and thus $\partial E_x / \partial y = 0$ for $z > 0$. By the symmetry of the plane, we can extend this condition to the interior of the plane and we are thus faced with the following simple first-order equation

$$\frac{d^2 E_x}{dz} = j\omega\mu\sigma E_x \tag{5.119}$$

The above equation is readily solved by substituting the trial solution $E_0 \exp(\alpha x)$

$$\alpha^2 E_x = j\omega\mu\sigma E_x \tag{5.120}$$

And thus

$$\alpha = \pm\sqrt{j\omega\mu\sigma} = \pm\sqrt{j}\sqrt{\omega\mu\sigma} \tag{5.121}$$

Since the fields must be finite as $z \to -\infty$, we keep the positive root. Working with polar coordinates, verify that the first term simplifies to $\sqrt{j} = (1 + j)/\sqrt{2}$. Let us define the constant

$$\delta \equiv \sqrt{\frac{2}{\omega\mu\sigma}} \tag{5.122}$$

With this definition we have the following form for the electric field inside the material at depth $z < 0$

$$E_x(z) = E_0 e^{z/\delta} e^{jz/\delta} \tag{5.123}$$

We call δ the *penetration depth* parameter or more commonly the *skin depth*. The motivation for this is clear since δ has units of length (m) and represents the decay rate of the field into the conductor. This is illustrated in Fig. 5.23. Since $\mathbf{J} = \sigma\mathbf{E}$ it is easy to see that the current density decays in a similar manner to the electric field. It can be shown that the magnetic field also diminishes exponentially.

Table 5.1 *The conductivity and penetration depth of various conductors at room temperature*

Material	Conductivity (S/m)	Skin Depth (μ)		
		100 MHz	1 GHz	10 GHz
Copper	5.80×10^7	6.6	2.1	0.66
Aluminum	3.72×10^7	8.2	2.6	0.82
Gold	4.44×10^7	7.6	2.4	0.76
Silver	6.17×10^7	6.4	2.0	0.64
Brass	1.57×10^7	12.7	4.0	1.27

For static fields it is easy to show that $E \equiv 0$ inside perfect conductors. Now we see more generally that AC fields decay to zero in *good* conductors. The rate of decay is quite rapid since σ is large for good conductors. In Table 5.1 we quantify δ for typical materials and some typical frequencies. For instance, for Cu at 1 GHz, the skin depth is only 2.1 μ. For this reason, δ is also called the *skin depth* since current flows along the outer skin of good conductors.

It is interesting to note that the phase term in Eq. (5.123) causes the electric field to flip direction inside the material at depth $z = \pi \delta$, thus causing the current to travel "backwards"! What causes this funny behavior? It is the magnetically induced currents. To observe this, note that time-varying conduction current creates a time-varying magnetic field which in turn induces a time-varying current opposing the original current. In the steady state, therefore, the current distribution takes on the form we have described.

It is reasonable to assume that the above solution can be applied to the *interior* of a *finite* rectangular conductor of width w and thickness t as long as the thickness is much greater than the skin depth. We do not expect the solution to apply to the edges of the rectangular conductor and in fact we shall later see that the field behavior at the edges is complicated. A round conductor of large radius (compared with the skin depth) should also have approximately the same field distribution, and, due to the absence of the edge discontinuity, the results should be even more applicable. In the next section we shall have occasion to make this comparison quantitative.

Since the current distribution is non-uniform, we employ Eq. (5.111) to obtain the internal impedance of the rectangular conductor. For the path of integration, we take the path along the surface at $z = 0$, where $\mathbf{E} = E_x \hat{x}$

$$Z_{int} = \frac{1}{I} \int_0^L E_x dx = \frac{E_x L}{I} \tag{5.124}$$

so the internal impedance per unit length, or the *surface impedance* is given by

$$Z_s = \frac{W}{L} Z_{int} = W \frac{E_x}{I} \tag{5.125}$$

The total current is given by integrating $\mathbf{J}$ along the depth of the conductor

$$I = W \int_{-\infty}^0 J(z) dz = W \sigma E_x \int_0^\infty e^{-(1+j)z/\delta} = W \delta \sigma E_x / (1 + j) \tag{5.126}$$

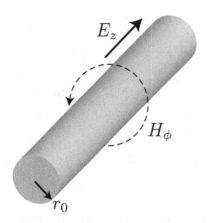

Figure 5.24 A round wire of infinite length has cylindrical symmetry.

Substituting in Eq. (5.125) we have

$$Z_s = \frac{(1+j)}{\sigma\delta} = (1+j)R_s \tag{5.127}$$

Does the above equation look familiar? The real part looks like the *sheet resistance* ($\Omega/\square$) of a material of conductivity σ and thickness δ. In other words, at frequency ω our rectangular slab appears the same as a slab at zero frequency with uniform current flowing through a thickness δ. Thus the resistance of the slab increases $\sqrt{\omega}$ as a function of frequency since $\delta \propto \sqrt{\omega}$. We can also observe from the imaginary part of Z_s that the *internal inductance* decreases at high frequency since the magnetic strength is reduced in the volume of the conductor.

Impedance of round wires

For a long round wire, the current J_z is invariant with z and the angle θ as shown in Fig. 5.24. Therefore, the Helmholtz equation simplifies

$$\nabla^2 \mathbf{J} = j\omega\mu\sigma\mathbf{J} = \tau^2\mathbf{J} \tag{5.128}$$

$$\frac{\partial^2 J_z}{\partial r^2} + \frac{1}{r}\frac{\partial J_z}{\partial r} + \tau^2 J_z = 0 \tag{5.129}$$

Two linearity independent solutions are the Bessel function and the Hankel function of the first kind

$$J_z = A J_0(\tau r) + B H_0^{(1)}(\tau r) \tag{5.130}$$

Since the Hankel function has a singularity at $r = 0$, it cannot be a solution. Normalizing to the current at the surface of the wire

$$J_z = \frac{\sigma E_0}{J_0(\tau r_0)} J_0(\tau r) \tag{5.131}$$

A plot of the current density in the wire is shown in Fig. 5.25. In the plot, a Cu wire with 1 mm diameter is used. Note that at low frequencies the current is essentially

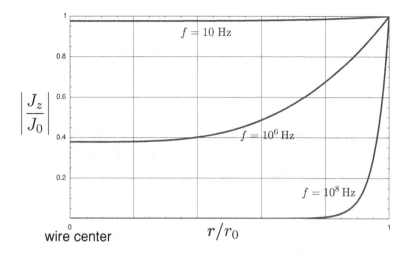

Figure 5.25 The current density inside of a round wire as a function of the radius. The origin represents the center of the wire. The current is uniform at low frequency but concentrated at the skin at high frequency.

uniform. At high frequency, though, the current decays exponentially as we penetrate the conductor.

In the limit that the radius is large, or equivalently $r_0/\delta \gg 1$, then the wire should behave like our plane conductor. In fact

$$\left| \frac{J_z}{\sigma E_0} \right| \approx e^{-(r_0-r)/\delta} \tag{5.132}$$

The impedance of a round wire can be computed by noting that only E_z and H_ϕ are present. Furthermore, we have

$$\oint \mathbf{H} \cdot \mathbf{dl} = I = 2\pi r_0 H_\phi \tag{5.133}$$

By $\nabla \times \mathbf{E} = -j\omega\mu\mathbf{H}$, it is easy to show that

$$H_\phi = \frac{1}{j\omega\mu} \frac{\partial E_z}{\partial r} \tag{5.134}$$

Using $E_z = J_z/\sigma = E_0 J_0(\tau r)/J_0(\tau r_0)$, the magnetic field is given by

$$H_\phi = \frac{E_0 \tau}{j\omega\mu} \frac{J_0'(\tau r)}{J_0(\tau r_0)} \tag{5.135}$$

Recall that $J_0'(x) = -J_1(x)$. Solving for the current

$$I = \frac{2\pi r_0 \sigma E_0}{\tau} \frac{J_1(\tau r_0)}{J_0(\tau r_0)} \tag{5.136}$$

Finally, we can write the internal impedance of the wire

$$Z_i = \frac{E_z(r_0)}{I} = \frac{\tau J_0(\tau r_0)}{2\pi r_0 \sigma J_1(\tau r_0)} \tag{5.137}$$

The above expression involves the Bessel function and its derivative of complex argument. While these functions have been amply tabulated, any modern computation environment such as `matlab` or `mathematica` can be used to evaluate the above expression.

In the low-frequency limit, the internal impedance of the wire reduces to

$$Z_i \approx \frac{1}{\pi r_0^2 \sigma}\left[1 + \frac{1}{48}\left(\frac{r_0}{\delta}\right)^2\right] + j\omega\frac{\mu}{8\pi} \tag{5.138}$$

The real part corresponds to a correction to the DC resistance of the wire (per unit length). The imaginary term corresponds exactly to the static internal inductance of the wire. As expected, the high-frequency limit matches the analysis for the semi-infinite plane

$$Z_i = \frac{(1+j)R_s}{2\pi r_0} \tag{5.139}$$

This is because at high frequency the perimeter of the conductor is much larger than the penetration depth δ and thus the plane conductor analysis becomes valid.

Approximate impedance of rectangular wires

The resistance of a rectangular wire as a function of frequency can be obtained numerically and fitted to the following equation [25]

$$R = \frac{l}{\sigma wt}\left[\frac{0.43093 x_w}{1 + 0.041\left(\frac{w}{t}\right)^{1.19}} + \frac{1.1147 + 1.2868 x_w}{1.2296 + 1.287 x_w^3} + 0.0035\left(\frac{w}{t} - 1\right)^{1.8}\right] \tag{5.140}$$

where $x_w = \sqrt{2 f\sigma\mu wt} \geq 2.5$ and for $x_w < 2.5$

$$R = \frac{l}{\sigma wt}\left[1 + 0.0122 x_w^{3+0.01 x_w^2}\right] \tag{5.141}$$

The above expression has an accuracy between 3% and 5% in the range of $w/t < 12$ and $x_w < 20$. It is interesting to compare the above result with the surface impedance calculation to see how much the edge effects alter the resistance.

Lossless inductors

The above discussion only applies to materials with finite conductivity. For lossless materials, the fields diminish in the volume of conductors and an infinitely dense current flows on the surface of conductors. The tangential electric field is zero inside and just outside of the conductors. The tangential magnetic field H_t is also zero inside of the volume of conductors but has a jump discontinuity equal to the surface current density J_s just outside the space of conductors

$$H_t = J_s \tag{5.142}$$

Thus the flux is independent of frequency and we do not expect the same variation in inductance due to the skin effect. On the other hand, the effective inductance will vary due to capacitive effects.

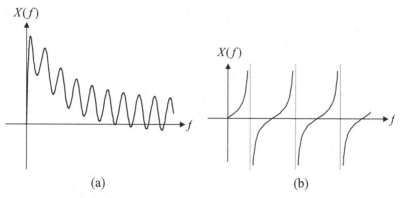

Figure 5.26 (a) The reactance of a lossy passive device as a function of frequency. (b) The reactance of a lossless passive device as a function of frequency.

Consider the complex power flowing into a black box shown in Fig. 5.9. If we integrate the Poynting vector $\mathbf{S} = \mathbf{E} \times \mathbf{H}^*$ across a surface S enclosing the black box

$$P = \frac{1}{2} \oint_S \mathbf{E} \times \mathbf{H}^* \cdot d\mathbf{S} = P_l + 2j\omega(W_m - W_e) \tag{5.143}$$

From electromagnetic theory [52] we have written the imaginary part of the above expression as the difference between the magnetic and electric stored energy. The reactance of the black box is thus related to the energy stored as follows

$$X = \frac{4\omega(W_m - W_e)}{II^*} \tag{5.144}$$

For an inductor $W_m > W_e$ and ideally $W_m \gg W_e$ so that the reactance of an inductor is positive. Furthermore it can be shown [52] that

$$\frac{dX}{d\omega} = \frac{4(W_m + W_e)}{II^*} \tag{5.145}$$

Please note that the above result is not a simple derivative of Eq. (5.144) (notice the change in sign). This shows that the derivative of the reactance is positive. This is a surprising result since it says that a lossless inductor cannot have a frequency-dependent profile such as shown in Fig. 5.26a, but it must resemble Fig. 5.26b. In other words, the reactance of an inductor is positive and *always* increasing as a function of frequency. The inductance as a function of frequency

$$L(\omega) = \frac{X}{\omega} \tag{5.146}$$

has slope

$$\frac{dL}{d\omega} = \frac{1}{\omega} \frac{dX}{d\omega} - \frac{X}{\omega^2} \tag{5.147}$$

Notice that every term in the above equation is positive for $\omega > 0$. From circuit theory, at low frequencies we expect

$$\frac{dX}{d\omega} = \frac{X}{\omega} \tag{5.148}$$

so that $dL/d\omega = 0$, implying a constant inductance. If the slope of the reactance is ever smaller than the reactance divided by frequency, the inductance will have a negative slope. But comparing Eq. (5.144) to Eq. (5.145), it is clear that $\frac{dX}{d\omega} \geq \frac{X}{\omega}$ so that

$$\frac{dL}{d\omega} \geq 0 \tag{5.149}$$

It is important to qualify the above result. First, we have assumed lossless materials in the construction of the inductor. Second, we have defined inductance by (5.146). Earlier we defined the inductance in terms of the magnetic energy alone. We should really call this the *effective* inductance since at high frequencies it will contain both magnetic and electrical energy. It is nevertheless interesting to notice that the effective inductance has a positive slope as a function of frequency somewhat counterintuitively. This result, though, is consistent with circuit theory. If we calculate the effective inductance of an LC tank circuit, we have

$$L_{eff} = \frac{L}{1 - LC\omega^2} \tag{5.150}$$

and for $\omega < \omega_0 \equiv 1/\sqrt{LC}$ the inductance is positive and increasing. As $\omega \to \omega_0$, the effective inductance takes on arbitrarily high values.

5.8 Quality factor of inductors

Using the fundamental definition of the Q factor as the ratio of energy storage to energy loss in a practical inductor over a cycle of excitation

$$Q \equiv \frac{\text{energy stored}}{\text{energy loss}} \tag{5.151}$$

For a physical inductor that has resistance R at frequency ω, in one cycle the inductor dissipates

$$E_{loss} = \frac{1}{2}|I|^2 RT = \frac{1}{2}|I|^2 R \frac{2\pi}{\omega} \tag{5.152}$$

Similarly, the energy stored in an inductor is

$$E_{stored} = \Im \left[\frac{1}{2}IV^*T \right] = \frac{1}{2}|I|^2 \omega L \frac{2\pi}{\omega} = \frac{1}{2}|I|^2 L 2\pi \tag{5.153}$$

and thus

$$Q = \frac{\omega L}{R} \tag{5.154}$$

The quality factor of inductors is important in many applications. For instance, a resonant tank exchanges energy between the magnetic and electric fields every cycle. If a fraction $1/Q$ of the stored energy is lost every cycle, then the system can sustain itself for about Q cycles. This determines, for instance, the power that must be replenished to the tank to sustain oscillations. While active devices such as transistors can be used to do this, these devices always add undesirable noise to the system.

To improve an inductor, therefore, we must minimize the various loss mechanisms associated with the inductor.

Inductor loss mechanisms

Conductive losses

At low frequencies, the resistance of an inductor is simply related to the DC resistance of the windings that make up the inductor. Since the resistance is an increasing function of temperature, the on-chip temperature should be estimated in determining the losses.

Internally induced losses

At high frequencies the effective resistance of an inductor is complicated by non-uniform current distribution in the interior of conductors. We have already seen that the *skin effect* tends to decrease the cross-sectional area of the device and thus we expect a resistance increase proportional to $\sqrt{f}$ due to this effect alone.

The physical asymmetry in the geometry of the structure will only exacerbate this. For instance, we observed that in a loop inductor the current tends to crowd to the inner radius of the conductor.

Externally induced losses

The proximity of nearby conductors also influences loss in an inductor. Due to the time-varying flux, *eddy currents* are induced to flow in conductors. As shown in Fig. 5.27, the effective resistance of these eddy current loops appears in series with the inductor. If we assume that the magnetic field of the induced currents interacts weakly with the magnetic field of the inductor, then we can neglect the change in inductance due to these currents and calculate the eddy currents directly as $j\omega\mathbf{A}$, where $\mathbf{A}$ is determined by the inductor currents alone. Otherwise, we must compute $\mathbf{A}$ by including the inductor current as well as the induced currents and this is a complicated situation.

Magnetization losses

Hitherto we have implicitly assumed that the material properties are linear, isotropic, and real. For most non-magnetic or weakly magnetic materials these are good assumptions. For magnetic materials, though, the magnetic permeability μ is non-linear and frequency dependent and dissipative. The exact physical cause of this is beyond the scope of this chapter

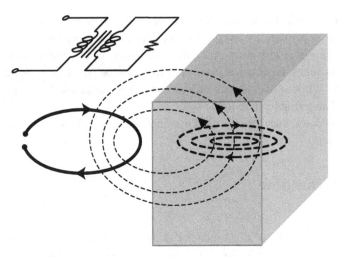

Figure 5.27 Model for eddy current losses as effective series losses.

and requires analysis of the quantum mechanics and thermodynamics of the constituent atoms. We can sweep such details under the rug and assume that $\mu = \mu' + j\mu''$ is a complex frequency-dependent quantity due to losses. The Q factor is thus bounded by μ'/μ'' due to the lossy part of the volume integral of $\mathbf{B} \cdot \mathbf{H}$ over all space.

Displacement current losses

Inevitably every real inductor stores some electric energy. We have up to now neglected the small displacement current $\mathbf{J}_d = j\omega\epsilon\mathbf{E}$ that flows in the volume of the problem space. If the material in question is lossy, then the permittivity has an imaginary component $\epsilon = \epsilon' - j\epsilon''$ as well. Physically, the electrically induced polarization in the material drains energy from the field and thus acts like a loss mechanism. The ratio of the imaginary part of the permittivity to the real part is known as the *loss tangent*

$$\tan\delta = \frac{\epsilon''}{\epsilon'} \tag{5.155}$$

and this is specified for various materials at various frequencies. We will defer a detailed discussion of the loss tangent. For now we will merely point out that the loss tangent acts as an additional loss mechanism when an inductor stores electric energy in addition to magnetic energy.

Radiative losses

For inductors that are large relative to the wavelength $\lambda = c/f$, additional losses can occur. For instance, consider breaking the inductor into many small sections and employing the partial inductance concept, the induced field at segment i due to current in segment j is simply $j\omega A_j$. When segments i and j are in close proximity, the vector $\mathbf{A}$ is purely real and so the induced field only stores energy. For points far removed, though, there is a propagation delay between these two points which adds an additional phase to $\mathbf{A_j}$ when evaluated at segment i. This additional phase delay causes the induced fields to have a *real* component, which appears as additional resistance. Physically, energy is lost to *radiation*, an electromagnetic wave which travels into space.

Overall quality factor due to multiple loss mechanisms

Consider the basic definition of the Q factor as the ratio of energy storage to energy loss. If we expand the various loss mechanisms we have

$$Q = \frac{W_{stored}}{W_{loss}} = \frac{W_{stored}}{W_{loss,1} + W_{loss,2} + \cdots} \tag{5.156}$$

Taking the reciprocal of the above we have

$$\frac{1}{Q} = \frac{W_{loss,1} + W_{loss,2} + \cdots}{W_{stored}} = \frac{1}{Q_1} + \frac{1}{Q_2} + \cdots \tag{5.157}$$

And thus the overall Q can be computed as the "parallel" combination of the individual Qs. The overall Q factor is therefore limited by the smallest Q factor and is always less than this quantity. As expected, when one loss mechanism dominates, the Q factor is approximately determined by this loss term. For instance, for an inductor constructed over an insulating substrate (such as GaAs) the displacement current losses may be orders of magnitude smaller than the ohmic losses and can be safely neglected. This is in stark contrast to Si, where the conductive substrate contributes significantly to the loss at high frequency.

5.9 Inductors and switching circuits

Ideal inductors are DC shorts. With the absence of resistive loss, the voltage drop along the path of an inductor is equal to the time rate of change of flux, or

$$V_L(t) = \frac{d\psi}{dt} = L\frac{dI}{dt}$$

which is zero for static fields. More interesting, though, is that this implies that the average voltage across an inductor is almost always zero. To see this, consider the Laplace Transform of the voltage. The average value can be obtained by taking the limit as $s \to 0$. Assuming the Laplace Transform of the current does not have any zeros we have

$$V_L(s \to 0) = \lim_{s \to 0} I_L s L = 0 \tag{5.158}$$

A similar conclusion applies for period excitation if we replace $s = j\omega$. Thus the average voltage across an inductor is zero. Therefore, for most periodic waveforms across an inductor, the average value of the steady-state voltage inductor must be zero.

As an application of this concept, consider Fig. 5.28a where an inductor terminal is periodically shorted to ground by an ideal switch with a duty cycle of say 50%. What is the steady-state voltage at node v_1 as a function of time? For 50% of the period, the voltage v_1 is zero due to the ideal switch applying the full rail voltage across the inductor terminals.

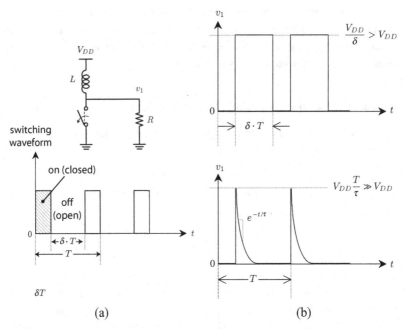

Figure 5.28 (a) Simple LR circuit switching circuit. (b) Approximate voltage waveforms at node v_1 for the case of $\tau \gg 1$ (top) and $\tau \ll 1$ (bottom) where $\tau = L/R$ is the characteristic time constant of the circuit.

In order to maintain an average value of zero, then the voltage v_1 must increase to at least a voltage of $2 \times V_{dd}$ for the remaining off cycle, a voltage above the supply! In fact, if we assume that the inductor is sufficiently large such that $\tau = L/R \gg 1$, then we will observe a steady-state voltage waveform similar to the top plot of Fig. 5.28b. In this case the energy stored in the inductor is sufficiently large that the energy drain during the off-period does not significantly reduce the magnetic energy of the inductor. Stated differently, the inductor is able to maintain a constant current.

On the other hand, if the inductor is small such that $\tau = L/R \ll 1$, then the magnetic energy in the inductor will exponentially drain and the steady-state voltage waveform will resemble the bottom graph of Fig. 5.28b.

Since the output voltage is larger than the supply voltage, this simple circuit principle can be used to build a voltage multiplier. In fact, as the duty cycle of the on period is increased, the voltage during the off-period has to take on correspondingly larger values to sustain a zero average voltage. By using feedback, we can generate a constant voltage by regulating the duty cycle of the waveform to maintain a constant DC voltage over the changing loads.

While at DC, inductors are shorts, at AC they in fact look like "bad" current sources. Just as capacitors at high frequency look like bad voltage sources, inductors have the dual behavior. A constant current drawn from an inductor produces no voltage drop. Inductors present increasingly more reactance to any attempts at changing the current. In many high-frequency circuits this behavior is exploited and large inductors, or RF chokes, are employed to deliver bias currents without disturbing the high-frequency behavior of circuits.

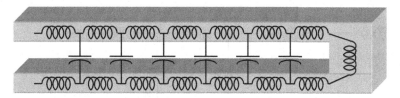

Figure 5.29 The capacitive coupling in a simple loop inductor.

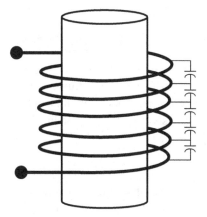

Figure 5.30 The capacitive coupling in a coil inductor.

5.10 Preview: how inductors mutate into capacitors

In Section 5.7 we observed that a structure appears inductive as long as the stored magnetic energy dominates over the electrically stored energy, $W_m > W_e$. Consider the loop inductor shown in Fig. 5.29. Schematically we have shown two paths between the terminals, a *conductive* path and a *displacement* path. The displacement path occurs by the time-varying electric field between the conductors. This electric field only exists if there is a net charge buildup on the two "plates" of the capacitors. It is important to observe that the actual displacement and conductive current paths are distributed and occur side by side. Thus our structure is really both an inductor and a capacitor. In Fig. 5.30 we have shown a common coil inductor and some representative capacitive elements that couple energy electrically from turn to turn of the inductor. While the coupling between adjacent turns is shown, every turn couples to every other turn in the coil.

Define the inductance per unit length as L' and the capacitance per unit length as C'. At low frequencies the loop presents a low impedance since $\omega L'$ is small, whereas $1/(\omega C')$ is large and thus the bulk of the current flows through the loop. Overall the behavior appears as inductive. At high frequencies, though, the inductive reactance is increasing, whereas the capacitive reactance is decreasing, so that there will be a frequency point when $1/(\omega C') = \omega L'$. At this point we may guess that the structure will no longer behave as an inductor and will in fact mutate into a capacitor. While the overall conclusion is correct, the frequency when this occurs is predicted incorrectly due to distributed effects.

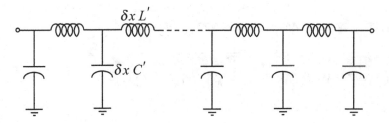

Figure 5.31 A distributed LC line.

But at what frequency does the structure *resonate*? At resonance the energy stored in the magnetic field is equal to the energy stored in the electric field. The answer to the question comes from *distributed* analysis of the structure. For instance, the loop inductor of Fig. 5.31 can be analyzed with transmission line theory (Chapter 9). The structure shown can be modeled as a shorted transmission line. Defining the inductance per unit length L' and the capacitance per unit length C', the impedance of a shorted transmission line is

$$Z_s = j Z_0 \tan(\omega l / c) \tag{5.159}$$

where l is the length of the line, $c = \frac{1}{\sqrt{L'C'}}$ is the propagation velocity, and $Z_0 = \sqrt{\frac{L'}{C'}}$ is the characteristic impedance of the line. Thus the impedance of the line is inductive until $\omega l / c < \pi/2$. But the tangent function is periodic and thus there are an infinite number of frequency ranges where the above device appears inductive. In terms of the inductance and capacitance per unit length, the circuit resonates at frequency ω if

$$n\frac{\pi}{2} = \beta l = \omega\sqrt{L'C'}l = \omega\sqrt{LC} \tag{5.160}$$

for n an integer. Solving for ω

$$\omega = n\frac{\pi}{2}\sqrt{\frac{1}{LC}} \tag{5.161}$$

The above result contradicts circuit theory. Circuit theory predicts a single resonant frequency at a frequency $2/\pi$ lower than distributed analysis. This is not surprising since circuit theory neglects any spacial variation in the fields. We shall return to this subject in Chapter 11. Since any physical inductor stores some parasitic capacitive energy, no matter how small, there is a frequency limit to the inductive behavior.

5.11   References

The ideas for this chapter come from working in the field for many years and so it is hard to trace the material back to the source. I know many of the ideas were inspired by Feynman's Lectures [18], and a careful reading of *Fields and Waves in Communication Electronics* [52]. In fact, the idea of writing this book is partly a result of reading Chapter 4 of this book and wanting to turn the chapter into an entire book.

6 Passive device design and layout

In this chapter we focus on the design and layout of passive devices. In particular we will emphasize the realization of inductors and capacitors in a integrated circuit process. We begin the discussion with a simple loop inductor. Since the loop is a core building block for more complicated structures, we will spend more time on this structure. By using loops in series and parallel, we can build spirals, solenoids, and other more complex structures. The design and layout of transformers for ICs builds on this background and is covered in detail in Chapter 10. Finally, we cover the design of MIM capacitors in an IC process.

6.1 Ring inductor

A loop or ring inductor is a natural inductor structure because we can cut the loop at any point to introduce the leads into the inductor, as shown in Fig. 6.1. Since the structure is planar, it can be easily incorporated onto a PCB or integrated circuit. A circular loop is approximated in an IC process with a polygon.[1] In an IC process allowing only "Manhattan" geometries, or orthogonal interconnect angles, then a rectangular loop inductor is a good choice. Intuitively, the inductance is a function of the loop area A since by the mean value theorem

$$\psi = \int_A \mathbf{B} \cdot \mathbf{dA} = B_{av} \cdot A$$

The resistance, though, depends on the perimeter of the loop. Thus, to maximize the Q of the structure, we should utilize a shape with maximum area to perimeter. It is well known that such a shape is indeed a circle. But a rectangular loop has only slightly smaller Q factors and is more convenient for layout.

The external inductance of a loop is computed by assuming a filamental current flows at the center of the loop, while computing the flux crossing the inner area enclosed by the loop.

The self-resonant frequency of the coil inductor can be very high, since there is little capacitive coupling in the structure. The lead capacitance will determine the self-resonant frequency for an isolated ring in free space, as shown in Fig. 6.2a. Unfortunately, if the

---

[1] A polygon with 64 sides is practically a circle. Some IC fabrications only allow 45° angles limiting us to an octagon. See Fig. 6.12.

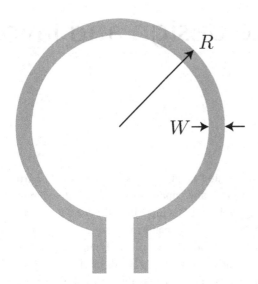

Figure 6.1 A single loop ring inductor structure.

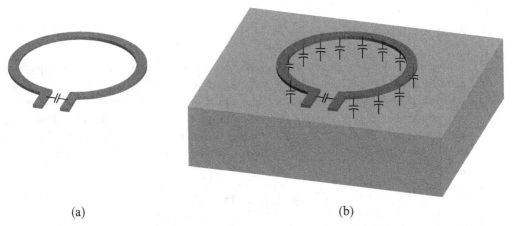

(a)   (b)

Figure 6.2 (a) In an isolated ring the self-capacitance, especially around the leads, sets the self-resonant frequency. (b) When the ring is fabricated on top of a conductive substrate, the distributed capacitance to the ground dominates the self-resonant frequency.

structure is implemented on-chip in Si, there is significant capacitance to the substrate due to the close proximity of the conductive substrate, as shown in Fig. 6.2b. This capacitance can be estimated by the parallel plate portion (proportional to area) and the fringing component (proportional to perimeter).

Skin effect

The resistance of a loop can be lowered by employing wider conductor traces in the loop. Naturally, though, at high frequencies the AC currents take the path of least impedance, as

Figure 6.3 Current constriction in a square ring inductor.

Figure 6.4 A loop inductor employing two symmetric traces in parallel to minimize current constriction.

illustrated in the current density plot of Fig. 6.3. The current flow is non-uniform and tends to flow along the inner side, or lower inductance path. This is easily modeled as two loops in parallel. Since the loop inductance of the inner loop is smaller, more current will take this path. As we split the loop into more and more coupled loops, we can in fact predict this current distribution more accurately.

One way to mitigate this non-uniform current flow is illustrated in Fig. 6.4, where the trace is split into two parallel paths. Each path is balanced as it spends an equal amount of time on the inside and outside of the loop. Thus the current should split uniformly between the two paths, leading to less current crowding in the structure.

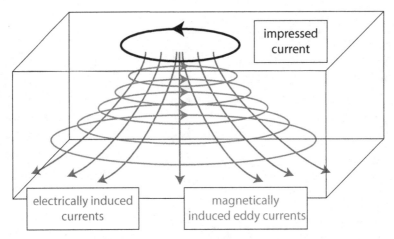

Figure 6.5 The electrically induced current flow in a typical inductor fabricated over a conductive substrate. The magnetically induced eddy currents flow in loops.

Substrate losses

Since the substrate is conductive, the substrate coupling in an inductor introduces loss. The substrate resistance depends strongly on the placement of substrate contacts, the ground connection at the substrate back plane, and the frequency of operation. At low frequencies the current spreads and can reach the back side. At high frequencies, the nearby substrate contacts dominate as current sinks. The flow of capacitive substrate current is shown in Fig. 6.5. It is important to distinguish between the capacitive substrate currents, arising from displacement current, from the substrate-induced eddy currents, flowing due to changing magnetic fields. Since the time-varying magnetic field penetrates the substrate, as shown in Fig. 6.5, the induced emf in the substrate leads to current flow. This current flow causes additional power loss to occur and appears as an effective increase in the series resistance of the ring inductor. For heavily conductive substrates, with $\rho < 0.01\Omega$-cm, this substrate loss can be a significant fraction of the overall losses. For moderately conductive substrate, though, $\rho > 1\Omega$-cm, the induced eddy currents are small and can be neglected at moderately high frequencies ($L < 20$ nH, $f < 10$ GHz).

Observe that there is a worst-case value of substrate resistance that results in the optimal power transfer from the ring to the substrate. Such a value is simply equal to the capacitor reactance (by the maximum power transfer theorem). The only way to eliminate capacitive substrate loss is to employ a substrate with resistivity of zero or infinity. An infinitely resistive substrate, or an insulating substrate, is ideal, but not widely available in Si IC processes. An infinitely conductive substrate, on the other hand, is undesirable since the induced eddy currents would render very low (practically zero) inductance from the ring. This is because the eddy currents generate a magnetic flux that opposes the inductor flux. So in practice, with finite non-zero substrate resistivity, we should strive to place our substrate contacts in such a way to move away from the "optimal" substrate resistance value.

One such approach is to attempt to move the substrate resistance to zero by employing a shield. The shield cannot be solid since the eddy currents would again reduce the inductance value, but a patterned shield can minimize the effect of the eddy currents without disturbing

(a)

(b)

Figure 6.6 (a) A patterned shield is employed to prevent electrical fields from penetrating the substrate while preventing eddy current (closed loops) flow over a large area. (b) A co-planar ground shield improves the quality factor of the ring inductor by intercepting a fraction of the fringing fields from reaching the lossy ground.

the displacement current flow, as shown in Fig. 6.6a. The shield can be constructed with a polysilicon or metal layer and patterned in such a way as to minimize any "loops" in the shield.

The main drawback of the shield is that it increases the substrate capacitance of the structure, leading to a lower self-resonant frequency. Furthermore, it cannot shield the devices from magnetically induced substrate losses. On the other hand, the shield can potentially reduce the amount of energy that is injected into or captured from the substrate, leading to better isolation. This will be discussed in more depth in Chapter 13.

A compromise to the shield structure is the "co-planar" ground shown in Fig. 6.6b. Here a ground plane surrounds the shield and captures most of the fringing fields from reaching the substrate. Since the ground does not form a closed loop around the ring, though, the ground plane does not form a transmission line with the inductor. In fact, if such a transmission line were formed, our ring would turn into a short section of a co-planar transmission and the magnetic flux would shift from the inside of the loop to the outside of the loop, between the current and return current. This would lower the inductance but lead to improved magnetic isolation, since the "monopole" currents of the loop are turned into dipole currents. This technique is very effective at high frequencies, where the desired inductance value allows the application of transmission lines.

6.2 The classic coil

The main limitation of a ring inductor is the achievable value of inductance. To realize higher inductance for a fixed area, we need to add more loops to the structure. By adding more loops in parallel, we can boost the value of inductance. This is accomplished by a spiral staircase style structure of a solenoid inductor, shown in Fig. 6.7. If the turns are wound in very close proximity, then each loop of the structure has a magnetic flux N times

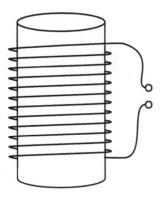

Figure 6.7 A solenoid "coil" inductor structure.

larger than a single loop. Since we traverse N loops, we expect to see an N^2 increase in the inductance. The winding resistance, on the other hand, will only increase like N. So overall the Q is boosted by N.

We can make this simple argument more quantitative as follows. A tightly wound very long solenoid has $B = 0$ outside and $B_x = 0$ inside, so that by Ampère's Law

$$B_y \ell = N I \mu_0 \tag{6.1}$$

where N is the number of current loops crossing the surface of the path. The vertical magnetic field is therefore constant

$$B_y = \frac{N I \mu_0}{\ell} = \mu_0 I n \tag{6.2}$$

The flux per turn is therefore simply given by

$$\Psi_{\text{turn}} = \pi a^2 B_y \tag{6.3}$$

The total flux through N turns is thus

$$\Psi = N \Psi_{\text{turn}} = N \pi a^2 B_y \tag{6.4}$$

$$\Psi = \mu_0 \frac{N^2 \pi a^2}{\ell} I \tag{6.5}$$

The solenoid inductance is thus

$$L = \frac{\Psi}{I} = \frac{\mu_0 N^2 \pi a^2}{\ell} \tag{6.6}$$

The above analysis is very simple and there are many more detailed and sophisticated calculations for the inductance of a coil. The interested reader is encouraged to consult some of the classic references [60] [22].

Rectangular coils

To implement a coil in an IC process, we can create a rectangular coil as shown in Fig. 6.8. Such a structure suffers from a very small cross-sectional area due to the small thickness of

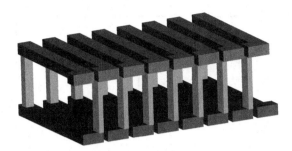

Figure 6.8 A rectangular coil inductor fabricated in a standard IC process. The dimensions of the vias have been exaggerated for illustration purposes.

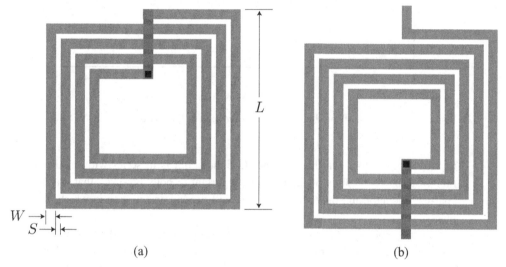

(a)   (b)

Figure 6.9 Square spiral inductor layout with (a) joined terminals, and (b) separated terminals.

the oxide layers. One advantage of such a structure, though, is that the core magnetic field lines are parallel to the substrate. Only the fringing fields penetrate the substrate and thus the eddy currents generated in the substrate are much smaller. More elaborate structures can be constructed with post processing. MEMS post processing has been demonstrated by several researchers [33] [67], allowing us to integrate high-quality inductors into an integrated circuit environment.

6.3 Spirals

Square spiral inductors are the most popular planar inductor geometries. Spiral inductors are wound on the top metal layer and a second layer is used to connect the innermost winding to the outside. As shown in Fig. 6.9, the inner turn can lead to several different points, depending on the application of the inductor. The layout of Fig. 6.9a forms a closed loop and is practical when both terminals of the inductor are accessed (say to a parallel capacitor

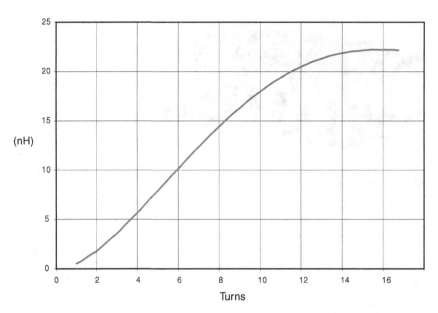

Figure 6.10 Variation of spiral inductance as a function of the number of turns N.

to form a resonator). On the other hand, if the spiral is in series with another element, the layout of Fig. 6.9b is preferred. It is important to note, however, that this layout is a "partial" inductor and the total inductance will depend on the loop in which the spiral is placed.

Due to the planar constraint, instead of adding a loop in the third dimension to spiral up (as in a solenoid), we spiral inward by adding loops inside each successive loop to form a spiral. Since the radius of each spiral is decreasing, the effective inductance does not increase like N^2, but like N^p with $p < 2$. A typical variation of inductance with turns is shown in Fig. 6.10. As long as $p > 1$, then the quality factor increases with increasing turns. In practice, though, the inner turns of a spiral suffer from increased losses due to enhanced skin/proximity effects. For this reason, to realize the highest Q inductor, it is better to use a single turn inductor with larger area. The advantage of a spiral over a ring inductor is that much higher inductance values can be generated in the same area.

Very similar comments with regards to the substrate losses of a loop inductor also apply to the spiral inductor, since it is essentially a series connection of loops. There are additional capacitive and loss currents, owing to the close proximity of the turns. For most IC structures, though, the substrate capacitance tends to dominate over the distributed winding capacitance.

The design of an optimal spiral is highly frequency and process dependent. This is because of the multiple loss mechanisms that arise from the distributed effects in the structure. In general, for a fixed area, we can employ many different values of trace width W, spacing S, and turns N, to achieve the same value of inductance. As we increase W for instance, the resistance drops. This drop is proportional to $1/W$ at low frequencies but much more gradual at higher frequencies, due to the skin effect, especially with $W \gg \delta$, where δ is

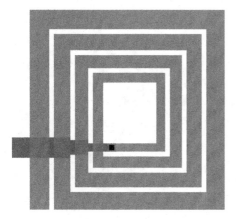

Figure 6.11 A square spiral inductor with tapered trace width. Inner turns do not need to be as wide due to enhanced current constriction.

the effective skin "width" (not depth). The substrate losses, though, tend to increase with W, since this increases the capacitance of the structure. At the same time, we observe that structures with more turns, N, tend to have higher resistive losses in the inner turns. For this reason "hollow" spirals are preferred [28]. The origin of these enhanced losses is due to the fact that the magnetic flux increases as we move towards the center of the spiral, due to the additive nature of the flux from each successive loop of the spiral. If the width of the structure is tapered, as shown in Fig. 6.11, then the performance of the spiral can be improved. Since the wide inner turns do not lower the resistance (due to current constriction), it is better to transfer the width to the outer turns, while keeping the total area of the spiral constant.

Finally, for low values of S, which can be extremely small in a modern IC process, we would expect excessive winding capacitance. For large values of S we find that the reduced proximity effects lead to more uniform current distributions, similar to an isolated conductor. But higher values of S lead to lower inductance, due to the reduction in coupling from turn-to-turn, and due to the smaller area of inner loops. There is thus an optimal value of S that allows a fair balance to occur.

Other variations on the layout include a rectangular layout, as opposed to a square outer area. A rectangular layout has lower Q than a square layout, but other constraints may impose a rectangular "opening" over which the spiral must be wound. Circular and polygon structures, shown in Fig. 6.12, have marginally higher Q factors and are preferred for this reason. The reason for the higher-quality factor follows the same line of argument for the ring inductor given previously.

6.4 Symmetric inductors

The main limitation of a spiral inductor is the asymmetry between the outer and inner lead of the structure when $N > 1$. This is evident in the Y parameters of the device which

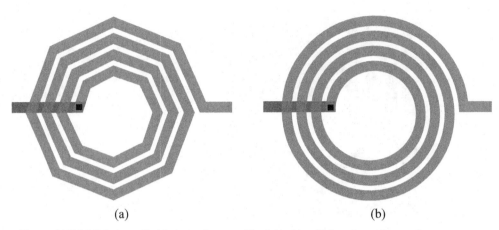

(a)   (b)

Figure 6.12 (a) Polygon spiral inductor layout with eight sides. This polygon has performance very similar to a (b) 64 sided polygon, which very accurately approaches the performance of a circular spiral.

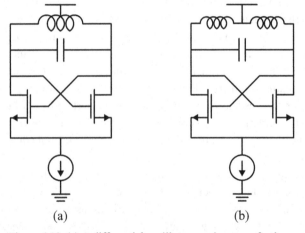

(a)   (b)

Figure 6.13 (a) A differential oscillator requires a perfectly symmetric inductor. The center "tap" of this inductor is used to deliver the bias voltage and current. (b) With a conventional spiral, two spirals must be connected in series for a fully symmetric layout.

shows that $Y_{11} \neq Y_{22}$. Since the outer-most turn interfaces to an open area, it tends to have more fringing capacitance. The inner turns, though, fringe much less, since the spiral turns are at approximately the same potential, and thus the fields have more "even" mode than "odd" mode. This asymmetry in the inductor is undesirable for fully balanced differential circuits. Furthermore, the electrical "center" of the spiral is hard to define. This center point is important to identify because if the structure is excited with a fully balanced signal, this point will not move electrically thus forming a virtual ground. We can thus use this point to connect a bias voltage without incurring any change into the operation of the inductor. In a differential oscillator, shown in Fig. 6.13a, for example, we can use this point to deliver the V_{DD} or ground connection. One symmetric layout is shown in Fig. 6.14a. Here the geometric center of the structure is also the electrical center, and thus this point can be used

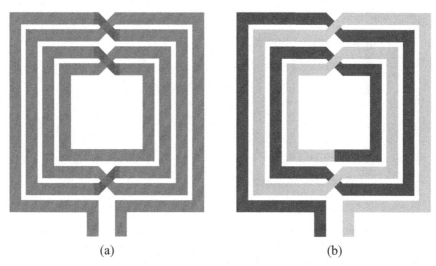

Figure 6.14 (a) A more symmetric inductor layout can be (b) decomposed into two coupled inductors.

as a "center tap" for the inductor. As shown in Fig. 6.14b, we can decompose this structure into two closely wound coupled inductors connected in series. In practice the coupling factor $k \sim 0.7$.

A center tapped inductor has a big advantage over using two series connected inductors shown in Fig. 6.13b. Since two separate inductors are only weakly coupled, the total inductance is simply $L + L + 2M \approx 2L$. If we wind the inductors together, on the other hand, a coupling factor of $k \sim 0.7$ can be utilized and the inductance is boosted to $2L(1 + k)$. Thus for equal inductance values, the area of the inductors can be substantially reduced leading to a higher-quality factor. In fact, at higher frequencies when the quality factor is dominated by the substrate, the two interwound inductors can have substantially higher Q values.

6.5 Multilayer inductors

Most modern IC processes offer many metal layers that can be employed to design more complicated three-dimensional structures. For instance, two or more spirals on different layers can be placed in series to increase the inductance, as shown in Fig. 6.15 and in three dimensions in Fig. 6.16. Note that the spirals must be wound correctly in order for the magnetic flux to be additive. Thus as the spiral winds inward on top, it should wind outward on the second layer, and then wind inward again on the third layer. For three equal inductance spirals, the inductance is increased to

$$L_t = L + M_2 + M_3 + L + M_2 + M_2 + L + M_2 + M_3 = 3L + 4M_2 + 2M_3 \qquad (6.7)$$

where M_2 is the mutual inductance between adjacent layers and M_3 is the mutual inductance between two spirals separated by another spiral in between. Since the coupling factors $k_2 \sim 0.8$ and $k_3 \sim 0.6$ are significant (due to the small separation of the layers), we can approach the upper bound of $L_t \leq 9L$, a N^2 boost in inductance value.

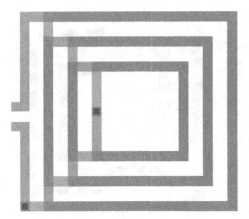

Figure 6.15 A multi-layer square spiral inductor is formed by connecting a spiral from the top layer to a spiral on the next level. Each spiral is wound so as to reinforce the magnetic field of the other. This is easily generalized to three or more layers.

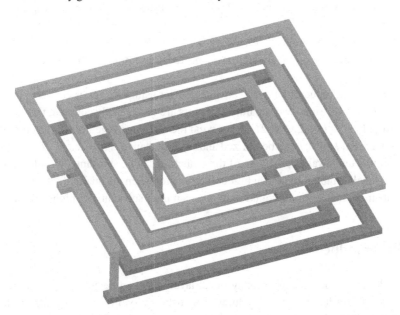

Figure 6.16 A three-dimensional view of the multi-layer series connected spiral shown in Fig. 6.15. The via thickness is exaggerated for illustration.

Even though very large and compact inductors can be realized with this approach, the self-resonant frequency suffers from the large spiral-to-spiral winding capacitance. The successive layers can be staggered somewhat to reduce the capacitance, but, overall, while these inductors can save area over a single layer structure, due to the closer proximity to the substrate, the quality factor of such structures suffers.

Another more obvious application of multiple layers is to simply connect the spirals in parallel, in order to realize an effectively thicker conductor. Many vias are employed to strap the layers together. This will indeed lower the resistance with a small impact on the

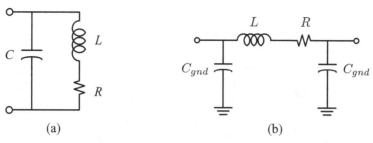

Figure 6.17 (a) A simple air-core inductor is modeled by a simple LCR network. (b) An inductor near a ground plane has a lower self-resonant frequency due to the extra capacitance.

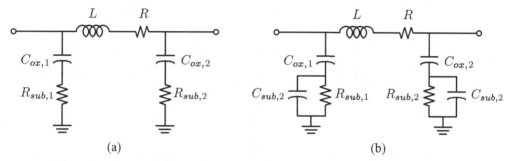

Figure 6.18 (a) An inductor fabricated over a lossy substrate can be modeled in a narrow band of frequencies with a Π model. (b) Additional capacitors can model displacement current in the substrate.

inductance value (since $k \sim 0.7$), but the lower layers are closer to the substrate and thus the improvement in the lower series resistance is eventually outweighed by the increased shunt losses through the substrate.

6.6 Inductor equivalent circuit models

The behavior of inductors can be modeled in practice with equivalent circuits consisting of magnetic and electric storage elements as well as loss elements. For instance, a simple air wound inductor is modeled as shown in Fig. 6.17a, where the series winding resistance is modeled by R and the winding capacitance is modeled by C. Even though the actual capacitance is a distributed across the windings, in practice the lumped value of C is adjusted to capture the first resonant frequency of the structure. The frequency dependence in the resistance R is simple to model in AC simulations, but difficult to model for time-domain simulations. For structures fabricated over an ideal ground plane, the capacitance to substrate can be modeled as shown in Fig. 6.17b. This is a simple model that can capture the behavior up to the first resonant frequency.

If the substrate is lossy, resistors in series with the capacitors can model the electrical substrate loss, as shown in Fig. 6.18a. In fact, it is easy to show that such a Π model is a one-to-one equivalent circuit for a two-port passive network

$$R_x + j\omega L = -\frac{1}{Y_{12}} \qquad (6.8)$$

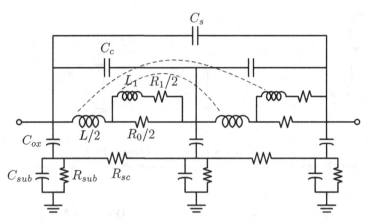

Figure 6.19 A multi-section model can be used to fit the inductor behavior over a wide frequency range.

and

$$(R_x + j\omega L) \| \left(R_{s1} + \frac{1}{j\omega C_{s1}} \right) = \frac{1}{Y_{11}} \tag{6.9}$$

$$(R_x + j\omega L) \| \left(R_{s2} + \frac{1}{j\omega C_{s2}} \right) = \frac{1}{Y_{22}} \tag{6.10}$$

It should be noted that in practice the values of the equivalent circuit elements are derived from optimization rather than from physical calculations. The physical values of the elements can be used as a starting point in the optimization. But slight adjustments in these component values can be used to obtain a better fit over a wide frequency range. For a broader band model, multiple Π sections can be cascaded. In the limit, the cascaded Π sections form an artificial transmission line that closely mimics the behavior of the inductor.

For moderately conductive substrates, the displacement current $\propto \omega \epsilon$ may be of the order of the conductive currents $\propto \sigma$. In this case, the model of Fig. 6.18b can be used for more accuracy. To model the frequency dependence of the series resistance due to skin and proximity effects, the model of Fig. 6.19 can be used [3]. Since at low frequencies the inductor windings are short circuits, the DC resistance of the structure is simply $R_0 \| R_1$. At high frequency, though, the resistance asymptotically approaches R_0. Thus the series resistance of the model increases as a function of frequency. An extra degree of freedom is added by adding mutual inductance between the winding inductor and the skin effect modeling inductor.

6.7 Integrated capacitors

High-quality capacitors are critical in RF circuits. As shown in the schematic of Fig. 6.20, these capacitors are employed in matching networks, as AC coupling capacitors (to isolate the bias of an amplifier from the driver), and as bypass capacitors to provide low inductance

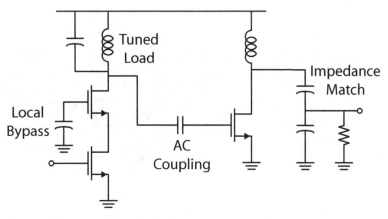

Figure 6.20 Integrated capacitors in a typical RF amplifier.

return to ground. While some specialized IC processes offer "MIM" (or metal-insulator-metal) capacitors as a special high-density capacitor device, many processes do not have this option. Otherwise, this option adds extra expense due to the required masking layers. But due to the high density of metalization in a modern IC, we can construct capacitors employing the standard metalization layers.

Metal-insulator-metal (MIM) capacitors

If a MIM capacitor is offered in a process, it is usually constructed from a parallel plate capacitor employing a thin oxide with a high dielectric constant. While the standard metals can be used to form a MIM capacitor, the capacitor density of such structures is quite low. For instance, if the metals are spaced at $1\,\mu$m apart, then capacitor density is only

$$C_{ox} = \frac{\epsilon_{ox}}{t} \approx \frac{3.9 \times 8.854 \times 10^{-12}\,\text{F/m}}{1 \times 10^{-6}\,\text{m}} = 35\,\text{aF}/\mu^2 \tag{6.11}$$

To lay out a 1 pF capacitor requires an area of approximately $170\,\mu \times 170\,\mu$. Alternatively, if we exploit the small lateral spacing of the process, we can realize much higher capacitor densities. For instance, a "finger" capacitor structure, shown in Fig. 6.21, utilizes the lateral fringing capacitance between two fingers. The spacing between two conductors is now of the order of 100 nm in a modern CMOS process. Since multiple layers are available, the capacitance increases with the number of layers without increasing the area. Moreover, a dense array of vias can create a virtual "wall," which boosts the capacitance greatly. For simplicity, assume that the process has six metal layers that can be routed with minimum spacing of 100 nm. Assume that vias require at least 200 nm of metal width. Thus each finger requires 300 nm of length, with the average spacing between fingers about $D_{eff} = 300$ nm. The capacitance per unit length between each finger is given by

$$C_f/L = T_{eff} \times \frac{\epsilon_{ox}}{D_{eff}} \approx 6.5\,\mu \cdot \frac{3.9 \times 8.854 \times 10^{-12}\,\text{F/m}}{0.3 \times 10^{-6}\,\text{m}} = 0.75\,\text{fF}/\mu \tag{6.12}$$

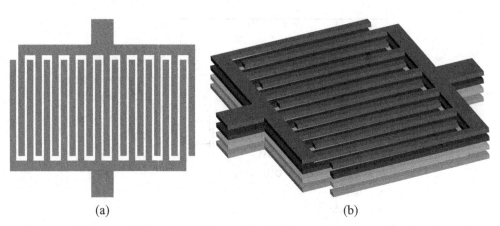

Figure 6.21 (a) Integrated "finger" capacitors exploit the high metal density to realize lateral capacitor structures. (b) Three-dimensional view of multi-layer finger capacitor structure.

where we assume that all six layers are employed, and each layer is $0.25\,\mu$ thick and spaced at $1\,\mu$ apart, giving $T_{eff} = 6.5\,\mu$. Thus the capacitance density is approximately

$$C_f/(W \cdot L) = 2.2\,\text{fF}/\mu/(0.2\,\mu + 0.1\,\mu) = 7.5\,\text{fF}/\mu^2 \qquad (6.13)$$

With such a lateral capacitor structure, the 1 pF capacitor can now be realized in an area of only $67\,\mu \times 67\,\mu$.

Capacitor Q

This high density allows relatively compact capacitors to be integrated into the RF chip. The finger length L must be chosen carefully to avoid de-Q'ing the capacitor with series resistance. Since the conductors have a resistivity $\rho \sim 100\,\text{m}\Omega/\square$, one finger has a resistance per length of $0.5\,\Omega/\mu$. But due to the distributed nature of the current flow (see Chapter 11), the actual resistance is only $1/3$ of the DC resistance. Intuitively this occurs since the full current does not flow through the entire finger. In fact, the current drops to zero by the time we reach the end of the finger.

Many short fingers in parallel must be employed to retain a high-quality factor. But in general we find it much easier to design a high-quality factor capacitor than a high Q inductor. To see this, consider a capacitor at 1 GHz. If we use $100\,\mu$ long fingers, the resistance per finger is $0.5\,\Omega \times 100\,\mu \times 2 \times \frac{1}{3}$ and the capacitance per finger is 75 fF, corresponding to a quality factor of nearly 64. As more fingers are added in parallel to realize a higher capacitance value, the resistance drops and the Q remains the same. So the Q factor is to first order only dependent on the length of the fingers.

The quality factor also depends on the quality of the dielectric layers. Native SiO_2 is an excellent dielectric with a very small loss tangent. Values as low as 1×10^{-6} have been reported. With such a small loss tangent, the series resistance would continue to dominate the quality factor of integrated capacitors. To see this note that the losses due to the parallel plate capacitance are given by

$$P_{Ld} = \frac{1}{4} \int_V |E|^2 \omega \epsilon'' dV \tag{6.14}$$

where ϵ'' is the imaginer component of the dielectric constant $\epsilon = \epsilon' + j\epsilon''$ and $\epsilon'' < 0$. For a uniform field the above expression simplifies

$$P_{Ld} = \frac{\omega \epsilon''}{4} E_0^2 W \times L \times t \tag{6.15}$$

where the applied voltage $V_0 = E_0 \times t$. Substituting we have

$$P_{Ld} = \frac{\omega \epsilon''}{4} \frac{V_0^2}{t} W \times L = \frac{\omega C}{4} \frac{\epsilon''}{\epsilon'} |V_0|^2 = \frac{G|V_0|^2}{4} \tag{6.16}$$

where $G = \omega C \frac{\epsilon''}{\epsilon'} = \omega C \tan \delta$.

The energy storage in the capacitor, though, is given by

$$E_C = \frac{1}{4} C |V_0|^2 \tag{6.17}$$

where the AC current $I = j\omega C V_0$ flows through the device. The losses through the capacitor plates are given by

$$P_{Lc} = 2 \times \frac{1}{4} |I|^2 \frac{R_p}{3} = \frac{1}{6} C^2 |V_0|^2 \omega^2 R_p \tag{6.18}$$

where the factor of $1/3$ accounts for the distributed nature of the current flow and the leading factor of 2 accounts for the presence of two plates of equal resistance.

Given the two loss mechanisms, we can compute the Q as follows

$$Q_C = \frac{\omega \frac{1}{4} C |V_0|^2}{\frac{1}{6} R_p C^2 \omega^2 R_p + \frac{1}{4} \omega C \tan \delta |V_0|^2} = \frac{1}{\frac{2}{3} \omega C R_p + \tan \delta} \tag{6.19}$$

For $\tan \delta \approx 10^{-6}$, the Q is bounded by 10^6, and so the plate resistance dominates the losses.

6.8 Calculation by means of the vector potential

Partial inductance

In the previous chapter, we introduced the concept of partial inductance, or the inductance of an "open" path. Physically, we know that inductance can only be associated with a closed path, so, when we speak of a partial inductance, we must exercise caution. A classic example of a confusing statement is the "inductance of a bond wire," which is a meaningless concept unless the return current is carefully considered. This is shown schematically in Fig. 6.22, where the inductance of the first path is defined by the current flow through the ground plane, whereas the inductance of the second path is defined by a loop that does not flow through the ground plane. The ground plane will lower the inductance of the second path due the image current flows (see Section 6.8).

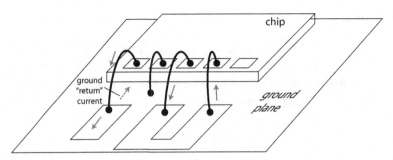

Figure 6.22 The "inductance" of a bondwire depends on the complete closed loop of the current flow.

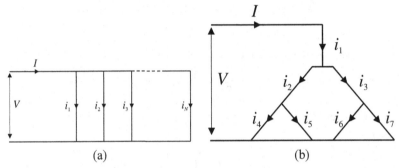

Figure 6.23 (a) A geometric structure with multiple loops but only two nodes. (b) A tree-like geometric structure.

But there is definite utility in the partial inductance concept when we look at efficient ways to calculate the inductance of practical circuits. For instance, consider the inductance of an arbitrary structure consisting of several loops and one pair of terminals, as shown in Fig. 6.23a. The structure has many mesh currents but only one node pair. Thus the voltage across every segment is equal and by KCL the sum of the currents must equal the terminal currents. The following expression, then, is the inverse of inductance of the entire structure

$$L^{-1} = \frac{I}{V} = \frac{1}{V} \sum_{i} I_i = \sum_{ij} [M^{-1}]_{ij} \tag{6.20}$$

where an inversion of the partial inductance matrix M is required.

As a counter example, Fig. 6.23b grows in a tree-like fashion and the number of meshes is less than the number of nodes. Setting up this system of equations is a bit more complicated using the partial inductance concept. To calculate the inductance of the above structure, we only have to note that if a current I is applied to the terminals of the structure, the voltage at the terminals is given by Eq. (5.100) and the currents are related by KCL

$$I = i_1 = i_2 + i_3 \tag{6.21}$$

$$i_2 = i_4 + i_5 \tag{6.22}$$

$$i_3 = i_6 + i_7 \tag{6.23}$$

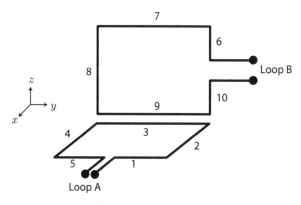

Figure 6.24 Two loops with several orthogonal sections.

and the voltages are related by KVL

$$V = v_1 + v_2 + v_4 \tag{6.24}$$

$$V = v_1 + v_2 + v_5 \tag{6.25}$$

$$V = v_1 + v_3 + v_6 \tag{6.26}$$

$$V = v_1 + v_3 + v_7 \tag{6.27}$$

$$\tag{6.28}$$

The above linear system of equations, along with Eq. (5.101), can be solved and expressed compactly in matrix notation

$$\begin{pmatrix} 1 & -1 & -1 & 0 & 0 & 0 & 0 \\ 0 & 1 & 0 & -1 & -1 & 0 & 0 \\ 0 & 0 & 1 & 0 & 0 & -1 & -1 \end{pmatrix} i = Bi = \begin{pmatrix} 0 \\ 0 \\ 0 \end{pmatrix} \tag{6.29}$$

and

$$\begin{pmatrix} 1 & 1 & 0 & 1 & 0 & 0 & 0 \\ 1 & 1 & 0 & 0 & 1 & 0 & 0 \\ 1 & 0 & 1 & 0 & 0 & 1 & 0 \\ 1 & 0 & 1 & 0 & 0 & 0 & 1 \end{pmatrix} v = Av = V \begin{pmatrix} 1 \\ 1 \\ 1 \\ 1 \end{pmatrix} \tag{6.30}$$

It is not too hard to show that

$$L = \hat{e}_1^T A Z B^{-1} \hat{e}_1 \tag{6.31}$$

where $\hat{e}_1^T = (1 0 \ldots 0)$ and the impedance matrix $Z = j\omega M$. This problem can also be solved using a mesh equation formulation.

In addition to serving as a computation aid, the partial inductance concept also adds a new vantage point in designing inductors. For instance, consider the mutual inductance between two orthogonal loops shown in Fig 6.24. Using the partial mutual inductance concepts we can immediately ignore several coupling terms since the vectors are orthogonal and $\mathbf{dl}_1 \cdot \mathbf{dl}_2 = 0$. Also, since several mutual coupling terms are identical we can quickly estimate M. This is not so obvious from the flux or energy concepts. As an example, consider the partial mutual coupling between segment 3 of loop A and loop B (ignoring the leads)

$$M_{3,B} = M_{3,6} + M_{3,7} + M_{3,8} + M_{3,9} + M_{3,10}$$

Dropping all the perpendicular segments ($M_{3,6} = M_{3,8} = M_{3,10} = 0$), we are left with

$$M_{3,B} = M_{3,7} + M_{3,9}$$

Magnetic vector integral and differential equations

In theory we can compute the self or mutual inductance of any structure as long as we calculate vector potential given by Eq. (5.87) and then follow with a line integral about the path of interest. In practice, though, the integrand of Eq. (5.87) is unknown since the high-frequency current distribution is unknown. How is the high-frequency current distribution obtained? From the magnetic vector potential

$$\mathbf{J} = \sigma \mathbf{E} = j\omega\sigma \mathbf{A} + \sigma \nabla \phi \tag{6.32}$$

In other words, at any given frequency to solve for $\mathbf{A}$ we must solve the following integral equation[2]

$$\mathbf{A}(r) = \mu \int_V' \frac{j\omega\sigma \mathbf{A}(\mathbf{r}')}{4\pi R} dV' \tag{6.33}$$

For simplicity we have dropped the electric potential ϕ from the above equation for the magnetostatic case but in general we must solve a pair of integral equations: one for the electrostatic potential and one for the magnetostatic potential. Current continuity can be used to relate charge and current. By the definition of magnetic vector potential and from Maxwell's equations we have

$$\nabla \times \nabla \times \mathbf{A} = \nabla \times \mathbf{B} = \mu \mathbf{J} + j\omega\mu\epsilon \mathbf{E} \tag{6.34}$$

Employing the vector identity

$$\nabla \times \nabla \times \mathbf{A} \equiv \nabla\nabla \cdot \mathbf{A} - \nabla^2 \mathbf{A} \tag{6.35}$$

and the simple Coulomb gauge $\nabla \cdot \mathbf{A} \equiv 0$[3] we have

$$-\nabla^2 \mathbf{A} = j\omega\mu\epsilon(-j\omega \mathbf{A} - \nabla \phi) + \mu \mathbf{J} \tag{6.36}$$

For magnetostatic problems we neglect the first two terms since we take the limit as $\omega \to 0$ and we are left with the spatial differential equation

$$\nabla^2 \mathbf{A} = -\mu \mathbf{J} \tag{6.37}$$

In Cartesian coordinates, the above equation is identical to three scalar Poisson's equations. In principle we can solve these differential equations to solve any magnetostatic

---

[2]   Note that $\mathbf{A}$ appears on both sides of the *integral equation*. Solving integral equations is analogous to (and as difficult as) solving differential equations.

[3]   Recall that we are free to choose the divergence of $\mathbf{A}$ as we see fit.

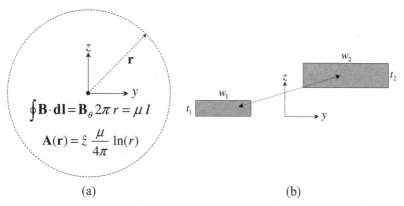

(a)   (b)

Figure 6.25 (a) The magnetic potential of a filament in two dimensions. (b) The mutual inductance of two rectangular conductors in two dimensions.

problem. In reality, the solution process is complex and requires numerical computations. In the following sections we focus on problems that admit analytic solutions.

Filamental calculations

For filamental circuits, the current distribution is known (trivially) and the volume integral of Eq. (5.87) simplifies to a line integral over the path of the current

$$\mathbf{A}(\mathbf{r}) = I\mu \int_C \frac{dl}{R} \tag{6.38}$$

There are no unknowns in the above integral, but there is a problem with the $1/R$ discontinuity. If we try to integrate the above over the path of the current, $R = 0$. However, as noted in Section 5.4 in deriving Neumann's formula, the above integral can be evaluated without any difficulty when computing the mutual inductance.

The most important case that we shall consider is the mutual inductance between two filaments in free space. To begin, consider an infinitely long filament. The magnetic vector potential in the vicinity of such a filament is readily computed

$$A(r) = \frac{\mu_0 I}{4\pi} \ln(r) \tag{6.39}$$

The magnetic vector potential falls off slowly as the natural log of distance from the filament. Therefore, the mutual inductance between two filaments separated by a radial distance r is thus

$$M(r) = \frac{\mu_0}{4\pi} \ln(r) \tag{6.40}$$

Using the above result, let us derive the mutual inductance between two conductors of the rectangular cross section shown in Fig. 6.25b. Each conductor has width w_i and thickness t_i and is located within the specified Cartesian coordinates. Since we have a solution for any two filaments in the cross sections, we can average $M(r)$ from above over every pair of constituent filaments in the cross sections to obtain the mutual inductance

$$M = \frac{\mu_0}{4\pi} \frac{1}{w_1 w_2 t_1 t_2} \int \int_{A_1} \int \int_{A_2} \ln\sqrt{(y-y')^2 + (z-z')^2} dy' dz' dy dz \qquad (6.41)$$

where $A_1 = w_1 t_1$ and $A_2 = w_2 t_2$ are the cross-sectional areas of the rectangular conductors. The above quadruple integral is difficult to evaluate, but fortunately has a closed-form solution given by [66]

$$-\frac{1}{2} F(y-y', z-z') \Big|_{y_1}^{y_2} \Big|_{y_3}^{y_4} \Big|_{z_1}^{z_2} \Big|_{z_3}^{z_4} - \frac{25}{12} w_1 w_2 t_1 t_2 \qquad (6.42)$$

where the function F is given by

$$F(y, z) = \frac{y^4 - 6y^2 z^2 + z^4}{24} \ln(y^2 + z^2) - \frac{yz}{3}\left(y^2 \tan^{-1}\frac{z}{y} + z^2 \tan^{-1}\frac{y}{z}\right) \qquad (6.43)$$

As is usually the case, it is not too difficult to verify that the above is the solution to the integral since

$$\frac{\partial^4 F(y-y', z-z')}{\partial y \partial y' \partial z \partial z'} = \ln((y-y')^2 + (z-z')^2) - \frac{25}{6} \qquad (6.44)$$

From the definition of F, the mutual inductance between two filaments can be readily calculated (preferably by a computer). What about the self inductance of a rectangular conductor? If we carefully take limits to let $y_3 \to y_1$ and $y_4 \to y_2$ we arrive at

$$L(w, t) = \frac{\mu_0}{4\pi}\left(\frac{1}{12 w^2 t^2} 8 t w^3 \tan^{-1}\left(\frac{t}{w}\right) + 8 t^3 w \tan^{-1}\left(\frac{w}{t}\right)\right.$$
$$\left. + t^4 \ln(t^2) + w^4 \ln(w^2) + (t^4 - 6 t^2 w^2 + w^4) \ln(t^2 + w^2)\right) - \frac{25}{12} \qquad (6.45)$$

Self inductance of conductors

Using the above techniques we can derive the self inductance of a flat conductor (zero-thickness) of length l and width w [46]

$$L = \frac{\mu l}{2\pi}\left(\sinh^{-1}\left(\frac{l}{w}\right) + \frac{l}{w}\sinh^{-1}\left(\frac{w}{l}\right) + \frac{w}{3l} - \frac{1}{3}\left(\frac{l}{w}\right)^2\left[\left(1 + \left(\frac{w}{l}\right)^2\right)^{\frac{3}{2}} - 1\right]\right) \qquad (6.46)$$

If the conductor has thickness t much less than w, then the above equation can be employed where w is replaced by $w + t$.

Geometric mean distance and the Grover/Greenhouse methods

Even for the simple two-dimensional geometry shown above, the computations are exorbitantly complicated. While three-dimensional expressions are even more complicated, we can often employ numerical integration techniques. One practical difficulty, though, is due to the singularity in the self-inductance computation. In this section we will discuss the *geometric mean distance* approximation. We shall see this technique is essentially a refinement of the two-dimensional approximation.

First, consider the mutual inductance between two parallel filaments of length l. It is easy to show that the following integral expression must be evaluated for the vector potential at (x, y, z) due to a filament at the origin

$$\mathbf{A}(x, y, z) = \frac{I\mu_0}{4\pi} \int_{-l/2}^{+l/2} \frac{1}{\sqrt{x^2 + y^2 + (z - z')^2}} dz' \tag{6.47}$$

Evaluating the integral

$$A_z(x, y, z) = \hat{z} \frac{\mu_0 I}{4\pi} \ln \left[\frac{(z + l/2) + \sqrt{x^2 + y^2 + (z + l/2)^2}}{(z - l/2) + \sqrt{x^2 + y^2 + (z - l/2)^2}} \right] \tag{6.48}$$

Integrating the above expression along a parallel path at $x = d$

$$M(l, d) = \frac{1}{I} \int_{-l/2}^{l/2} A_z(d, 0, z') dz' \tag{6.49}$$

This gives the following geometric expression for the mutual inductance

$$M(l, d) = \frac{\mu_0}{4\pi} l \left[\ln \left(\sqrt{1 + \left(\frac{l}{d} \right)^2} + \frac{l}{d} \right) - \sqrt{1 + \left(\frac{d}{l} \right)^2} + \frac{d}{l} \right] \tag{6.50}$$

Using an approach similar to the previous section, we can average the above expression over a cross-sectional area of two rectangular conductors to obtain the *partial* mutual inductance between two rectangular slabs

$$M_{12} = \frac{1}{A_1 A_2} \int_{A_1} \int_{A_2} M(L, r) dA_1 dA_2 \tag{6.51}$$

where $r = \sqrt{(x_1 - x_2)^2 + (y_1 - y_2)^2 + (z_1 - z_2)^2}$. For the case of self inductance, we can set $A_1 = A_2$ and perform the integration. This will result in integrable singularity in the calculation.

If the filaments under consideration are of length such that $L \gg d$, then the results for $M(L, d)$ can be simplified if we neglect the term $\left(\frac{d}{L} \right)^2$

$$M(L, d) \approx \frac{\mu_0}{4\pi} L \left(\frac{d}{L} - \ln d + \ln 2L - 1 \right) \tag{6.52}$$

Integration of the above expression across the cross section yields a simpler calculation. The only difficult term to integrate is the $\ln(d)$. We have already met this integral in the two-dimensional case. This term, the *geometric mean distance* (or GMD) is the average value of the logarithm of the distance between all points between the conductors. The integral also contains a simpler term, the *arithmetic mean distance* (AMD), simply the average distance between the conductors.

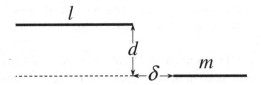

Figure 6.26 Two parallel filaments of unequal length.

In general, for any two parallel conductors of an arbitrary cross section, the GMD approximation gives the following result for the mutual inductance

$$M(L, d) \approx \frac{\mu L}{4\pi} (\text{AMD}/L - \text{GMD} + \ln(2L) - 1) \qquad (6.53)$$

as long as $\left(\frac{d}{L}\right)^2 \ll 1$.

In many practical situations the conductors are isolated so that $d \gg l$. The GMD approximation does not hold in these cases but the simple filamental approximation is valid even up to high frequencies. This is plausible since at considerable distance from the source, the exact current distribution is immaterial and an equivalent filament serves as an excellent approximation. Since we have seen that at high frequencies only the current distribution changes, the approximation is good even at AC frequencies.

General orthogonal geometries

The partial mutual inductance between unequal parallel filaments shown in Fig. 6.26 is given by the following expression [22]

$$M = \frac{\mu_0}{4\pi} \left[\alpha \sinh^{-1} \frac{\alpha}{d} - \beta \sinh^{-1} \frac{\beta}{d} - -\gamma \sinh^{-1} \frac{\gamma}{d} + \delta \sinh^{-1} \frac{\delta}{d} \right.$$
$$\left. - \sqrt{\alpha^2 + d^2} + \sqrt{\beta^2 + d^2} + \sqrt{\gamma^2 + d^2} - \sqrt{\delta^2 + d^2} \right] \qquad (6.54)$$

where

$$\alpha = l + m + \delta, \ \beta = l + \delta, \ \gamma = m + \delta \qquad (6.55)$$

The distance δ can go negative if the filaments overlap. The above result is generated by integrating the magnetic vector potential given by Eq. (6.48) over the shown path.

Using some clever symmetry arguments, we can also derive this result without doing any integrals. Consider the case shown in Fig. 6.27a, where we begin with two equal length parallel segments of length ℓ and break one segment to a length m. With reference to Fig. 6.27b, the mutual inductance for the equal segments can be written

$$M(l) = M(m) + M(l - m) + 2M_x \qquad (6.56)$$

The mutual inductance M_x is shown in Fig. 6.27c. The notation $M(p)$ is the filamental mutual inductance between two parallel equal length filaments of length p. But we can also express M_x in terms of our desired mutual inductance M

$$M_x = M - M(m) \qquad (6.57)$$

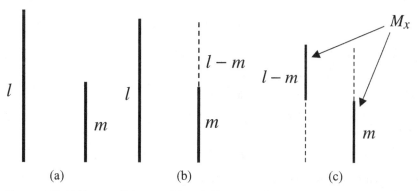

Figure 6.27 (a) Two parallel unequal length filaments with ends at equal positions. (b) An equivalent system of real and fictitious segments used in computing the mutual inductance. (c) The mutual inductance M_x can be expressed in terms of the desired result (see text).

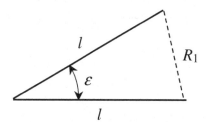

Figure 6.28 Two equal length filaments meeting at a point.

Substituting we have an expression for the desired mutual inductance

$$M = \frac{1}{2}\left(M(l) + M(m) - M(l - m)\right) \tag{6.58}$$

where each term in the above expression is given by Eq. (6.52). In the problems (online) you are asked to use such symmetry arguments to compute the mutual inductance of various filamental configurations. Using the GMD argument, you can substitute the GMD for l and generalize the above calculations to appropriate cases.

Arbitrary geometry

The filamental partial mutual inductance between any arbitrary arrangements of filaments can be computed using the above recipe. Compute the magnetic vector potential for the first filament (Eq. (6.48)). Next, integrate this expression over the path of the second filament. In theory these computations are easy but in practice they are tedious. The results are from Grover and repeated here since this source is out of print [22].

Consider the simple case of equal filaments meeting at a point (Fig. 6.28).

$$M = \frac{\mu_0}{\pi} l \cos \epsilon \, \tanh^{-1} \frac{l}{l + R_1} \tag{6.59}$$

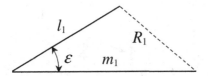

Figure 6.29 Two unequal length filaments meeting at a point.

where

$$R_1^2 = 2l^2(1 - \cos \epsilon) \tag{6.60}$$

For unequal filaments meeting at a point use

$$M = \frac{\mu_0}{2\pi} \cos \epsilon \left[l_1 \tanh^{-1} \frac{m_1}{l_1 + R} + m_1 \tanh^{-1} \frac{l_1}{1 + R} \right] \tag{6.61}$$

As shown in Fig. 6.29, the filament lengths are l_1 and m_1 and subtend an angle ϵ. The distance between the filament ends is R. The following relations can be used to substitute for either R or ϵ

$$\cos \epsilon = \frac{l_1^2 + m_1^2 - R^2}{2l_1 m_1} \tag{6.62}$$

and

$$\frac{R^2}{l_1^2} = 1 + \frac{m_1^2}{l_1^2} - 2\frac{m_1}{l_1} \cos \epsilon \tag{6.63}$$

Generalizing further we present the formula for unequal filaments in the same plane, not meeting. As shown in Fig. 6.30 the filaments lengths are l and m and the following geometric relations apply

$$2 \cos \epsilon = \frac{\alpha^2}{lm} \tag{6.64}$$

$$\alpha^2 = R_4^2 - R_3^2 + R_2^2 - R_1^2 \tag{6.65}$$

$$\mu = \frac{l \left[2m^2 \left(R_2^2 - R_3^2 - l^2 \right) + \alpha^2 \left(R_4^2 - R_3^2 - m^2 \right) \right]}{4l^2 m^2 - \alpha^4} \tag{6.66}$$

$$\nu = \frac{m \left[2l^2 \left(R_4^2 - R_3^2 - m^2 \right) + \alpha^2 \left(R_2^2 - R_3^2 - l^2 \right) \right]}{4l^2 m^2 - \alpha^4} \tag{6.67}$$

$$R_1^2 = (\mu + l)^2 + (\nu + m)^2 - 2(\mu + l)(\nu + m) \cos \epsilon \tag{6.68}$$

$$R_2^2 = (\mu + l)^2 + \nu^2 - 2\nu(\mu + 1) \cos \epsilon \tag{6.69}$$

$$R_3^2 = \mu^2 + \nu^2 - 2\mu\nu \cos \epsilon \tag{6.70}$$

$$R_4^2 = \mu^2 + (\nu + m)^2 - 2\mu(\nu + m) \cos \epsilon \tag{6.71}$$

And at last, with the preliminaries out of the way, the partial mutual inductance is given by

$$\frac{M}{2 \cos \epsilon} = \frac{\mu_0}{4\pi} \left[(\mu + l) \tanh^{-1} \frac{m}{R_1 + R_2} + (\nu + m) \tanh^{-1} \frac{l}{R_1 + R_4} \right.$$
$$\left. - \mu \tanh^{-1} \frac{m}{R_3 + R_4} - \nu \tanh^{-1} \frac{l}{R_2 + R_3} \right] \tag{6.72}$$

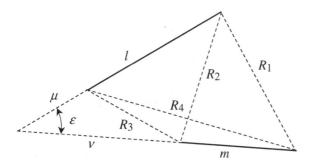

Figure 6.30 Two arbitrary filaments lying in a common plane.

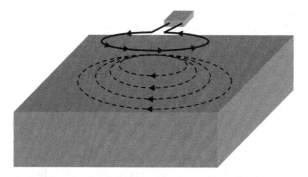

Figure 6.31 A loop inductor near an ideal ground plane.

Using symmetry arguments and Eq. (6.59) we have

$$M_{lm} = (M_{\mu+l,\nu+m} + M_{\nu\mu}) - (M_{\mu+l,\nu} + M_{\nu+m,\mu}) \qquad (6.73)$$

The most general result, the partial mutual inductance of two straight filaments placed in any desired position, is presented in Appendix 6.10. You may wonder why we present the above results when we could have simply quoted this most general case from the outset. The reason is obvious when we write the equation down – it is a *long* formula. Thus it is wise to use the simplest equation possible, not only to save computation time but also to enhance numerical stability. Using the above formulas, the inductance of any piecewise linear struc-ture can be approximated. The self inductance for each section can be computed directly and the mutual inductance can be approximated from the above filamental equations. Given enough segments and a modern workstation, any smooth curve can also be approximated.

Effect of an ideal ground plane

An ideal ground plane is defined as an infinitely conductive plane. If an inductive circuit resides *near* the ground plane, then from our experience thus far we can see that induced currents will flow in the ground plane in such a way as to oppose the impinging magnetic field. This is shown schematically in Fig. 6.31. Since the induced currents *reduce* the magnetic flux, the effective inductance seen at the terminals of the inductor is lowered.

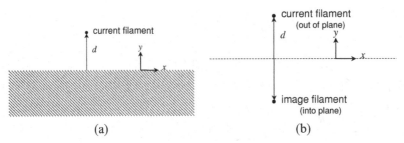

Figure 6.32 (a) An infinitely long filament residing parallel to a ground plane. (b) An equivalent system of an image filament without the ground plane.

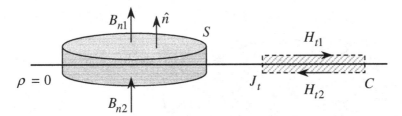

Figure 6.33 A surface and curve enclosing regions inside and outside of the conductive region.

Often, magnetically induced currents are referred to as *eddy currents*. Since magnetic fields arise due to currents, we must include eddy currents in Eq. (5.87) when computing the magnetic vector potential or the magnetic field. Thus we must either solve the resulting integral equation (5.87) or the differential Eq. (6.37).

For now, we will present a powerful alternative method of solution that does not require solving any integral or differential equations, the *method of images*. For many symmetric situations, we can invoke this approximation to include the induced ground currents. For instance, consider an infinitely long filamental current at a distance d over an infinitely conductive ground plane shown in Fig. 6.32a. Now consider an equivalent system of two conductors separated by a distance $2d$ shown in Fig. 6.32b, where the introduced current filament at $z = -d$ has a current flowing in a direction opposite to the original filament. We would like to show that the field configurations of the two systems are identical for $z > 0$.

Since the ground plane is infinitely conductive, the fields decay to zero infinitely fast and thus $\mathbf{E} = 0$ and $\mathbf{H} = 0$ inside the conductor. The only induced currents, then, are on the surface of the conductor. The vector fields in Fig. 6.32a are uniquely determined by the sources (the filamental current) and the boundary conditions. Let us take the boundary to be the surface of the ground plane. Imposing boundary conditions on the normal component of the magnetic field, consider the surface integral of $\mathbf{B}$ over the pillbox surface at the boundary shown in Fig. 6.33a. Let the height of the cylinder go to zero so the only contribution to the integral is the normal component of the field at the top and bottom surface

$$\int_S \mathbf{B} \cdot d\mathbf{S} = B_{n1} - B_{n2} = 0 \tag{6.74}$$

The right-hand side is zero by Maxwell's equations. Since B_{n2} is zero inside the conductive region, $B_{n1} = 0$ as well. Note that the current filaments of Fig. 6.32b produce a zero

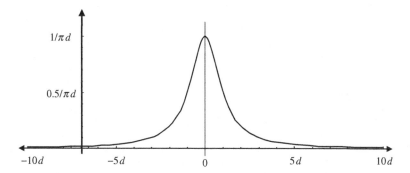

Figure 6.34 The current density at the surface of the ground plane.

normal field at $z = 0$. By using the Biot-Savart Law or the magnetic vector potential of a filament we see that the magnetic field is tangential at the surface. The tangential component is given by

$$B_t = B_x(x, 0) = \frac{\mu I}{2\pi} \left[\frac{y - d}{x^2 + (y - d)^2} - \frac{y + d}{x^2 + (y + d)^2} \right]_{y=0} \tag{6.75}$$

If the systems are equivalent, then we can calculate the functional form of the current density in the ground plane from the above expression

$$J_t = H_t = \frac{1}{\mu} B_t = \frac{I}{\pi} \frac{d}{x^2 + d^2} \tag{6.76}$$

We have plotted this surface current in Fig. 6.34. As expected, the current peaks near the filamental source. The peak value is $I/(\pi d)$ and increases inversely as the filament is moved closer to the ground plane.

Thus, the effect of the ground plane is to lower the self inductance by the mutual inductance between the source and its image. For a conductor with inductance L_0 in the absence of the ground plane, then, the effective inductance near the ground plane is reduced to $L_{eff} = L_0 - M$, where M is the mutual inductance between the conductor and its image. Or, alternatively, $L_{eff} = L(1 - k)$, where k is the coupling coefficient.

6.9 References

This chapter is an extension of a previous book I wrote as a graduate student at Berkeley [43]. Many of the equations come from the trusty reference by Grover [22] and Ternan [60]. The vector potential approach and partial inductance techniques were popularized by researchers such as Greenhouse [21] and Ruelli [54].

6.10 Appendix: Filamental partial mutual inductance

Here we present the partial mutual inductance of two straight filaments placed in any desired position. The geometry is shown in Fig. 6.35. Let the distances between the segment ends

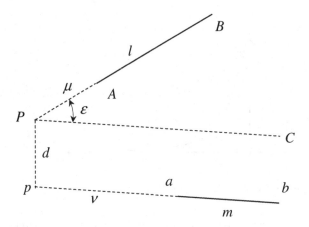

Figure 6.35 Two arbitrarily placed filaments.

be denoted by R_i

$$Bb = R_1 \tag{6.77}$$

$$Ba = R_2 \tag{6.78}$$

$$Aa = R_3 \tag{6.79}$$

$$Ab = R_4 \tag{6.80}$$

$$\frac{M}{\cos \epsilon} = \frac{\mu}{2\pi} \left((l + \mu) \tanh^{-1} \left(\frac{m}{R_1 + R_2} \right) - v \tanh^{-1} \left(\frac{l}{R_2 + R_3} \right) \right.$$
$$\left. + (m + v) \tanh^{-1} \left(\frac{l}{R_1 + R_4} \right) - \mu \tanh^{-1} \left(\frac{m}{R_3 + R_4} \right) \right) - d \, \Omega \, \csc(\epsilon) \tag{6.81}$$

where

$$\Omega = \tan^{-1} \left(\frac{d^2 \cos \epsilon + (\mu + l)(v + m) \sin^2 \epsilon}{d R_1 \sin \epsilon} \right) - \tan^{-1} \left(\frac{d^2 \cos \epsilon + (\mu + l)v \sin^2 \epsilon}{d R_2 \sin \epsilon} \right)$$
$$+ \tan^{-1} \left(\frac{d^2 \cos \epsilon + \mu v \sin^2 \epsilon}{d R_3 \sin \epsilon} \right) - \tan^{-1} \left(\frac{d^2 \cos \epsilon + \mu(v + m) \sin^2 \epsilon}{d R_4 \sin \epsilon} \right) \tag{6.82}$$

The various parameters can be written in terms of the geometric parameters

$$\cos \epsilon = \frac{\alpha^2}{2lm} \tag{6.83}$$

$$\alpha^2 = R_4^2 - R_3^2 + R_2^2 - R_1^2 \tag{6.84}$$

$$d^2 = R_3^2 - \mu^2 - v^2 + 2\mu v \cos \epsilon \tag{6.85}$$

$$\frac{\mu}{l} = \frac{2m^2 \left(R_2^2 - R_3^2 - l^2 \right) + \alpha^2 \left(R_4^2 - R_3^2 - m^2 \right)}{4l^2 m^2 - \alpha^4} \tag{6.86}$$

$$\frac{v}{m} = \frac{2l^2 \left(R_4^2 - R_3^2 - m^2 \right) + \alpha^2 \left(R_2^2 - R_3^2 - l^2 \right)}{4l^2 m^2 - \alpha^4} \tag{6.87}$$

Alternatively, the above equations can be inverted to obtain

$$R_1^2 = d^2 + (\mu + l)^2 + (v + m)^2 - 2(\mu + l)(v + m)\cos\epsilon \tag{6.88}$$

$$R_2^2 = d^2 + (\mu + l)^2 + v^2 - 2v(\mu + l)\cos\epsilon \tag{6.89}$$

$$R_3^2 = d^2 + \mu^2 + v^2 - 2\mu v \cos\epsilon \tag{6.90}$$

$$R_4^2 = d^2 + \mu^2 + (v + m)^2 - 2\mu(v + m)\cos\epsilon \tag{6.91}$$

7 Resonance and impedance matching

Many common circuits make use of inductors and capacitors in different ways to achieve their functionality. Filters, impedance matching circuits, resonators, and chokes are common examples. We study these circuits in detail and in particular we shall focus on the desirable properties of the passive components in such circuits.

7.1 Resonance

We begin with the textbook discussion of resonance of RLC circuits. These circuits are simple enough to allow full analysis, and yet rich enough to form the basis for most of the circuits we will study in this chapter.

Incidentally, simple second-order resonant circuits can also model a wide array of physical phenomena, such as pendulums, mass-spring mechanical resonators, molecular resonance, microwave cavities, sections of transmission lines, and even large-scale structures such as bridges. Understanding these circuits will afford a wide perspective into many physical situations.

Series RLC circuits

The RLC circuit shown in Fig. 7.1 is deceptively simple. The impedance seen by the source is simply given by

$$Z = j\omega L + \frac{1}{j\omega C} + R = R + j\omega L \left(1 - \frac{1}{\omega^2 LC}\right) \tag{7.1}$$

The impedance is purely real at the *resonant frequency* when $\Im(Z) = 0$, or $\omega = \pm\frac{1}{\sqrt{LC}}$. At resonance the impedance takes on a minimal value. It is worthwhile to investigate the cause of resonance, or the cancellation of the reactive components due to the inductor and capacitor. Since the inductor and capacitor voltages are always $180°$ out of phase, and one reactance is dropping while the other is increasing, there is clearly always a frequency when the magnitudes are equal. Thus resonance occurs when $\omega L = \frac{1}{\omega C}$. A phasor diagram, shown in Fig. 7.2, shows this in detail.

So what is the magic about this circuit? The first observation is that, at resonance, the voltage across the reactances can be larger, in fact much larger, than the voltage across the

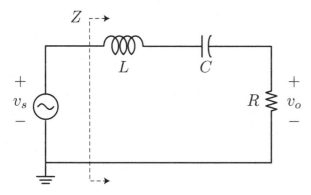

Figure 7.1 A series RLC circuit.

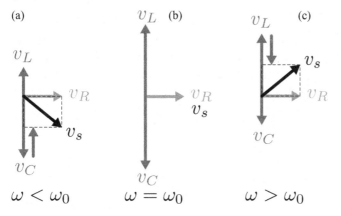

Figure 7.2 The phasor diagram of voltages in the series RLC circuit (a) below resonance, (b) at resonance, and (c) beyond resonance.

resistors, R. In other words, this circuit has voltage gain. Of course it does not have power gain, for it is a passive circuit. The voltage across the inductor is given by

$$v_L = j\omega_0 L i = j\omega_0 L \frac{v_s}{Z(j\omega_0)} = j\omega_0 L \frac{v_s}{R} = jQ \times v_s \tag{7.2}$$

where we have defined a circuit Q factor at resonance as

$$Q = \frac{\omega_0 L}{R} \tag{7.3}$$

It is easy to show that the same voltage multiplication occurs across the capacitor

$$v_C = \frac{1}{j\omega_0 C} i = \frac{1}{j\omega_0 C} \frac{v_s}{Z(j\omega_0)} = \frac{1}{j\omega_0 C} \frac{v_s}{R} = -jQ \times v_s \tag{7.4}$$

This voltage multiplication property is the key feature of the circuit that allows it to be used as an impedance transformer.

It is important to distinguish this Q factor from the intrinsic Q of the inductor and capacitor. For now, we assume the inductor and capacitor are ideal. We can rewrite the Q

factor in several equivalent forms, owing to the equality of the reactances at resonance

$$Q = \frac{\omega_0 L}{R} = \frac{1}{\omega_0 C} \frac{1}{R} = \frac{\sqrt{LC}}{C} \frac{1}{R} = \sqrt{\frac{L}{C}} \frac{1}{R} = \frac{Z_0}{R} \tag{7.5}$$

where we have defined the $Z_0 = \sqrt{\frac{L}{C}}$ as the characteristic impedance of the circuit.

Circuit transfer function

Let's now examine the transfer function of the circuit

$$H(j\omega) = \frac{v_o}{v_s} = \frac{R}{j\omega L + \frac{1}{j\omega C} + R} \tag{7.6}$$

$$H(j\omega) = \frac{j\omega RC}{1 - \omega^2 LC + j\omega RC} \tag{7.7}$$

Obviously, the circuit cannot conduct DC current, so there is a zero in the transfer function. The denominator is a quadratic polynomial. It's worthwhile to put it into a standard form that quickly reveals important circuit parameters

$$H(j\omega) = \frac{j\omega \frac{R}{L}}{\frac{1}{LC} + (j\omega)^2 + j\omega \frac{R}{L}} \tag{7.8}$$

Using the definition of Q and ω_0 for the circuit

$$H(j\omega) = \frac{j\omega \frac{\omega_0}{Q}}{\omega_0^2 + (j\omega)^2 + j\frac{\omega\omega_0}{Q}} \tag{7.9}$$

Factoring the denominator with the assumption that $Q > \frac{1}{2}$ gives us the complex poles of the circuit

$$s^{\pm} = -\frac{\omega_0}{2Q} \pm j\omega_0 \sqrt{1 - \frac{1}{4Q^2}} \tag{7.10}$$

The poles have a constant magnitude equal to the resonant frequency

$$|s| = \sqrt{\frac{\omega_0^2}{4Q^2} + \omega_0^2 \left(1 - \frac{1}{4Q^2}\right)} = \omega_0 \tag{7.11}$$

A root-locus plot of the poles as a function of Q appears in Fig. 7.3. As $Q \to \infty$, the poles move to the imaginary axis. In fact, the real part of the poles is inversely related to the Q factor.

Circuit bandwidth

As shown in Fig. 7.4, when we plot the magnitude of the transfer function, we see that the selectivity of the circuit is also related inversely to the Q factor. In the limit that $Q \to \infty$, the circuit is infinitely selective and only allows signals at resonance ω_0 to travel to the load.

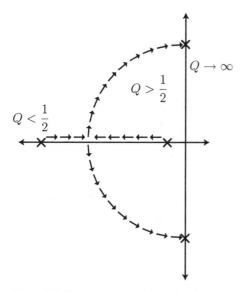

Figure 7.3 The root locus of the poles for a second-order transfer function parameterized by Q. The poles begin on the real axis for $Q < \frac{1}{2}$ and become complex, tracing a semi-circle for increasing Q.

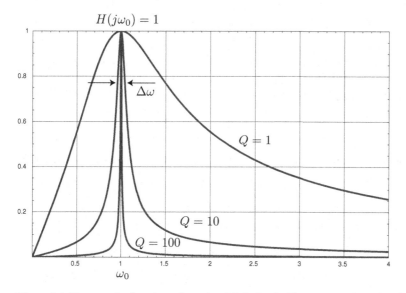

Figure 7.4 The transfer function of a series RLC circuit. The output voltage is taken at the resistor terminals. Increasing Q leads to a more peaky response.

Note that the peak gain in the circuit is always unity, regardless of Q, since at resonance the L and C together disappear and effectively all the source voltage appears across the load.

The selectivity of the circuit lends itself well to filter applications (Section 7.5). To characterize the peakiness, let's compute the frequency when the magnitude squared of the

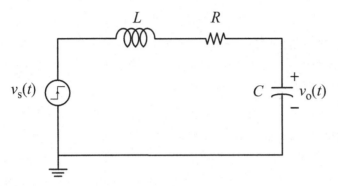

Figure 7.5 A step function applied to a series RLC circuit with output taken across the capacitor.

transfer function drops by half

$$|H(j\omega)|^2 = \frac{\left(\omega\frac{\omega_0}{Q}\right)^2}{\left(\omega_0^2 - \omega^2\right)^2 + \left(\omega\frac{\omega_0}{Q}\right)^2} = \frac{1}{2} \tag{7.12}$$

This happens when

$$\left(\frac{\omega_0^2 - \omega^2}{\omega_0\omega/Q}\right)^2 = 1 \tag{7.13}$$

Solving the above equation yields four solutions, corresponding to two positive and two negative frequencies. The peakiness is characterized by the difference between these frequencies, or the bandwidth, given by

$$\Delta\omega = \omega_+ - \omega_- = \frac{\omega_0}{Q} \tag{7.14}$$

which shows that the normalized bandwidth is inversely proportional to the circuit Q

$$\frac{\Delta\omega}{\omega_0} = \frac{1}{Q} \tag{7.15}$$

You can also show that the resonance frequency is the geometric mean frequency of the 3 dB frequencies

$$\omega_0 = \sqrt{\omega_+\omega_-} \tag{7.16}$$

Circuit damping factor

So far we have characterized the second-order series RLC circuit by its frequency response, leading to the concept of the resonant frequency ω_0 and quality factor Q. An equivalent characterization involves the time domain response, particularly the step response of the circuit. This will lead to the concept of the damping factor. As shown in Fig. 7.5, we will now consider the output as the voltage across the capacitor. This corresponds to the common situation in digital circuits, where the "load" is a capacitive reactance of a gate.

First let's recall the familiar step response of an RC circuit, or the limit as $L \to 0$ in the series RLC circuit, the circuit response to a step function is a rising exponential function that asymptotes towards the source voltage with a time scale $\tau = 1/RC$.

When the circuit has inductance, it's described by a second-order differential equation. Applying KVL, we can write an equation for the voltage $v_C(t)$

$$v_s(t) = v_C(t) + RC\frac{dv_C}{dt} + LC\frac{d^2v_C}{dt^2} \tag{7.17}$$

with initial conditions

$$v_0(t) = v_C(t) = 0\text{V} \tag{7.18}$$

and

$$i(0) = i_L(0) = 0\text{A} \tag{7.19}$$

For $t > 0$, the source voltage switches to V_{dd}. Thus Eq. (7.17) has a constant source for $t > 0$

$$V_{dd} = v_C(t) + RC\frac{dv_C}{dt} + LC\frac{d^2v_C}{dt^2} \tag{7.20}$$

In the steady state, $\frac{d}{dt} \to 0$, leading to

$$V_{dd} = v_C(\infty) \tag{7.21}$$

which implies that the entire voltage of the source appears across the capacitor, driving the current in the circuit to zero. Subtracting this steady-state voltage from the solution, we can simplify our differential equation

$$v_C(t) = V_{dd} + v(t) \tag{7.22}$$

$$V_{dd} = V_{dd} + v(t) + RC\frac{dv}{dt} + LC\frac{d^2v}{dt^2} \tag{7.23}$$

or, more simply, the homogeneous equation

$$0 = v(t) + RC\frac{dv}{dt} + LC\frac{d^2v}{dt^2} \tag{7.24}$$

The solution of which is nothing but a complex exponential $v(t) = Ae^{st}$, leading to the characteristic equation

$$0 = 1 + RCs + LCs^2 = 1 + (s\tau)2\zeta + (s\tau)^2 \tag{7.25}$$

where we have defined

$$\tau = \frac{1}{\omega_0} \tag{7.26}$$

and

$$\zeta = \frac{1}{2Q} \tag{7.27}$$

which has solutions

$$st = -\zeta \pm \sqrt{\zeta^2 - 1} \tag{7.28}$$

This equation can be characterized by the value ζ, leading to the following three important cases

$$\begin{aligned} \zeta < 1 \quad &\text{underdamped} \\ \zeta = 1 \quad &\text{critically damped} \\ \zeta > 1 \quad &\text{overdamped} \end{aligned} \tag{7.29}$$

The terminology "damped" will become obvious momentarily. When we plot the general time-domain solution, we have

$$v_C(t) = V_{dd} + A \exp(s_1 t) + B \exp(s_2 t) \tag{7.30}$$

which must satisfy the following boundary conditions

$$v_C(0) = V_{dd} + A + B = 0 \tag{7.31}$$

and

$$i(0) = C \frac{dv_C(t)}{dt}\bigg|_{t=0} = 0 \tag{7.32}$$

or the following set of equations

$$As_1 + Bs_2 = 0 \tag{7.33}$$
$$A + B = -V_{dd} \tag{7.34}$$

which has the following solution

$$A = \frac{-V_{dd}}{1 - \sigma} \tag{7.35}$$

$$B = \frac{\sigma V_{dd}}{1 - \sigma} \tag{7.36}$$

where $\sigma = \frac{s_1}{s_2}$. So we have the following equation

$$v_C(t) = V_{dd} \left(1 - \frac{1}{1 - \sigma}(e^{s_1 t} - \sigma e^{s_2 t}) \right) \tag{7.37}$$

If the circuit is overdamped, $\zeta > 1$, the poles are real and negative

$$s = \frac{1}{\tau}(-\zeta \pm \sqrt{\zeta^2 - 1}) = \begin{cases} s_1 \\ s_2 \end{cases} < 0 \tag{7.38}$$

The normalized response, with $\zeta = 2, 1, 0.5, 0.3$, is shown in Fig. 7.6. Qualitatively, we see that the $\zeta = 2$ circuit response resembles the familiar RC circuit. This is true until $\zeta = 1$, at which point the circuit is said to be *critically damped*. The circuit has two equal poles at

$$s = \frac{1}{\tau}(-\zeta \pm \sqrt{\zeta^2 - 1}) = -\frac{1}{\tau} \tag{7.39}$$

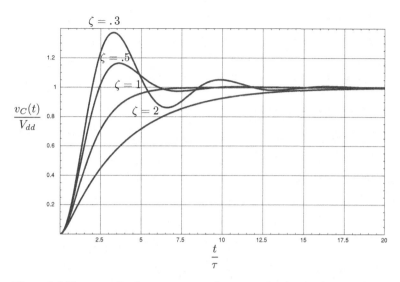

Figure 7.6 The normalized step response of an RLC circuit. The damping factor ζ is varied to cover three important regions: an overdamped response with $\zeta > 1$, a critically damped response with $\zeta = 1$, and an underdamped response with $\zeta < 1$.

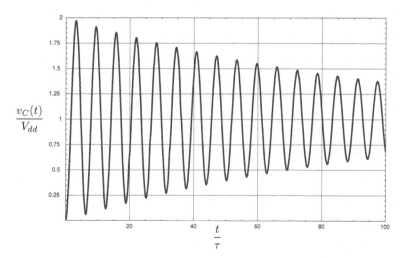

Figure 7.7 The step response of an RLC circuit with low damping ($\zeta = 0.01$).

In which case the time response is given by

$$\lim_{\zeta \to 1} v_C(t) = V_{dd}\left(1 - e^{-t/\tau} - \frac{t}{\tau}e^{-t/\tau}\right) \tag{7.40}$$

As seen in Fig. 7.6, the circuit step response is faster with increasing $\zeta \leq 1$, but still similar to the RC circuit. The *underdamped* case, $\zeta < 1$, is the most interesting case here, as the response is now markedly different. The circuit *overshoots* the mark and settles down to the steady-state value after oscillating. The amount of oscillation increases as the damping factor ζ is reduced. The response with very low damping, $\zeta = .01$, is shown in Fig. 7.7.

The underdamped case is characterized by two complex conjugate poles

$$s\tau = -\zeta \pm j\sqrt{1 - \zeta^2} = a \pm jb \tag{7.41}$$

Note the complex conjugate of the A coefficient is simply

$$A^* = \frac{-V_{dd}}{1 - \sigma^*} = \frac{V_{dd}}{\sigma^{-1} - 1} = \frac{\sigma V_{dd}}{1 - \sigma} = B \tag{7.42}$$

which allows us to write the voltage across the capacitor as

$$v_C(t) = V_{dd} + e^{at/\tau}(Ae^{jbt} + A^*e^{-jbt}) \tag{7.43}$$

or more simply

$$v_C(t) = V_{dd} + e^{at/\tau}2|A|\cos(\omega t + \phi) \tag{7.44}$$

where

$$|A| = \frac{V_{dd}}{|1 + \sigma|} \tag{7.45}$$

and

$$\phi = \angle\frac{V_{dd}}{1 + \sigma} \tag{7.46}$$

The oscillating frequency is determined by the imaginary part of the poles, $\omega = b/\tau = \omega_0\sqrt{1 - \zeta^2}$, which approaches the natural resonance frequency as the damping is reduced. The decay per period, or the envelope damping of the waveform, is determined by a, which equals $\zeta\omega_0$. This is why the damping factor controls the amount of overshoot and oscillation in the waveform.

Energy storage in RLC "tank"

Let's compute the ebb and flow of the energy at resonance. To begin, let's assume that there is negligible loss in the circuit. The energy across the inductor is given by

$$w_L = \frac{1}{2}Li^2(t) = \frac{1}{2}LI_M^2\cos^2\omega_0 t \tag{7.47}$$

Likewise, the energy stored in the capacitor is given by

$$w_C = \frac{1}{2}Cv_C^2(t) = \frac{1}{2}C\left(\frac{1}{C}\int i(\tau)d\tau\right)^2 \tag{7.48}$$

Performing the integral leads to

$$w_C = \frac{1}{2}\frac{I_M^2}{\omega_0^2 C}\sin^2\omega_0 t \tag{7.49}$$

The total energy *stored* in the circuit is the sum of these terms

$$w_s = w_L + w_C = \frac{1}{2}I_M^2\left(L\cos^2\omega_0 t + \frac{1}{\omega_0^2 C}\sin^2\omega_0 t\right) = \frac{1}{2}I_M^2 L \tag{7.50}$$

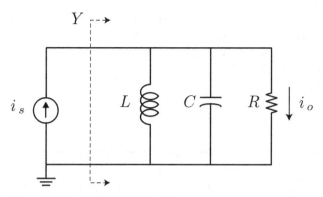

Figure 7.8 A parallel RLC circuit.

which is a constant! This means that the reactive stored energy in the circuit does not change and simply moves between capacitive energy and inductive energy. When the current is maximum across the inductor, all the energy is in fact stored in the inductor

$$w_{L,\max} = w_s = \frac{1}{2} I_M^2 L \tag{7.51}$$

Likewise, the peak energy in the capacitor occurs when the current in the circuit drops to zero

$$w_{C,\max} = w_s = \frac{1}{2} V_M^2 C \tag{7.52}$$

Now let's re-introduce loss in the circuit. In each cycle, a resistor R will dissipate energy

$$w_d = P \cdot T = \frac{1}{2} I_M^2 R \cdot \frac{2\pi}{\omega_0} \tag{7.53}$$

The ratio of the energy stored to the energy dissipated is thus

$$\frac{w_s}{w_d} = \frac{\frac{1}{2} L I_M^2}{\frac{1}{2} I_M^2 R \frac{2\pi}{\omega_0}} = \frac{\omega_0 L}{R} \frac{1}{2\pi} = \frac{Q}{2\pi} \tag{7.54}$$

This gives us the physical interpretation of the *quality factor* Q as 2π times the ratio of energy stored per cycle to energy dissipated per cycle in an RLC circuit

$$Q = 2\pi \frac{w_s}{w_d} \tag{7.55}$$

We can now see that if $Q \gg 1$, then an initial energy in the tank tends to slosh back and forth for many cycles. In fact, we can see that in roughly Q cycles, the energy of the tank is depleted.

Parallel RLC circuits

The parallel RLC circuit shown in Fig. 7.8 is the dual of the series circuit. By "dual" we mean that the role of voltage and current are interchanged. Hence the circuit is most naturally probed with a current source i_s. In other words, the circuit has current gain as opposed to voltage gain, and the admittance minimizes at resonance as opposed to the impedance. Finally, the role of capacitance and inductance are also interchanged. In principle, therefore, we do not have to repeat all the detailed calculations we just performed for the series case, but in practice it is a worthwhile exercise.

The admittance of the circuit is given by

$$Y = j\omega C + \frac{1}{j\omega L} + G = G + j\omega C \left(1 - \frac{1}{\omega^2 LC}\right) \tag{7.56}$$

which has the same form as Eq. (7.1). The resonant frequency also occurs when $\Im(Y) = 0$, or when $\omega = \omega_0 = \pm\frac{1}{\sqrt{LC}}$. Likewise, at resonance the admittance takes on a minimal value. Equivalently, the impedance at resonance is maximum. This property makes the parallel RLC circuit an important element in tuned amplifier loads (see Section 8). It is also easy to show that at resonance the circuit has a current gain of Q

$$i_C = j\omega_0 C v_o = j\omega_0 C \frac{i_s}{Y(j\omega_0)} = j\omega_0 C \frac{i_s}{G} = jQ \times i_s \tag{7.57}$$

where we have defined the circuit Q factor at resonance by

$$Q = \frac{\omega_0 C}{G} \tag{7.58}$$

in complete analogy with Eq. (7.3). Likewise, the current gain through the inductor is also easily derived

$$i_L = -jQ \times i_s \tag{7.59}$$

The equivalent expressions for the circuit Q factor are given by the inverse of the relations of Eq. (7.5)

$$Q = \frac{\omega_0 C}{G} = \frac{R}{\omega_0 L} = \frac{R}{\frac{1}{\sqrt{LC}}L} = \frac{R}{\sqrt{\frac{L}{C}}} = \frac{R}{Z_0} \tag{7.60}$$

The phase response of a resonant circuit is also related to the Q factor. For the parallel RLC circuit, the phase of the admittance is given by

$$\angle Y(j\omega) = \tan^{-1}\left(\frac{\omega C \left(1 - \frac{1}{\omega^2 LC}\right)}{G}\right) \tag{7.61}$$

The rate of change of phase at resonance is given by

$$\left.\frac{d\angle Y(j\omega)}{d\omega}\right|_{\omega_0} = \frac{2Q}{\omega_0} \tag{7.62}$$

A plot of the admittance phase as a function of frequency and Q is shown in Fig. 7.9.

Circuit transfer function

Given the duality of the series and parallel RLC circuits, it is easy to deduce the behavior of the circuit. Whereas the series RLC circuit acted as a filter and was only sensitive to

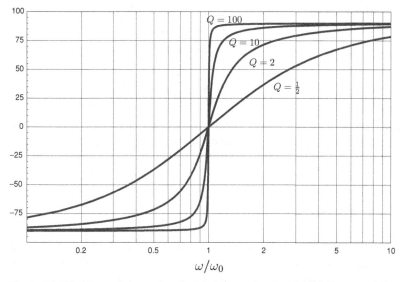

Figure 7.9 The phase of a second-order admittance as a function of frequency. The rate of change of phase at resonance is proportional to the Q factor.

voltages near resonance ω_0, likewise the parallel RLC circuit is only sensitive to currents near resonance

$$H(j\omega) = \frac{i_o}{i_s} = \frac{v_o G}{v_o Y(j\omega)} = \frac{G}{j\omega C + \frac{1}{j\omega L} + G} \tag{7.63}$$

which can be put into the same canonical form as before

$$H(j\omega) = \frac{j\omega \frac{\omega_0}{Q}}{\omega_0^2 + (j\omega)^2 + j\frac{\omega\omega_0}{Q}} \tag{7.64}$$

where we have appropriately re-defined the circuit Q to correspond the parallel RLC circuit. Notice that the impedance of the circuit takes on the same form

$$Z(j\omega) = \frac{1}{Y(j\omega)} = \frac{1}{j\omega C + \frac{1}{j\omega L} + G} \tag{7.65}$$

which can be simplified to

$$Z(j\omega) = \frac{j\frac{\omega}{\omega_0}\frac{1}{GQ}}{1 + \left(\frac{j\omega}{\omega_0}\right)^2 + j\frac{\omega}{\omega_0 Q}} \tag{7.66}$$

At resonance, the real terms in the denominator cancel

$$Z(j\omega_0) = \frac{j\frac{R}{Q}}{\underbrace{1 + \left(\frac{j\omega_0}{\omega_0}\right)^2}_{=0} + j\frac{1}{Q}} = R \tag{7.67}$$

It is not hard to see that this circuit has the same half power bandwidth as the series RLC circuit, since the denominator has the same functional form

$$\frac{\Delta\omega}{\omega_0} = \frac{1}{Q} \tag{7.68}$$

A plot of this impedance versus frequency has the same form as Fig. 7.4 multiplied by the resistance R.

Energy storage in a parallel RLC circuit is completely analogous to the series RLC case and in fact the general equation relating circuit Q to energy storage and dissipation also holds in the parallel RLC circuit.

7.2 The many faces of Q

As we have seen, in RLC circuits the most important parameter is the circuit Q and resonance frequency ω_0. Not only do these parameters describe the circuit in a general way, but they also give us immediate insight into the circuit behavior.

The Q factor can be computed several ways, depending on the application. For instance, if the circuit is designed as a filter, then the most important Q relation is the half-power bandwidth

$$Q = \frac{\omega_0}{\Delta\omega} \tag{7.69}$$

We shall also find many applications where the phase selectivity of these circuits is of importance. An example is a resonant oscillator where the noise of the system is rejected by the tank based on the phase selectivity. In an oscillator any "excess phase" in the loop tends to move the oscillator away from the natural resonant frequency. It is therefore desirable to maximize the rate of change of phase of the circuit impedance as a function of frequency. For the parallel RLC circuit we derived the phase of the admittance (Eq. (7.62)), which gives us another way to interpret and compute Q

$$Q = \frac{\omega_0}{2} \frac{d\angle Y(j\omega)}{d\omega} \tag{7.70}$$

For applications where the circuit is used as a voltage or current multiplier, the ratio of reactive voltage (current) to real voltage (current) is most relevant. As we shall see, in RFID systems (Section 10.10) this is an important application of the circuit. For a series case we found

$$Q = \frac{v_L}{v_R} = \frac{v_C}{v_R} \tag{7.71}$$

and for the parallel case

$$Q = \frac{i_L}{i_R} = \frac{i_C}{i_R} \tag{7.72}$$

Finally, when the step response or time domain transient response is of interest, the circuit Q or equivalently, the damping factor ζ, describes the behavior of the circuit, with $\zeta = 1$ the boundary between damped behavior and under-damped oscillatory response.

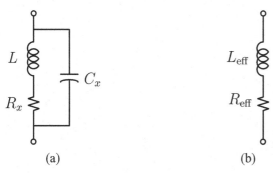

Figure 7.10 (a) A simple model for an inductor with winding capacitance C_x and winding resistance R_x. (b) A simplified equivalent circuit for the non-ideal inductor model.

The last and one of the most important interpretations of Q is in the definition of energy, relating the energy storage and losses in a RLC "tank" circuit. We can define the Q of a circuit at frequency ω as the energy stored in the tank (W) divided by the rate of energy loss

$$Q = W \left/ \frac{dW}{d\phi} \right. = \omega W \left/ \frac{dW}{dt} \right.$$

Practical issues with resonators

Inductor equivalent circuit

The previous discussion was oversimplified since any real RLC circuit will consist of non-ideal and lossy Ls and Cs. For example, consider the simple equivalent circuit for the inductor shown in Fig. 7.10a, where C_x accounts for the winding capacitance and R_x is the series winding resistance. This simple model is sufficiently accurate over a narrow range of frequencies for inductors realized on non-lossy substrates or an air coil inductor. In many situations we would like to model the inductor as an equivalent ideal inductor with series loss, as shown in Fig. 7.10b. This is done by simply equating the impedance at a given frequency resulting in

$$R_{\text{eff}} \approx R \left(1 + \left(\frac{\omega}{\omega_0} \right)^2 \right) \tag{7.73}$$

$$L_{\text{eff}} \approx L \left(1 + \left(\frac{\omega}{\omega_0} \right)^2 \right) \tag{7.74}$$

The above approximation holds when the circuit is operated far below the inductor self-resonant frequency (SRF) $\omega \ll \omega_0$. Note that the second time constant in the circuit RC is at an even higher frequency than the resonant frequency for $Q > 1$.

For more complex structures, for example an inductor modeled on a lossy substrate such as silicon, one approach is to simply model each element with an equivalent circuit, as shown in Fig. 7.11a. Here we assume that the capacitor and resistor are nearly ideal, but the inductor has parasitic capacitance and resistance, modeling the self-resonance and the

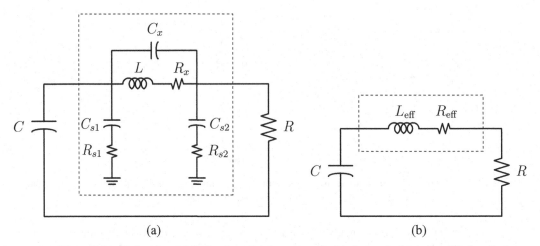

Figure 7.11 (a) A series RLC circuit containing a non-ideal inductor L. (b) A simplified equivalent circuit for the series RLC circuit.

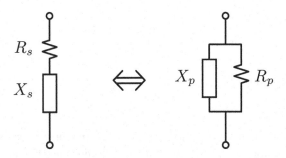

Figure 7.12 A series to parallel transformation for an arbitrary impedance Z.

conductor and substrate losses. Analyzing such a circuit is trivial but messy, leading to higher-order equations containing many poles/zeros and it is unlikely to give us insight into the circuit behavior. An alternative approach is to observe that no matter how complicated the inductor model, at a fixed frequency below the component resonance frequency, the net impedance of the inductor can be represented as $Z_L = R_{eff} + j\omega L_{eff}$. While this is obviously true, we will derive some simple formulas to quickly derive this equivalent impedance from a complicated circuit model.

Shunt-series transformation

The key calculation aid is the series to parallel transformation. Consider the impedance shown in Fig. 7.12, which we wish to represent as a parallel impedance. We can do this at a single frequency as long as the impedance of the series network equals the impedance of the shunt network

$$R_s + jX_s = \frac{1}{\frac{1}{R_p} + \frac{1}{jX_p}} \tag{7.75}$$

Equating the real and imaginary parts

$$R_s = \frac{R_p X_p^2}{R_p^2 + X_p^2} \tag{7.76}$$

$$X_s = \frac{R_p^2 X_p}{R_p^2 + X_p^2} \tag{7.77}$$

which can be simplified by using the definition of Q

$$Q_s = \frac{X_s}{R_s} = \frac{R_p^2 X_p}{R_p X_p^2} = \frac{R_p}{X_p} = Q_p = Q \tag{7.78}$$

which shows that

$$R_p = R_s(1 + Q^2) \tag{7.79}$$

and

$$X_p = X_s(1 + Q^{-2}) \approx X_s \tag{7.80}$$

where the approximation applies under high Q conditions.

Simplifying practical RLC resonators

The simplified RLC circuit appears in Fig. 7.11b, which has the same equivalent form as a second-order RLC circuit. Here the effective inductance L_{eff} of the circuit is different from L. If the circuit is operated far below the SRF of the inductor, then $L_{\text{eff}} \approx L$. The loss in the circuit now includes the effective series resistance and equivalent shunt substrate losses of the inductor. Again, at frequencies far below the SRF of the inductor, the resistance is dominated by the conductive loss of the inductor winding. We can thus conclude that for a practical RLC circuit employing non-ideal RLC elements, we can treat the circuit as a simple series RLC if all elements are used below their self-resonant frequency and represented as equivalent series impedances. The self-resonant frequency of the circuit is then determined by the net capacitance and inductance in the loop, and the Q factor is determined by the total series resistance in the loop. For this reason, series RLC circuits are sensitive to any series resistance in the circuit and the layout must be done properly in order to minimize any contact or interconnect resistance.

It is thus clear that the circuit Q is necessarily lower than the component Q_x, where Q_x is defined in the "series" sense. If the net impedance of a component is $Z = R + jX$, then

$$Q_x = \frac{X}{R} \tag{7.81}$$

At resonance the circuit Q is given by

$$Q = \frac{X'}{R + R_{x,L} + R_{x,C}} \tag{7.82}$$

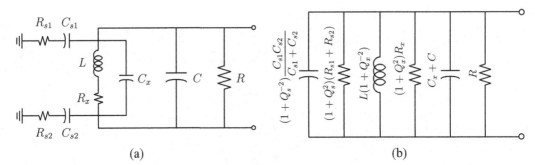

Figure 7.13 (a) A parallel RLC circuit containing a non-ideal inductor L. (b) A simplified equivalent circuit for the parallel RLC circuit.

where X' is the net capacitive reactance at resonance and $R_{x,L|C}$ represent the series losses in the reactive components. Given that typically $X' \approx X$, we can re-write the above as

$$\frac{1}{Q} \approx \frac{R + R_{x,1} + R_{x,2}}{X} = \frac{1}{Q_{id}} + \frac{1}{Q_L} + \frac{1}{Q_C} \tag{7.83}$$

where Q_{id} is the Q of the circuit using "ideal" components. This means that the circuit Q is approximately the parallel combination of the Qs

$$Q = Q_{id}||Q_L||Q_C < Q_{id} \tag{7.84}$$

Example 11

Design a series RLC circuit with $R = 5\,\Omega$, using an inductor with $L = 5$ nH and component $Q_L = 30$ and a capacitor with component $Q_C = 200$ to resonate at 1 GHz. We begin by calculating the equivalent series resistance of each component.

$$R_{x,L} = \frac{X_L}{Q_L} = \frac{31.416}{30} \approx 1.05\,\Omega$$

$$R_{x,C} = \frac{X_C}{Q_C} = \frac{31.416}{200} \approx 0.16\,\Omega$$

The total series resistance of the circuit is thus $R_T = 6.21\,\Omega$, resulting in a net $Q = 5$. In comparison, the unloaded $Q = 6.3$.

The exact same considerations apply for a parallel RLC circuit with the exception that now *shunt* parasitics de-Q the circuit. To see this, consider a non-ideal RLC circuit shown in Fig. 7.13. Now we convert each non-ideal element into an ideal admittance $Y = G_x + jB'$. The net conductance of the circuit is given by

$$G_T = G_{x,L} + G_{x,C} + G \tag{7.85}$$

The resulting equivalent circuit is shown in Fig. 7.13. For operation below the SRF of the components, therefore, Eq. (7.84) applies.

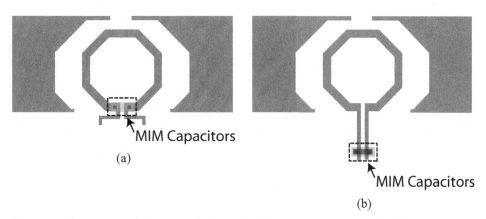

Figure 7.14 (a) An LC tank layout employing a ring inductor and two series connected MIM capacitors. (b) An inferior LC tank layout due to extra lead inductance and loss.

LC tanks

In many applications, such as oscillators, we use an LC tank as a frequency reference. In an oscillator, for instance, the frequency of oscillation is determined by the frequency where the phase shift through the loop is a $0°$ (any multiple of 2π). As the phase selectivity of an RLC tank is determined by the Q, we maximize the "tank" Q by using the highest-quality inductance and capacitance available. Furthermore, we eliminate or minimize the loading resistance so we omit R to obtain an LC tank. Any residual R in the parallel RLC tank is due to losses (component Q_c) or unwanted loading.

The tank impedance at resonance is therefore real and only limited by the realizable quality factor Q

$$Z(\omega_0) = R_p = \omega_0 L \times Q \tag{7.86}$$

The layout of an integrated LC tank circuit appears in Fig. 7.14a. In the layout of the LC tank in an IC environment, it is often easy to forget that every wire interconnect adds resistance and potentially couples to the substrate and thus it can de-Q the circuit. Here a small ring inductor resonates with MIM capacitors. An electric ground shield surrounds the ring without forming a closed loop. Notice that the MIM capacitors are placed directly underneath the leads of the inductor. The measured impedance of this structure is shown in Fig. 7.15. As expected, the circuit behaves as an LC circuit with the ring inductor losses dominating the Q. Despite the high-frequency resonance of 56 GHz, the ring tank behaves extraordinarily like a simple lumped LC circuit. This follows because the dimensions of the structure are much smaller than the wavelength over the operating range.

It is important to note that the layout of Fig. 7.14b is sub-optimal (but commonly used) as it adds unnecessary lead inductance and resistance to the parallel tank. Every effort should be made in the layout to put the capacitor leads as close to the inductor leads to minimize losses.

Notice that the layout in Fig. 7.14a uses two series MIM capacitors. This is done for two important reasons. Even though the net capacitance is the same as a single capacitor of half the size, larger capacitors have better matching and are more accurate. Furthermore,

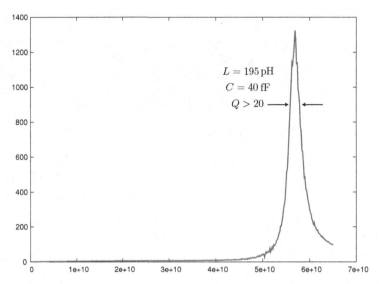

Figure 7.15 The on-chip measured LC tank impedance.

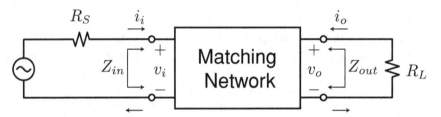

Figure 7.16 A generic matching network as a black box.

due to the differential excitation, the capacitor common node is not excited and acts as an AC ground. This is beneficial since the bottom plate is closer to the substrate and thus is a lossier node.

7.3 Impedance matching

In the words of one experienced designer, "RF design is all about impedance matching." In this section we would like to show how inductors and capacitors are handy elements at impedance matching. Viewed as a black box shown in Fig. 7.16, an impedance matcher changes a given load resistance R_L to a source resistance R_S. Without loss of generality, assume $R_S > R_L$, and a power match factor of $m = R_S/R_L$ is desired. In fact any matching network that boosts the resistance by some factor can be flipped over to do the opposite matching.

Since $R_L = v_o/i_o$ and $R_S = v_i/i_i$, we can see that this transformation can be achieved by a voltage gain, $v_i = kv_o$. Assuming the black box is realized with passive elements without memory, power conservation implies

$$i_i v_i = i_o v_o \tag{7.87}$$

thus the current must drop by the same factor, $i_i = k^{-1} i_o$, resulting in

$$Z_{in} = \frac{v_i}{i_i} = \frac{k v_o}{k^{-1} i_o} = k^2 \frac{v_o}{i_o} = k^2 R_L \tag{7.88}$$

which means that $k = \sqrt{m}$ to achieve an impedance match. There are many ways to realize such a circuit block and in this section we will explore techniques employing inductors and capacitors. In Chapter 10 we will explore transformer-based impedance matching.

Why play the matchmaker?

Optimal power transfer

Perhaps the most important reason for matching is to maximize the power transfer from a source to a load. Recall that the maximum power available from a fixed voltage source with impedance Z_S is obtained when the load is the complex conjugate impedance, or $Z_L = Z_S^*$. To see that such an optimum impedance must exist notice that the power delivered is zero when an open or short is attached to the source and non-zero for any practical value of the load. By continuity, we can see that the curve of P_L has a maximum.

Optimal noise figure

There are other reasons to match impedances and we will mention them in passing. Another important case is to minimize the noise figure of an amplifier driven for a given source impedance. It can be shown that the noise figure of a two-port (e.g. a single transistor or a multi-stage amplifier) takes on the following general form

$$F = F_{min} + \frac{R_n}{G_g} |Y_s - Y_{s,opt}|^2 \tag{7.89}$$

The parameters R_n, G_g, and $Y_{s,opt}$ are properties of the amplifier at a particular bias point. The noise figure is thus minimized when the source impedance is equal to $Y_{s,opt}$. So another function for a matching network is to perform this transformation. In fact, in a modern integrated circuit we have control over the physical dimension of the transistor and thus we can size a transistor appropriately so that a noise match corresponds to an optimal power gain match.

Minimum reflections in transmission lines

Another important reason for impedance matching will become evident when we study switching transients on a transmission line (Chapter 12). Here impedance matching is more commonly known as "terminating" a transmission line. Without a proper termination, reflections can lead to inter-symbol interference. A practical example is a ghost of a television screen caused by reflections off the antenna and television amplifier. These reflections travel back and forth on the feed-line and produce weak secondary copies of the signals, which appear as ghosts on the screen.

Optimal efficiency

Power amplifiers present more reasons to match impedances. In particular, while an impedance match results in maximum power transfer, an entirely different impedance is needed to achieve *optimal efficiency*. To see this observe that the drain efficiency of an amplifier can be written as

$$\eta = \frac{P_L}{P_{dc}} = \frac{\frac{1}{2}v_o i_o}{I_Q V_{sup}} \tag{7.90}$$

where I_Q is the average current drawn from the supply voltage V_{sup} over one cycle and v_o and i_o are the output voltage and current swings in the amplifier. Rewriting the above equation into normalized form

$$\eta = \frac{1}{2}\hat{I}\hat{V} \tag{7.91}$$

where $\hat{I} = i_o/I_Q$ is the normalized current swing and $\hat{V} = v_o/V_{sup}$ is the normalized voltage swing. By conservation of energy, $\hat{I} \times \hat{V} \leq 2$. For a tuned amplifier, $\hat{V} \leq 1$ and $\hat{I} \leq 2$. Once these values have been specified to achieve a certain efficiency, the value of the load impedance is fixed by the ratio

$$R_{L,opt} = \frac{v_o}{i_o} = \frac{\hat{V}}{\hat{I}} \times \frac{V_{sup}}{I_Q} \tag{7.92}$$

Let us do an example. Say we would like to deliver 1 W of power in a modern CMOS process with $V_{sup} < 2$ V (due to breakdown). Since the maximum voltage swing cannot exceed the supply ($\hat{V} \leq 1$), the required current swing is found from the required output power

$$\frac{1}{2}i_o v_o = P_L \tag{7.93}$$

or $i_o \geq 2P_L/V_{sup}$, which works out to $i_o \geq 1$ A. The optimal load impedance is thus as low as $2V/1A = 2\,\Omega$, a very small resistance. The load in most communication systems is the system impedance Z_0, usually $Z_0 = 50\Omega$ (RF) or 75Ω (video), or even a higher impedance. Clearly, an impedance matching network is needed to to convert the load impedance Z_0 to 2Ω to achieve the required output power and optimal efficiency.

Capacitive and inductive dividers

Perhaps the simplest matching networks are simple voltage dividers. Consider the capacitive voltage divider shown in Fig. 7.17a. At RF frequencies, if $R_L \gg X_2$, then we can see that the circuit will work as advertised. Assuming that negligible current flows into R_L, the current flowing into the capacitors is given by

$$i = \frac{v_i}{j(X_1 + X_2)} \tag{7.94}$$

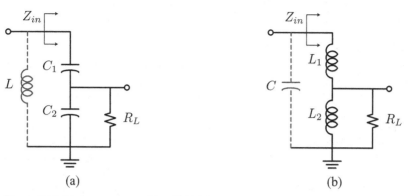

Figure 7.17 (a) A tapped capacitive divider impedance transformer. (b) A tapped inductor impedance transformer. The reactance in the structure can be resonated with an appropriate elements (shown in dashed line).

and the voltage is therefore

$$v_o = v_{C_2} = jX_2 \times i = v_i \frac{X_2}{X_1 + X_2} = v_i \frac{1}{1 + \frac{C_2}{C_1}} = k v_i \qquad (7.95)$$

which means that the load resistance is boosted by a factor of k^2

$$R_{in} \approx \left(1 + \frac{C_2}{C_1}\right)^2 R_L \qquad (7.96)$$

We can arrive at the same destination by using the shunt $\leftrightarrow$ series transformation twice. The final value of R_{in} is given by a $1 + Q_2^2$ reduction, followed by a $1 + Q_s^2$ enhancement

$$R_{in} = \frac{1 + Q_s^2}{1 + Q_2^2} R_L \qquad (7.97)$$

where $Q_2 = \frac{R_L}{X_2}$, $X_s = X_1 || X_2'$, and

$$Q_s = \frac{X_s}{R_L}(1 + Q_2^2) \qquad (7.98)$$

The final expression is derived after some algebra

$$R_{in} = \frac{R_L}{1 + Q_2^2} + \left(\frac{X_s}{R_L}\right)^2 + \left(\frac{X_s}{X_2}\right)^2 R_L \qquad (7.99)$$

Under the assumption that $X_2 \ll R_L$, the final term dominates

$$R_{in} \approx \left(\frac{X_s}{X_2}\right)^2 R_L = \left(1 + \frac{C_2}{C_1}\right)^2 R_L \qquad (7.100)$$

as expected. The reactance of the capacitive divider can be absorbed by a resonating inductance as shown in Fig. 7.17. In a similar vein, an inductive divider-matching circuit can be designed as shown in Fig. 7.17.

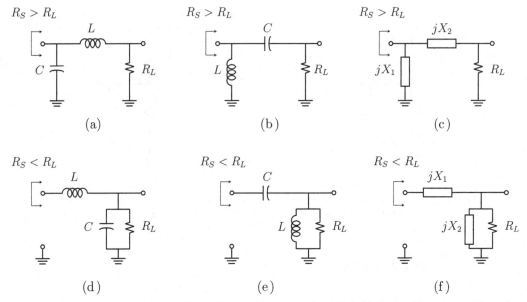

Figure 7.18 Several incarnations of L-matching networks. In (a)–(c) the load is connected in series with the reactance boosting the input resistance. In (d)–(f) the load is in shunt with the reactance, lowering the input resistance.

An L match

Consider the L-matching networks, shown in Fig. 7.18, named due to the topology of the network. We shall see that one direction of the L match boosts the load impedance (in series with load), whereas the other lowers the load impedance (in shunt with the load). Let's focus on the first two networks shown in Figs. 7.18a,b. Here, in absence of the source, we have a simple series RLC circuit. Recall that in resonance the voltage across the reactive elements is Q times larger than the voltage on the load! In essence, that is enough to perform the impedance transformation. Without doing any calculations, you can immediately guess that the impedance seen by the source is about Q^2 larger than R_L. Furthermore, since the circuit is operating in resonance, the net impedance seen by the source is purely real. To be sure, let's do the math.

A quick way to accomplish this feat is to begin with the series to parallel transformation, as shown in Fig. 7.19, where the load resistance in series with the inductor is converted to an equivalent parallel load equal to

$$R_p = (1 + Q^2)R_L \tag{7.101}$$

where $Q = X_L/R_L$, and $X'_L = X_L(1 + Q^{-2})$. The circuit is now nothing but a parallel RLC circuit and it is clear that at resonance the source will see only R_p, or a boosted value of R_L. The boosting factor is indeed equal to $Q^2 + 1$, very close to the value we guessed from the outset.

To gain insight into the operation of Figs. 7.18d–f, consider an Norton equivalent of the same circuit shown in Fig. 7.20. Now the circuit is easy to understand since it is simply a

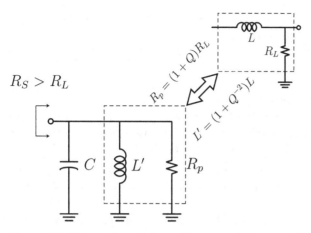

Figure 7.19 The transformed L-matching network into a parallel RLC equivalent circuit.

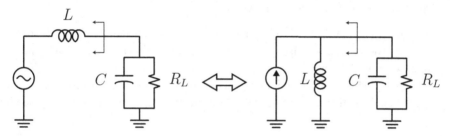

Figure 7.20 The source voltage driving the L-matching network can be transformed into an equivalent Norton current source.

parallel resonant circuit. We know that at resonance the current through the reactances is Q times larger than the current in the load. Since the current in the series element (L in Fig. 7.18d) is controlled by the source voltage, we can immediately see that $i_s = Qi_L$, thus providing the required current gain to lower the load resistance by a factor of Q^2.

As you may guess, the mathematics will yield a similar result. Simply do a parallel to series transformation of the load to obtain

$$R_s = \frac{R_p}{1 + Q^2} \tag{7.102}$$

$$X'_p = \frac{X_p}{1 + Q^{-2}} \tag{7.103}$$

The resulting circuit is a simple series RLC circuit. At resonance, the source will only see the reduced-series resistance R_s.

L-matching design equations

The following design procedure applies to an L match using the generic forms of Figs. 7.18c and f. The actual choice between Figs. 7.18a,d and Fig. 7.18b,e depends on the application. For instance, Figs. 7.18b,e provide AC coupling (DC isolation), which may be required in many applications. In other applications, a common DC voltage may be needed, making the networks of Figs. 7.18a,d the obvious choice.

Let $R_{hi} = \max(R_S, R_L)$ and $R_{lo} = \min(R_S, R_L)$. The L-matching networks shown in Fig. 7.18 are designed as follows:

1 Calculate the boosting factor $m = \frac{R_{hi}}{R_{lo}}$.
2 Compute the required circuit Q by $(1 + Q^2) = m$, or $Q = \sqrt{m - 1}$.
3 Pick the required reactance from the Q. If you are boosting the resistance, e.g. $R_S > R_L$, then $X_s = Q \cdot R_L$. If you are dropping the resistance, $X_p = \frac{R_L}{Q}$.
4 Compute the effective resonating reactance. If $R_S > R_L$, calculate $X_s' = X_s(1 + Q^{-2})$ and set the shunt reactance in order to resonate, $X_p = -X_s'$. If $R_S < R_L$, then calculate $X_p' = \frac{X_p}{1+Q^{-2}}$ and set the series reactance in order to resonate, $X_s = -X_p'$.
5 For a given frequency of operation, pick the value of L and C to satisfy these equations.

Insertion loss of an L-matching network

We would like to include the losses in our passive elements into the design of the matching network. The most detrimental effect of the component Q is the insertion loss, which reduces the power transfer from source to load.

Let's begin by using our intuition to derive an approximate expression for the loss. Note that the power delivered to the input of the matching network P_{in} can be divided into two components

$$P_{in} = P_L + P_{diss} \tag{7.104}$$

where P_L is the power delivered to the load and P_{diss} is the power dissipated by the non-ideal inductors and capacitors. The insertion loss is therefore given by

$$IL = \frac{P_L}{P_{in}} = \frac{P_L}{P_L + P_{diss}} = \frac{1}{1 + \frac{P_{diss}}{P_L}} \tag{7.105}$$

Recall that for the equivalent series RLC circuit in resonance, the voltages across the reactances are Q times larger than the voltage across R_L. We can show that the reactive power is also a factor of Q larger. For instance, the energy in the inductor is given by

$$W_m = \frac{1}{4}Li_s^2 = \frac{1}{4}\frac{v_s^2}{4R_S^2}L \tag{7.106}$$

or

$$\omega_0 \times W_m = \frac{1}{4}\frac{v_s^2}{4R_S}\frac{\omega_0 L}{R_S} = \frac{1}{2}\frac{v_s^2}{8R_S}Q = \frac{1}{2}P_L \times Q \tag{7.107}$$

where P_L is the power to the load at resonance

$$P_L = \frac{v_L^2}{2R_S} = \frac{v_s^2}{4 \cdot 2 \cdot R_S} = \frac{v_s^2}{8R_S} \tag{7.108}$$

The total reactive power is thus exactly Q times larger than the power in the load

$$\omega_0(W_m + W_e) = Q \times P_L \tag{7.109}$$

By the definition of the component Q_c factor, the power dissipated in the non-ideal elements of net quality factor Q_c is simply

$$P_{diss} = \frac{P_L \cdot Q}{Q_c} \tag{7.110}$$

which by using Eq. (7.105) immediately leads to the following expression for the insertion loss

$$IL = \frac{1}{1 + \frac{Q}{Q_c}} \tag{7.111}$$

The above equation is very simple and insightful. Note that using a higher network Q, e.g. a higher matching ratio, will incur more insertion loss with the simple single-stage matching network. Furthermore, the absolute component Q is not important but only the component Q_c normalized to the network Q. Thus, if a low-matching ratio is needed, the actual components can be moderately lossy without incurring too much insertion loss.

Also note that the actual inductors and capacitors in the circuit can be modeled with very complicated sub-circuits, with several parasitics to model distributed and skin effect, but in the end, at a given frequency, we can calculate the equivalent component Q_c factor and use it in the above equation.

Note that Q_c is the net quality factor of the passive elements. If one element dominates, such as a low-Q inductor, then Q_L can be used in its place. The exact analysis for a lossy inductor and capacitor is simple enough and yields an expression that is identical to Eq. (7.111) when only inductor losses are taken into account, but differs when both inductor and capacitors losses are included

$$IL = \frac{1}{1 + \frac{Q}{Q_L}} \frac{1}{1 + \frac{Q}{Q_c}} \tag{7.112}$$

which can be written as

$$IL = \frac{1}{1 + Q\left(Q_L^{-1} + Q_C^{-1}\right) + \frac{Q^2}{Q_L Q_c}} \tag{7.113}$$

which equals the general expression we derived under "high-Q" conditions, e.g. $Q_c \gg Q$.

When completing the design with real elements, it is also necessary to shift the component values slightly, due to the extra loss. While formulas for these perturbations can be calculated, a modern computer and optimizer really make this exercise unnecessary.

Reactance absorption

In most situations, the load and source impedances are often complex and our discussion so far only applies to real load and source impedances. An easy way to handle complex loads is to simply absorb them with reactive elements. For example, consider the complex load shown in Fig. 7.21. To apply an L-matching circuit, we can begin by simply resonating out the load reactance at the desired operating frequency. For instance, we add an inductance L_{res} in shunt with the capacitor to produce a real load. From here the design procedure is identical. Note that we can absorb the inductor L_{res} into the shunt L-matching element.

$R_S < R_L$

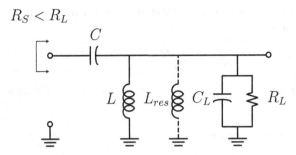

Figure 7.21 The complex load C_L in parallel with R_L is matched to a real source impedance by first applying a parallel inductor L_{res} to resonate out C_L. The load can now be matched using a standard matching network. In the final design, the resonating L_{res} can be simply absorbed into L.

$R_S > R_L$ $R_S > R_L$ $R_S > R_L$

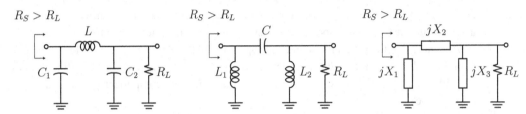

Figure 7.22 Several incarnations of a Π-matching network. The first is a low-pass structure, the second a high-pass structure. The third is a general Π network.

$R_S > R_i$ $R_i < R_L$

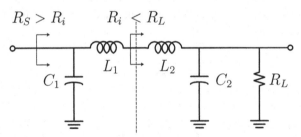

Figure 7.23 The Π network can be decomposed into a back-to-front cascade of two L-matching networks. The impedance is first reduced down to $R_i < R_L$, then increased back up to $R_S > R_L > R_i$.

From now onwards we can simply discuss the real matching problem, since a complex load or source can be handled in a similar fashion. Often there are multiple ways to perform the absorption with each choice yielding slightly different network properties such as Q (bandwidth), and different frequency selectivity (e.g. low-pass, high-pass, bandpass).

A Π match

The L match circuit is simple and elegant but is somewhat constrained. In particular, we cannot freely choose the Q of the circuit since it is fixed by the required matching factor m. This restriction is easily solved with the Π-matching circuit, also named from its topology, shown in Fig. 7.22. The idea behind the Π match can be easily understood by studying the cascade of two back-to-front L matches shown in Fig. 7.23. In this circuit the first L match

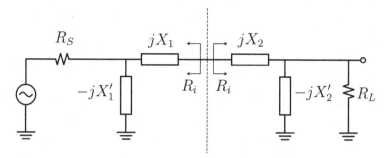

Figure 7.24 The reflected input and output impedance are both equal to R_i at the center of the Π network.

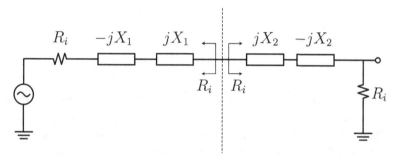

Figure 7.25 The L sections can be converted into series sections to produce one big LCR circuit.

will lower the load impedance to an intermediate value R_i

$$R_i = \frac{R_L}{1 + Q_1^2} \tag{7.114}$$

or

$$Q_1 = \sqrt{\frac{R_L}{R_i} - 1} \tag{7.115}$$

Since $R_i < R_L$, the second L match needs to boost the value of R_i up to R_s. The Q of the second L network is thus

$$Q_2 = \sqrt{\frac{R_S}{R_i} - 1} > \sqrt{\frac{R_S}{R_L} - 1} \tag{7.116}$$

When we combine the two L networks, we obtain a Π network with a higher Q than possible with a single-stage transformation. In general the Q, or equivalently the bandwidth $B = \frac{\omega_0}{Q}$, is a free parameter that can be chosen at will for a given application. Note that when the source is connected to the input, the circuit is symmetric about the center, as shown in Fig. 7.24. Now it is rather easy to compute the network Q by drawing a series equivalent circuit about the center of the structure, as shown in Fig. 7.25. If the capacitors and inductors in series are combined, the result is a simple RLC circuit with Q

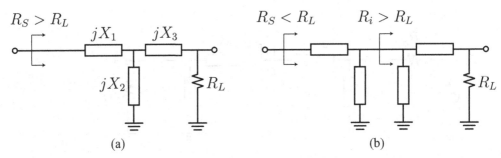

Figure 7.26 (a) The T-matching network can be decomposed into two front-to-back L sections. (b) The first L section boosts the resistance to a value of $R_i > R_L$ and the second L structure drops the impedance to $R_S < R_i$.

given by

$$Q = \frac{X_1 + X_2}{2R_i} = \frac{Q_1 + Q_2}{2} \tag{7.117}$$

It is important to note the inclusion of the source resistance when calculating the network Q as we are implicitly assuming a power match. In a power amplifier, the source impedance may be different and the above calculation should take that into consideration. For instance, if the PA is modeled as a high-impedance current source (Class A/B operation), then the factor of 2 disappears. The design procedure begins with the specification of the network Q. Eq. (7.117) is then used to find R_i, and from there the L-matching procedure outlined above takes over.

A T match

The T-matching network, shown in Fig. 7.26a, is the dual of the Π network. By now you can see that the names all correspond the physical topology of the circuit. The T network can also be decomposed into a cascade of two back-to-front L networks, as shown in Fig. 7.26b. The first L transforms the resistance up to some intermediate value $R_i > R_S$, and the second L transforms the resistance back down to R_S. Thus the net Q is higher than a single-stage match. The network Q can be derived in an analogous fashion and yields the same solution

$$Q = \frac{1}{2} \left(\sqrt{\frac{R_i}{R_L} - 1} + \sqrt{\frac{R_i}{R_S} - 1} \right) \tag{7.118}$$

Multi-section low Q matching

We have seen that the Π- and T- matching networks are essentially two-stage networks, which can boost the network Q. In many applications we actually would like to achieve the opposite effect, e.g. low network Q is desirable in broadband applications. Furthermore, a low Q design is less susceptible to process variations. Also, a lower Q network lowers the loss of the network, as is evident by examining Eq. (7.111).

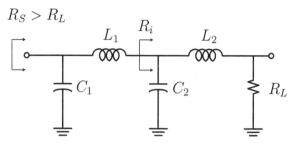

$R_S > R_L$

Figure 7.27 A two-stage low-pass L-matching network. The first stage steps up the intermediate resistance $R_S < R_i < R_L$, thus lowering the Q over a single-stage design.

To lower the Q of an L-matching network, we can employ more than one stage to change the impedance in smaller steps. Recall that $Q = \sqrt{m - 1}$, and a large m factor requires a high Q match. If we simply change the impedance by a factor $k < m$, the Q of the first L section is reduced. Likewise, a second L section will further change the resistance to the desired R_S with a step size $l < m$, where $l \cdot k = m$. An example two-stage network is shown in Fig. 7.27a. Reflecting all impedances to the center of the network, the real part of the impedance looking left or right is R_i at resonance. Thus the power dissipation is equal for both networks. The overall Q is thus given by

$$Q = \frac{\omega(W_{s1} + W_{s2})}{P_{d1} + P_{d2}} = \frac{\omega W_{s1}}{2P_d} + \frac{\omega W_{s2}}{2P_d} = \frac{Q_1 + Q_2}{2} \qquad (7.119)$$

$$Q = \frac{1}{2}\left(\sqrt{\frac{R_i}{R_L} - 1} + \sqrt{\frac{R_S}{R_i} - 1}\right) \qquad (7.120)$$

Note the difference between the above and Eq. (7.118). The R_i term appears once in the denominator and once in the numerator since it is an intermediate value. What is the lowest Q achievable? To find out, take the derivative of (7.120) with respect to R_i and solve for the minimum

$$R_{i,\text{opt}} = \sqrt{R_L R_S} \qquad (7.121)$$

which results in a Q approximately lower by a square root factor

$$Q_{opt} = \sqrt{\sqrt{\frac{R_S}{R_L}} - 1} \approx m^{1/4} \qquad (7.122)$$

It is clear that the above equations apply to the opposite case when $R_L > R_S$ by simply interchanging the role of the source and the load.

To achieve an even lower Q, we can keep adding sections as shown in Fig. 7.28. The optimally low Q value is obtained when the intermediate impedances are stepped in geometric progression

$$\frac{R_{i1}}{R_{lo}} = \frac{R_{i2}}{R_{i1}} = \frac{R_{i3}}{R_{i2}} = \cdots = \frac{R_{hi}}{R_{in}} = 1 + Q^2 \qquad (7.123)$$

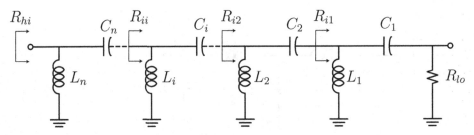

Figure 7.28 A high-pass multi-section L-matching network.

where $R_{hi} = \max(R_S, R_L)$ and $R_{lo} = \min(R_S, R_L)$. In the limit that $n \to \infty$, we take very small "baby" steps from R_{lo} to R_{hi} and the circuit starts to look like a tapered transmission line. Multiplying each term in the above equation

$$\frac{R_{i1}}{R_{lo}} \cdot \frac{R_{i2}}{R_{i1}} \cdot \frac{R_{i3}}{R_{i2}} \cdot \ldots \cdot \frac{R_{hi}}{R_{in}} = \frac{R_{hi}}{R_{lo}} = (1 + Q^2)^N \tag{7.124}$$

which results in the optimal Q factor for the overall network

$$Q = \sqrt{\left(\frac{R_{hi}}{R_{lo}}\right)^{1/N} - 1} \tag{7.125}$$

The loss in the optimal multi-section line can be calculated as follows. Using the same approach as Section 7.3, note that the total power dissipated in the matching network is given by

$$P_{diss} = \frac{NQP_L}{Q_u} \tag{7.126}$$

where N sections are used, each with equal Q due to the condition set forth by Eq. (7.123). This leads to the following expression

$$IL = \frac{1}{1 + N\frac{Q}{Q_u}} \tag{7.127}$$

or

$$IL = \frac{1}{1 + \frac{N}{Q_u}\sqrt{\left(\frac{R_{hi}}{R_{lo}}\right)^{1/N} - 1}} \tag{7.128}$$

It is interesting to observe that this expression has an optimum for a particular value of N. It is easy enough to plot IL for a few values of N to determine the optimal number of sections. Intuitively adding sections can decrease the insertion loss, since it also lowers the network Q factor. Adding too many sections, though, can counterbalance this benefit.

Example 12

Suppose a power amplifier delivering 100 W of power has an optimal load resistance of .5Ω, but needs to drive a 50Ω antenna. Design a matching network, assuming that the component Qs of 30 are available.

First note that a matching factor of $m = 50/.5 = 100$ is needed. The table below shows the network Q and insertion loss as a function of the number of sections N. Clearly three sections yield the optimal solution. But since a three-section filter is more expensive, and has only marginally better performance, a two-section matching network may be preferable.

N	Q (Eq. 7.125)	IL (dB) (Eq. 7.128)
1	9.95	−1.24
2	3	−0.79
3	1.91	−0.76
4	1.47	−0.78
5	1.23	−0.81
6	1.07	−0.85

7.4 Distributed matching networks

Matching circuits employing transmission lines are covered in Chapter 9. Quarter wave transmission lines, transmission line stubs, and other distributed structures will be covered. Transformer-based matching, including transmission line transformers, are covered in Chapter 10 and Chapter 11.

7.5 Filters

A filter is a circuit with a specified frequency response characteristic. Filters are key elements in high-speed communication circuits, since often our information is buried in a lot of noise and interference. A filter can attenuate out-of-band signals and thus reduce the required dynamic range of analog and digital circuits, which leads directly to power savings (e.g. lower resolution ADC and fewer bits in the DSP).

Matching networks and filters have much in common. We may view a matching network as a filter with different load and source impedances. In effect all the circuits we have met in this chapter are filters, with the important distinction that, while some filtering action was occurring, it was not a well-controlled part of the design.

LC filters grow naturally from a simple low-pass LC divider shown in Fig. 7.29a. At low frequencies, the inductor is a short and the capacitor is an open and so the input passes to the output unattenuated. At higher frequencies, though, the inductor tends to increase in reactance and the capacitance shunts the output to ground. How do we increase the out-of-band attenuation? Why not simply cascade another LC filter section as shown in Fig. 7.29b. In fact, we can continue to add sections to increase the out-of-band attenuation.

A high-pass version of this circuit is the dual where the capacitors and inductors switch places, as shown in Fig. 7.28. A bandpass version of the circuit is also easy if we observe that a series LC circuit is a short at resonance and a shunt LC circuit is an open at resonance.

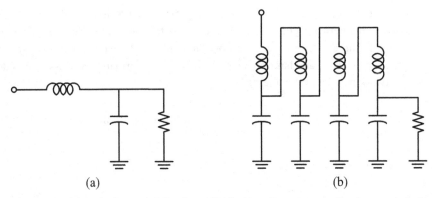

(a)   (b)

Figure 7.29 (a) A single-stage LC voltage divider has a low-pass transfer characteristic. Its topology is identical to an L-matching network. (b) Cascades of voltage dividers are equivalent to cascades of L sections leading to higher out-of-band attenuation.

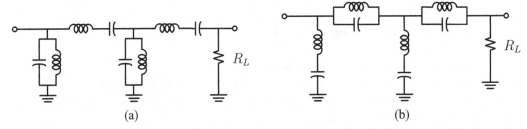

(a)   (b)

Figure 7.30 (a) A bandpass multi-section filter. (b) A multi-section notch filter.

Then the circuit of Fig. 7.30a acts like a bandpass filter. If we switch the locations of the series and parallel resonant tanks, we obtain the network shown in Fig. 7.30b, a notch filter. Now the circuit response is a null at resonance.

The design of prototype filters is a very well-developed discipline and many good sources exist [69] that tabulate filters into families with distinct characteristics (say minimum ripple or linear phase response).

7.6   References

This chapter has origins in lecture notes I used in several communication circuits courses taught at Berkeley. I learned this material from classic references, such as *Communication Circuits: Analysis and Design* [8] and *Solid State Radio* [31]. A more recent addition to the bookshelf, Tom Lee's, *The Design of CMOS RFIC* [34] also has an excellent chapter covering some of this material.

8 Small-signal high-speed amplifiers

The design and analysis of amplifiers is as old as the field of electronics. Since the invention of active devices, engineers have striven to maximize the performance of amplifiers to achieve higher gain, lower noise, wider bandwidths, and higher dynamic range. Over the years two distinct styles of amplifier design have emerged. Analog amplifier design styles dominate in the frequency range of DC to approximately 1 GHz. The upper frequency limit is really arbitrary and has been a strong function of technology f_T and improved integration and miniaturization. Most analog IC amplifiers are designed to drive high-impedance loads and/or mostly capacitive loads. Only the output stage may be designed to drive an external load resistance. Furthermore, since the signals tend to be baseband, the analysis of such amplifiers is based on the voltage or current gain transfer function and pole/zero analysis. Feedback is almost always employed since large gain can be extracted at low frequency and traded off to achieve higher speed, robustness against process tolerance, and linearity.

At the other extreme, we have microwave amplifier design based on two-port theory. The active element such as the transistor is treated as a two-port device and the device is carefully impedance matched to obtain the optimal performance. In the course of the design, extensive use of the Smith Chart enables the designer to trade off gain, noise, bandwidth, and matching. Usually the active device layout is fixed and characterized by measured small-signal parameters, and only the matching networks can be altered in the design of the amplifier.

This is in stark contrast to the analog design, where the device is modeled by a compact model[1] which depends on terminal voltages and device geometry. The compact model is valid over multiple operating regions and is developed from measurements. Occasionally, though, measurements are not available and so the compact model may be based on device/process simulations. Usually the device models are not as accurate as measured data since a compromise must be made in designing a model that is valid over a wide operating range of bias and geometry. Add process and temperature variation and we see that an analog circuit should be as insensitive to the active device characteristics as possible. Hence the preference for feedback amplifiers.

In this chapter we will briefly review some analog style amplifier designs, especially the topologies that can provide high-frequency performance. Then we will discuss some

---

[1] Most modern "compact" models are actually quite rotund, exceeding 10 000 lines of C code.

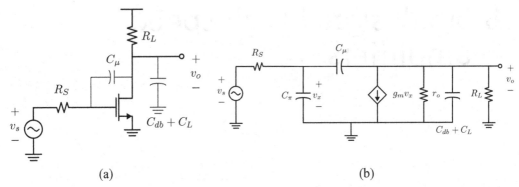

(a)   (b)

Figure 8.1 (a) A resistively loaded common-source (emitter) amplifier. (b) Equivalent circuit for common-source (emitter) amplifier.

classic RF tuned amplifiers. Finally, we will study the general properties of amplifiers as two-ports using classical network theory. In this way we hope to bridge the gap between analog amplifiers, RF amplifiers, and microwave amplifier design techniques.

8.1   Broadband amplifiers

Resistive load amplifiers

Consider the generic resistively loaded amplifier shown in Fig. 8.1a, where for simplicity only the AC schematic is shown. It is interesting to observe that the same equivalent circuit represents a bipolar amplifier at high frequencies, as C_π tends to dominate the input impedance at high frequencies

$$\frac{r_\pi}{X_\pi} = \frac{\beta_0/g_m}{1/\omega C_\pi} = \beta_0 \frac{\omega}{\omega_T} \tag{8.1}$$

where $\omega_T = g_m/C_\pi$ is the unity gain frequency of the device. If we identify high-frequency operation as $f_T/10$, and assuming $\beta_0 \sim 100$, then we have

$$r_\pi/X_\pi = 100/10 = 10 \tag{8.2}$$

It is well known that the amplifier frequency response suffers from the Miller feedback capacitance C_μ and a severe gain-bandwidth tradeoff is required. Under the Miller approximation, the input pole is given by

$$\omega_0^{-1} \approx R_S(C_\pi + |A_v|C_\mu) \tag{8.3}$$

Letting $\mu = \frac{C_\mu}{C_\pi}$ we have

$$\omega_0^{-1} = R_S C_\pi (1 + \mu|A_v|) \approx R_S C_\pi \mu |A_v| \tag{8.4}$$

In modern nano-scale CMOS technology nodes, $\mu \sim .3 - .5$ is not uncommon. In contrast, long channel transistors enjoyed a rather small drain–oxide overlap. Assuming the voltage

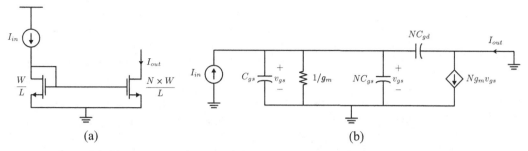

Figure 8.2 (a) A current mirror can be employed as a high-frequency current amplifier. (b) The simplified equivalent circuit for the current mirror.

gain is given by the low-frequency value of $-g_m R_L$, we have

$$\omega_0^{-1} = R_S C_\pi \mu g_m R_L = (g_m R_S)(g_m R_L)\frac{C_\pi}{g_m}\mu \tag{8.5}$$

or

$$\omega_0^{-1} = |A_v|^2 \frac{R_S}{R_L}\omega_T^{-1}\mu \tag{8.6}$$

The amplifier has a bandwidth reduction factor proportional to A_v^2

$$\omega_0 \times |A_v|^2 = \omega_T \times \left(\frac{R_L}{R_S}\right) \times \frac{1}{\mu} \tag{8.7}$$

Broadband amplifiers, in contrast, generally can achieve a gain-bandwidth tradeoff approaching the device f_T

$$G \times \omega = \omega_T \tag{8.8}$$

A simple example, shown in Fig. 8.2a, illustrates the point. A $1 \times N$ current mirror has broadband frequency response which can be illustrated with the equivalent circuit of Fig. 8.2b. The diode-connected device can be replaced with a conductance of value g_{m1} in shunt with the amplifier input capacitance C_π. If the current amplifier drives a low impedance load, the transfer function is given by

$$G_i = \frac{i_o}{i_s} = \frac{g_{m2}}{Y_{in}(s)} = \frac{g_{m2}}{g_{m1} + sC_\pi + N \times sC_\pi} \tag{8.9}$$

or

$$G_i = \frac{\frac{g_{m2}}{g_{m1}}}{1 + (N+1)\frac{s}{\omega_T}} \tag{8.10}$$

where $\omega_T = g_{m1}/C_\pi$ is the usual unity gain frequency of the transistor. Note that the transconductance of output device is N times larger, since it can be thought of as N devices in parallel. The complete transfer function is therefore

$$G_i = \frac{N}{1 + \frac{s}{\omega_T/(N+1)}} \tag{8.11}$$

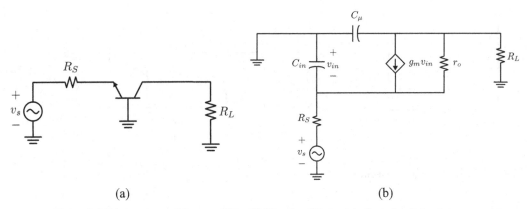

(a)   (b)

Figure 8.3 (a) A common-base amplifier. (b) The simplified equivalent circuit for the common-base (gate) amplifier.

and the gain-bandwidth product is given by

$$G_i \times \omega_{-3db} = \frac{N}{N+1}\omega_T \approx \omega_T \qquad (8.12)$$

It is important to note that the above analysis holds only if we assume the load impedance is extremely low, ideally a short. If we connect a physical resistor to the output, the Miller effect will produce a significant feedback current, which invalidates our assumptions. Also note that this amplifier is inherently linear since a non-linear v_{in} is generated to correctly produce a linear output current. In other words, the large signal transfer function of the circuit is also an ideal current multiplier (until the device leaves the forward active region).

The common-gate or common-base amplifier, shown in Fig. 8.3, is also broadband, as it does not suffer from the Miller effect, but it can only provide voltage gain. But the low input impedance is useful in high-frequency applications. It is easy to show that the voltage gain of the amplifier is given by

$$A_v = \frac{G_m R_L}{(1 + s(C_o + C_\mu)R_L)\left(1 + s\frac{C_\pi R_S}{1 + g_m R_S}\right)} \qquad (8.13)$$

where the degenerated transconductance is given by $G_m = \frac{g_m}{1 + g_m R_S}$. If the product $g_m R_S$ is made very large, the transfer function simplifies to

$$A_v = \frac{R_L/R_S}{(1 + s/\omega_T)(1 + s(C_o + C_\mu)R_L)} \qquad (8.14)$$

which contains two poles, one at ω_T, and another set by the output capacitance and the feedback capacitance C_μ. If every effort is taken to minimize the output capacitance (e.g. by placing a broadband buffer to drive the next stage), the second pole becomes $C_\mu R_L$. In an impedance matched environment, though, $R_S = 1/g_m$, so $g_m R_S = 1$, or (8.13) becomes

$$A_{vm} = \frac{\frac{1}{2}g_m R_L}{(1 + s/2 \cdot \omega_T)(1 + s(C_o + C_\mu)R_L)} \qquad (8.15)$$

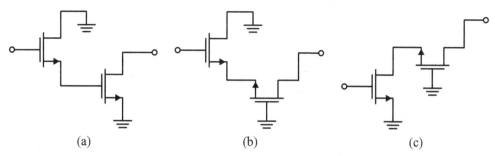

(a)   (b)   (c)

Figure 8.4 Wideband two-stage amplifiers. (a) A source follower driving a common-source amplifier. (b) A source follower driving a common-gate amplifier. (c) A common-source amplifier drives a common-gate amplifier, or a cascode amplifier.

which has two poles, one at twice ω_T, and another set by the output load. If the output is also impedance matched to the source, $g_m R_L = 1$

$$A_{vmm} = \frac{\frac{1}{2} g_m R_L}{(1 + s/2 \cdot \omega_T)(1 + s(C_o + C_\mu)/g_m)} \tag{8.16}$$

To see the limiting frequency of operation, assume C_o is minimized and $C_\mu = \mu C_\pi$, where $\mu < 1$. Then, if $\mu < \frac{1}{2}$, the first pole of the amplifier is set by twice ω_T.

The two-stage amplifiers shown in Fig. 8.4 suffer less from the Miller effect. In Fig. 8.4a, a source follower drives a common-source amplifier, thus lowering the source resistance seen by the Miller capacitor. In Fig. 8.4b, a source follower drives a common-gate amplifier, raising the input impedance of a basic common-gate amplifier without drastically altering the gain. This topology is also recognized as a differential amplifier driven by a single-ended input.

The third amplifier, shown in Fig. 8.4c, is a cascade of the common-source and common-gate amplifier, which is widely known as the cascode topology. The amplifier is simple and elegant as it provides both voltage and current gain. Since the devices can be stacked, the DC current is shared by the two stages, resulting in low-power amplifier block. The version shown in Fig. 8.5 includes more details. The voltage gain across the transconductance Miller capacitor can be made as small as desired by sizing the cascode transistor

$$A_{v,1} = -\frac{g_{m1}}{g_{m2}} \tag{8.17}$$

at the cost of loading the amplifier with a non-dominant pole. In a well-balanced design, the dominant pole is due to the output of the amplifier

$$\omega_{-3\,dB} = \frac{1}{R_L C_o} \tag{8.18}$$

where $C_o = C_{\mu,2} + C_{db,2} + C_L$. This capacitance is independent of the gain of the amplifier since the gate terminal of M2 is at a fixed AC potential. The cascode boosts the gain of the amplifier by allowing a larger load resistance for a given bandwidth. It also increases the output impedance of the amplifier to roughly $g_m r_o^2$, which decreases the self-loading.

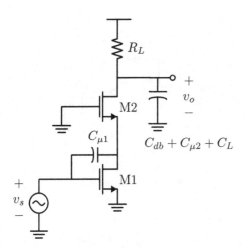

Figure 8.5 A resistively loaded cascode amplifier does not suffer from the Miller effect and is much more stable due to reduced feedback.

The gain-bandwidth product of the amplifier is thus bounded by

$$A_v \times \omega_{-3\,dB} = \frac{g_m}{C_{\mu,2} + C_{db,2} + C_L} \approx \frac{g_m}{C_L} \tag{8.19}$$

For stability reasons, we typically operate this amplifier at a frequency sufficiently lower than ω_T and thus C_L is made large enough to ensure this condition. In theory, therefore, this amplifier has a gain-bandwidth product approaching a significant fraction of the ω_T of the device.

There are a few problems with this amplifier. First, if we are greedy for gain, we have to pay with headroom since a larger load resistor R_L consumes larger DC headroom. This may lead to unreasonably high supply voltages. In most applications, we do not have control over the supply due to the intrinsic breakdown in a transistor. Higher f_T devices also have lower breakdown voltages, leading to a natural limit to the gain of the amplifier. For example, a 130 nm CMOS technology may limit the supply to 1.3 V. In analog circuits the voltage headroom is usually solved by employing active loads. Thus the upper limit of gain is set by the maximum current drive or power dissipation. A cascode amplifier operating with 10mA of bias current, therefore, cannot have a load resistor larger than about 130Ω.

A transistor in saturation can obtain an output impedance r_o much larger than V_{ds}/I_{ds}. Gain boosting a cascode load, for instance, can greatly enhance the output impedance. But an active load has several drawbacks. First it further limits the output swing of the amplifier since operation into the triode region must be avoided for both the load and the cascode transconductance device. In many applications a large swing is necessary (say in a LO buffer or PA driver amplifier). Furthermore, the non-linearity of the load degrades the linearity of the amplifier, leading to excess distortion.

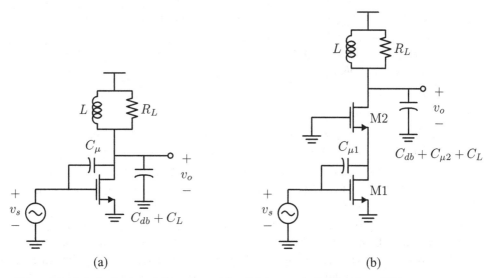

Figure 8.6 (a) A single stage LC tuned amplifier. (b) A cascode LC tuned amplifier.

Tuned amplifiers

The RLC loaded amplifier shown in Fig. 8.6a solves several of the voltage headroom problems. In Fig. 8.6a, a single transconductance device drives a shunt RLC load, which results in a voltage gain of

$$A_v = -g_m Z(j\omega) = \frac{-g_m}{Y(j\omega)} \tag{8.20}$$

The peak gain occurs at resonance

$$A_{v,\max} = -g_m R_{\text{eff}} \tag{8.21}$$

where R_{eff} is the *loaded* resistance of the tank

$$R_{\text{eff}} = R_L || r_o || R_{x,L} || R_{x,C} \tag{8.22}$$

In order to maximize the gain, we employ high-Q inductors and capacitors in the load and omit the explicit load resistance R_L. The peak gain is thus

$$A_{v,\max} \approx -g_m(R_{x,L} || R_{x,C}) \tag{8.23}$$

Assuming the Q factor is dominated by the inductor, a good assumption for monolithic IC inductors, we have

$$A_{v,\max} \approx -g_m R_{x,L} = -g_m Q_L \omega L \tag{8.24}$$

where we used the series to parallel transformation to calculate the effective shunt resistance due to the inductor

$$R_{\text{eff}} = \left(1 + Q_L^2\right) R_{x,L} \approx Q_L^2 R_{x,L} = Q_L^2 \frac{\omega L}{Q_L} = Q_L \times \omega L \tag{8.25}$$

The gain is maximized at a fixed bias current and frequency by maximizing the product $Q_L \times L$. So, in theory, there is no limit to the voltage gain of the amplifier as long as we can maximize the quality factor Q_L. Note that the capacitance of the circuit is not detrimental since it is resonated away with the shunt inductance. In other words L is chosen such that

$$L = \frac{1}{\omega_0^2 C_{\text{eff}}}$$

(8.26)

where $C_{\text{eff}} = C_{db} + (1 - |A_v^{-1}|)C_\mu + C_L$. The ability to tune out the parasitic capacitances in the circuit is a major advantage of the tuned amplifier. This is especially important as it allows low-power operation. Another important advantage of the circuit is that there is practically no DC voltage drop across the inductor, allowing very low-supply voltage operation.

Another less obvious advantage is the improved voltage swing at the output of the amplifier. Usually the voltage swing is limited by the supply voltage and the $V_{ds,sat}$ of the amplifier. In this case, though, the voltage can swing above the supply. Since the average DC voltage across an inductor is zero, the output voltage can swing around the DC operating point of V_{dd}. This is a major efficiency boost for the amplifier and is an indispensable tool in designing power amplifiers and buffers.

You may now wonder why we need the cascode amplifier shown in Fig. 8.6b. Besides the obvious advantage of boosting the output impedance, thus maximizing the Q of the load, the cascode device solves a major stability problem of the amplifier. We will show later that a feedback path can easily lead to unwanted oscillations in the amplifier. The cascode version of the tuned amplifier has virtually zero feedback and thus is much more stable. The loss in voltage headroom is a small price to pay considering the improved headroom afforded by the inductor.

It is interesting to note that the bandwidth of the circuit is still determined by the RC time constant at the load. The bandwidth is given by

$$BW = \frac{\omega_0}{Q} = \frac{\omega_0}{\omega_0 RC} = \frac{1}{RC}$$

(8.27)

The ultimate sacrifice for high-frequency operation in a tuned amplifier is that the amplifier is narrow-band with zero gain at DC. In fact, the larger the Q of the tank, the higher the gain and the lower the bandwidth. To win some of the bandwidth back requires other techniques, such as shunt peaking (Section 8.1) and distributed amplifiers (Section 11.4). You may naturally wonder, "How high can we go?" Based on the simple analysis thus far, it seems that for any given frequency, no matter how high, we can simply absorb the parasitic capacitance of the amplifier with an appropriately small inductor (say a short section of transmission line) and thus realize an amplifier at an arbitrary frequency. This is of course ludicrous and we will re-examine this question at the end of the chapter.

Feedback amplifiers

The application of feedback around amplifier gain blocks has many desirable qualities, such as broadband operation, stable gain over process and temperature, and reduction of distortion. The cost of feedback is reduced gain and potential instability. At high frequencies, this is doubly important since gain does not come cheaply (power consumption) and stability is a primary concern.[2] For this reason most RF amplifiers have consisted of tuned amplifiers employing a few stages of amplification. But this trend may change as technology progresses. Today transistors are low cost and extremely fast, but not very reliable. The case can be made in many applications that feedback amplifiers are actually good low-cost contenders.

Because of stability concerns, practical designs cannot have more than two stages of amplification before feedback is applied. This is because phase delay in amplifier stages is significant at RF, and stability problems will ensue. Using cascades of dominant pole amplifiers, though, is too conservative and leads to rapid bandwidth "shrinkage." For n stages we have a -3 dB bandwidth of

$$|G_n(j\omega)| = \left|\left(\frac{G_0}{1 + j\omega\tau}\right)^n\right| = \frac{1}{\sqrt{2}} \tag{8.28}$$

or

$$\omega_{-3\text{dB}} = \frac{1}{\tau}\sqrt{2^{1/n} - 1} \tag{8.29}$$

which shows a rapid bandwidth shrinkage with n. For $n = 3$, for instance, we have a bandwidth reduction of 50%. A more desirable situation is to use amplifiers with sharper roll-off characteristics. To see this, imagine the cascade of n "brick-wall" response amplifiers. Clearly, this cascade would not involve any bandwidth reduction. Feedback allows one to have a great amount of control over the poles of an amplifier. If we are smart, we can arrange for the poles to lie on a circle (rather than on the real axis) or ellipse to realize sharp cutoff characteristics.

Single stage feedback

Feedback in a single-stage amplifier is so common it is often not even considered feedback. Two robust topologies include series degeneration shown in Fig. 8.7a, shunt feedback shown in Fig. 8.7b, and a combination dual feedback shown in Fig. 8.7c. Series feedback stabilizes and linearizes the transconductance of an amplifier but increases the input impedance. The shunt feedback in turn lowers the input impedance. We will return to the popular shunt feedback amplifier in a later section.

Shunt peaking

The central idea of shunt peaking is simple and yet powerful. As shown in Fig. 8.8a, in the simplest form, a load consisting of a resistor and an inductor in series leads to a zero in the transfer function. This can be used to cancel the pole of the transfer function and within a band of frequencies create a flat-frequency response. With reference to Fig. 8.8b, let us

---

[2] You've probably heard the saying that "RF amplifiers are good oscillators."

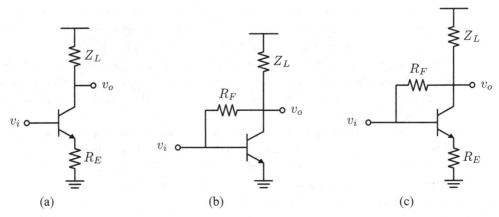

(a)   (b)   (c)

Figure 8.7 (a) Series feedback or "degeneration" results in a stable transconductance over process and temperature. (b) Shunt feedback lowers the input impedance of a common-emitter amplifier. (c) Combined series–shunt feedback allows a good tradeoff between matching and gain.

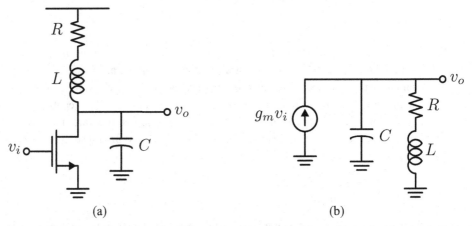

(a)   (b)

Figure 8.8 (a) A common-source amplifier with a shunt-peaking load. (b) Equivalent circuit for the shunt-peaking amplifier.

allow the equations to do the talking

$$Z(s) = (sL + R) || \frac{1}{sC} = \frac{R\left(1 + s\frac{L}{R}\right)}{1 + sRC + s^2LC} \tag{8.30}$$

$$|Z(j\omega)| = R\sqrt{\frac{1 + \left(\frac{\omega L}{R}\right)^2}{\left(1 - \omega^2 LC\right)^2 + (\omega RC)^2}} \tag{8.31}$$

These equations can be written in normalized form with m as the ratio of two time constants

$$m = \frac{RC}{L/R} \tag{8.32}$$

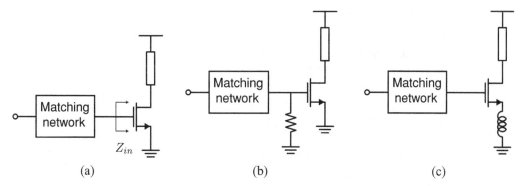

Figure 8.9 (a) Matching networks often must convert a predominantly imaginary load impedance to a real value. (b) A simple solution is to simply terminate the matching network with a physical resistor. (c) A more elegant solution uses a feedback synthesized resistor input match.

Letting $\tau = L/R$ we have

$$Z(s) = \frac{R(1 + s\tau)}{1 + s\tau m + s^2\tau^2 m} \tag{8.33}$$

$$|Z(j\omega)| = R\sqrt{\frac{1 + (\omega\tau)^2}{(1 - \omega^2\tau^2 m)^2 + (\omega\tau m)^2}} \tag{8.34}$$

Your grade school siblings can help you find the bandwidth by solving the above quadratic

$$\frac{\omega}{\omega_1} = \sqrt{\left(1 + m - \frac{m^2}{2}\right) + \sqrt{\left(1 + m - \frac{m^2}{2}\right)^2 + m^2}} \tag{8.35}$$

The maximum bandwidth obtainable occurs for $m = \sqrt{2}$, or a bandwidth boost of 85%. This comes at the expense of 20% peaking. A good compromise value occurs for $m = 2$, which leads to only 3% peaking and a bandwidth enhancement of 82%. If you simply cannot tolerate any peaking, let $m = 1 + \sqrt{2}$ and the bandwidth is still 72% larger. Finally, in a broadband application where a linear phase or flat delay response is desired, it can be shown that the optimal value of $m \approx 3.1$ is your choice, yielding a 57% bandwidth enhancement.

Reactive series feedback

In our discussion of matching networks, we considered networks that transform from a given source impedance to a given load impedance. Consider now the load shown in Fig. 8.9a, the input of an active device. At moderate frequencies the input impedance is dominated by C_{gs}. We need to somehow transform the input capacitance to a real load resistance. In Chapter 11 we shall see that any real MOS amplifier has a real component to the input impedance and thus there is a finite real component to the input impedance. If the transistor layout has ample fingers to minimize the physical polysilicon (or metal) gate resistance, the remaining gate-induced channel resistance is given by $1/5g_m$. Thus the Q factor of the

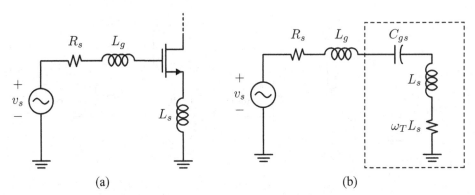

Figure 8.10 (a) The complete input-matching network requires a gate inductor L_g to resonate with the input capacitance C_{gs}. (b) The equivalent circuit for the input match is a series RLC circuit.

input of the MOS transistor is given by

$$Q_{gate} \approx \frac{5g_m}{\omega C_{gs}} = 5\frac{\omega_T}{\omega} \tag{8.36}$$

At moderate frequencies $\omega \ll \omega_T$, this is a high-Q input impedance. If we resonate out this capacitance with a shunt inductor, the resulting shunt resistance $Q^2 R_i$ is too large to match to the low-source resistance. On the other hand, if we use a series inductor, the input resistance is simply R_i at resonance, too small to match. So what do we do?

We could explicitly add a resistor to the gate, as shown in Fig. 8.9b, but this will add extra noise to the circuit. A more elegant non-obvious solution is to add an inductor to the source of the amplifier, shown in Fig. 8.9c. To see the benefit, consider the general case of a degeneration impedance Z connected to the source. It is easy to show that the input impedance becomes (neglecting C_{gd})

$$Z_{in} = Z + \frac{1}{j\omega C_{gs}} + \frac{g_m Z}{j\omega C_{gs}} \tag{8.37}$$

The action of the feedback produces the third term, which can be programmed appropriately. Note that if $Z = j\omega L_s$, the third term becomes purely real and independent of frequency

$$\Re(Z_{in}) = R_{in} = \frac{g_m L_s}{C_{gs}} = \omega_T L_s \tag{8.38}$$

By simply controlling the value of L_s, we can control the input impedance. We can also vary the ω_T of the device by placing extra capacitance in the shunt with C_{gs}.

It is interesting to observe that the source impedance in effect drives a series RLC circuit, shown in Fig. 8.10. The voltage gain can be calculated by noting that the voltage v_{gs} is the voltage across a series resonant capacitor, which means that it is Q times as large as the voltage across the source resistor. For an input match, $v_{R_s} = \frac{1}{2}v_s$, so the voltage gain at resonance is given by

$$v_o = -g_m R_L v_{gs} = -g_m R_L Q \times \frac{v_s}{2} \tag{8.39}$$

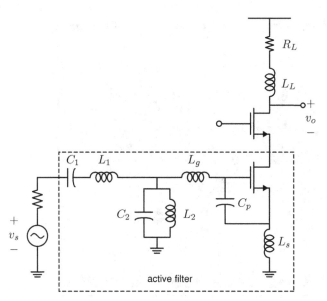

Figure 8.11 An ultra-wide band amplifier employing a three-section Chebychev active filter at input.
The series RLC network formed by the transconductance stage forms the third section of the filter.

or

$$A_v = -\frac{1}{2} g_m R_L Q \tag{8.40}$$

The bandwidth of the matching stage of the inductively degenerated amplifier is set by
the Q factor of the input. Since the source impedance is fixed, there is little freedom in
controlling the Q factor of the input stage. In most designs, the Q is fairly low and thus the
input stage is relatively wideband. But many applications require much larger bandwidth.
A good example is an ultra wideband (UWB) amplifier that needs to cover the 3–10 GHz
band.

The bandwidth of the amplifier can be extended by the observation that the series RLC
input impedance can be absorbed into the ladder filter structure, as shown in Fig. 7.30a. The
calculation of the amplifier gain is nearly identical to the simple inductively degenerated
amplifier. Note that the amplifier input impedance is given by

$$Z_{in} = \frac{R_S}{W(s)} \tag{8.41}$$

The function $W(s)$ is the filter transfer function. In band we have $|W(s)| \approx 1$, whereas out
of band $|W(s)| \approx 0$. The truth of these statements depends on the in-band ripple and the
steepness of the filter skirt. Thus the input current generates an output voltage

$$\frac{v_o}{v_{in}} = \frac{v_o}{i_{in}} \times \frac{i_{in}}{v_{in}} = \frac{g_m}{s C_{gs}} Z_L(s) \times \frac{1}{R_S W(s)} \tag{8.42}$$

The shunt-peaking load compensates for the current gain roll off. It can be shown that the
above amplifier can be optimized to generate optimal power gain and noise performance
by proper design. A broadband amplifier, shown in Fig. 8.11, employs these techniques to

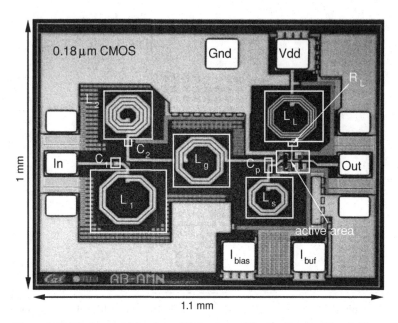

Figure 8.12 The layout of the ultra-wideband amplifier implemented in a $0.18\,\mu$m CMOS technology.

achieve wide bandwidth and matching. The amplifier uses a three-section Chebychev input filter and shunt peaking to achieve a bandwidth from 3–10 GHz [2]. The layout, shown in Fig. 8.12, is dominated by the area of the spiral inductors.

Practical issues with inductive degeneration

The placement of L_s in the source does raise an important and subtle question since the input of an amplifier often represents the interface between an external signal and the internal amplifier, a path which traces through the package and bond wires or solder bumps. Commonly, the chip is connected to the PCB through a package consisting of bondwires and leads, as shown in Fig. 8.13, which generate a lot of parasitic inductance.

As shown in Fig. 8.14a, since this current is generated by an off-chip source, the on-chip spiral inductor is in series with a parasitic inductance. The parasitic component is determined by the total perimeter of this loop. At low frequency, this parasitic is beneficial since it can contribute to the source and gate inductors. Bond wire inductors are high-quality inductors and are superior to on-chip inductors. The downside, though, is that their precise value is difficult to determine. For this reason, if the parasitic inductance is much smaller than the on-chip degeneration, it can be safely ignored. Otherwise careful design is required to set the tuning frequency and matching correctly. Note that an on-chip bypass capacitor does not help, since the current has to leave the circuit (see next section).

The concept of *partial* inductance is again particularly useful, since we can partition the input network into several partial paths consisting of the on-chip spiral, the bondwires, and the lead and board traces. The partial inductance from the source of the amplifier to the negative terminal of the source is responsible for the negative feedback. It is important

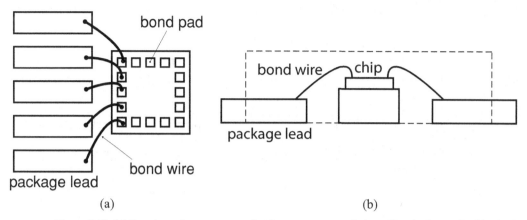

Figure 8.13 (a) Top view of a common packaging structure consisting of bond wires and solder leads that connect to the PCB. (b) Side view of structure.

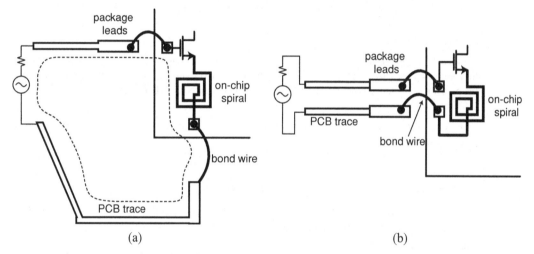

Figure 8.14 (a) The design of an inductively degenerated amplifier must account for the inductance of the entire path, including the spiral inductor, the on-chip lead inductance, the transistor, and the important off-chip parasitics, such as bond wires (or solder bumps), the package leads, and the PCB trace parasitics. (b) By placing the input bond pads closer, the parasitic component of the source inductance is minimized.

to remember that all these partial paths are mutually coupled. When we speak of the total inductance seen by the source, though, there is only one inductor and no need to consider mutual coupling. This is a rather confusing situation and leads to many blackboard arguments between engineers!

In practice, the mutual inductance between the bondwires and the package traces are the dominant terms. Putting bond wires and leads in parallel does not help as much as you may initially imagine, since the bond wires are coupled inductors. For instance, for two bond wires in parallel, the voltage V drop across the bond wires is given by

$$V = j\omega L_1 I_1 + j\omega M I_2 = j\omega L_2 I_2 + j\omega M I_1 \qquad (8.43)$$

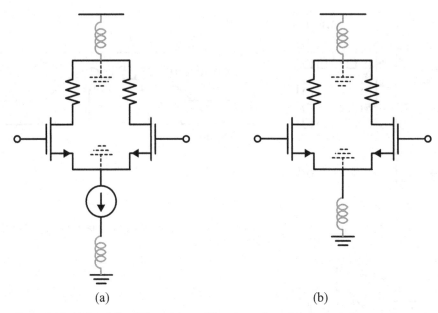

(a)   (b)

Figure 8.15 (a) In a fully differential amplifier, the package/bond wire inductors play a minor role as they are in series with a high-impedance current source. (b) Even in a pseudo-differential amplifier, the parasitic inductance is shorted by the virtual ground created by a balanced drive.

and since $I_1 + I_2 = I$, it is easy to show that the effective inductance of the parallel combination is given by

$$L_{eff} = \frac{L_1 L_2 - M^2}{L_1 + L_2 - 2M} \tag{8.44}$$

For two uncoupled inductors, $M = 0$, as expected the inductance drops to the parallel combination. But take the extreme case of $L = L_1 = L_2$ and $M = k\sqrt{L_1 L_2} = L$ for $k = 1$. Factoring the above equation we have

$$L_{eff} = \frac{L^2 - M^2}{2(L - M)} = \frac{(L + M)(L - M)}{2(L - M)} = \frac{1}{2}(L + M) = L \tag{8.45}$$

For this extreme case, the inductance does not drop at all. In a real package, adjacent bond wires can have $k \sim 0.7$, so the benefit in using multiple bondwires helps, but not as much as you may have expected.

One way to reduce the inductance of the degeneration is to place the input lead and the ground lead adjacent and in parallel, as shown in Fig. 8.14b. This minimizes the area of the loop driving the input transistor. Alternatively, we can see that the partial inductance of the bond wires from the source lead is reduced by the action of the negative mutual inductance from the gate lead (since the currents travel in opposite directions).

A differential input stage is thus preferable in that the degeneration inductor is completely on-chip, as seen in Fig. 8.15a. Even if the tail current is removed, the input impedance of the balanced circuit of Fig. 8.15b, does not depend on the bond wire inductance for balanced excitation. Accurate inductors can be synthesized on-chip but the bond wires represent a

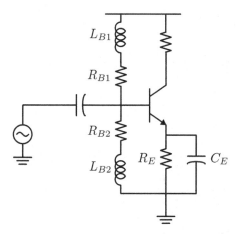

Figure 8.16 A discrete common emitter amplifier is biased using large inductor "chokes" and an emitter bypass capacitor. The bias resistors provide a stable and effective quiescent operation point. The inductor is effectively an open at AC frequencies, de-coupling the bias resistors from the amplifier. The resistor R_E stabilizes the bias point of the transistor, and the capacitor C_E *bypasses* the resistor to ground to boost the AC gain.

rather large and less predictable inductance. The differential circuit is therefore more robust and manufacturable.

RF chokes and bypass caps

A large inductor, often referred to as an RF choke, can be used to deliver bias voltages to a given node in the circuit without disturbing the AC performance. This is shown in Fig. 8.16, where an inductor biases the amplifier input with the proper voltage. An RF choke is often drawn with vertical parallel bars to indicate that it is acting as a choke. The bars indicate the presence of a magnetic core to boost the inductance. In the IC environment, though, any inductor layout with high enough inductance at a given frequency of operation can be treated as a choke.

At AC frequencies, the choke is an open circuit since ωL is a large reactance in comparison with other circuit impedances. Alternatively, in the time domain, we can think of the choke as a poor current source, analogous to a capacitor acting as a poor voltage source.

Often a large resistor can replace a choke, saving chip area. But if the DC node draws a varying current (say the base current of a BJT), the resistor is not useful, since the DC drop will vary with process and temperature. A further benefit of the choke is that it introduces less noise and loss into the circuit compared with a resistor.

A good structure for the realization of integrated circuit choke with inductance in excess of 100s of nH is a multi-layer series connected inductor, as shown in Fig. 6.16, but employing 3–4 inductors in series. In Fig. 8.17 it is evident that the total area taken up by the choke can be relatively small in an RF layout. One common concern for the choke is the SRF, which is undesirably low in this layout due to the metal-to-metal inter-winding capacitance. If the resonance occurs near the operating frequency, the AC operation may be compromised.

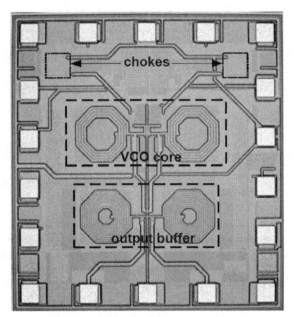

Figure 8.17 A VCO employing series-connected multi-layer inductor chokes. The chokes are relatively small and can be easily incorporated into empty space in the layout.

Large capacitors play the dual role as "bypass" capacitors or "coupling" capacitors. Coupling capacitors are useful as they isolate the DC voltage between two AC signals. Large "bypass" capacitors are also commonly employed to bypass biasing resistors in a circuit. For instance, the emitter resistor R_E shown in Fig. 8.16 is convenient for biasing as the negative feedback provides stability. But it hampers the AC gain, so a big bypass capacitor C_E effectively removes it from the circuit at the frequency of operation.[3]

Capacitors in general are essential in high-frequency circuits as they act as good AC voltage sources since $1/(\omega C)$ is small. This is important at high frequency since lead inductance can isolate a battery or supply from an RF circuit, leading to stability and coupling problems. An example is shown in Fig. 8.18a where a two-stage amplifier is redrawn in Fig. 8.18b where the lead inductance is essentially an open at high frequency. The second stage now acts like a shunt-feedback amplifier and may oscillate. A large bypass capacitor can create an on-chip or local ground, eliminating the feedback path.

Often layouts of chips are littered with large bypass capacitors. In high-frequency designs it is much better to distribute the bypass capacitor throughout the chip, rather than placing it all in one place. In fact, the bypass capacitors should be placed as close as possible to critical bias points in transistors. Take, for instance, a cascode amplifier shown in Fig. 8.19. For illustration we have placed the bypass capacitor a distance away from the gate, creating a small lead inductance L_g between the capacitor and the gate. The input impedance looking into the gate of the cascode device can have a negative real part. We have already

---

[3]   By proper design, the zero created by C_E can be exploited to maximize the bandwidth of the amplifier.

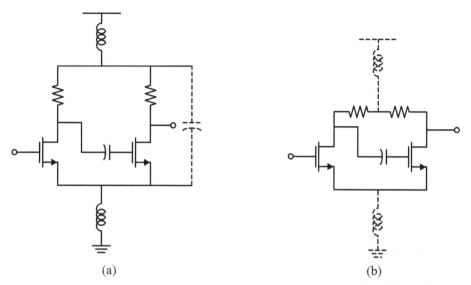

Figure 8.18 (a) The two-stage amplifier at low frequency, where the parasitic lead inductances are effectively shorts, and the on-chip bypass capacitor is an open. (b) The same circuit at high frequency without the bypass capacitors. The load resistors form a feedback path leading to instability.

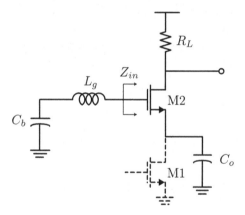

Figure 8.19 The cascode stage needs to be bypassed carefully due to the negative input resistance looking into the gate of the cascode stage. The negative resistance is generated by the capacitive degeneration due to the transconductance stage output impedance at high frequency.

calculated the effect of general degeneration impedance in Eq. (8.37). At high frequencies, the cascode device is degenerated by the output capacitance of the transconductance stage M1. A capacitive degeneration produces an input impedance approximately equal to

$$Z_{in} = \frac{1}{j\omega C_o} + \frac{1}{j\omega C_{gs}} - \frac{g_m}{\omega^2 C_o C_{gs}} = -jX_{in} - R_{in} \qquad (8.46)$$

The danger is now clearly visible. If the resonant circuit composed of L_g and the input capacitance produced by X_{in} occur at a frequency where the R_{in} is larger than the equivalent tanks series losses, the circuit will oscillate.

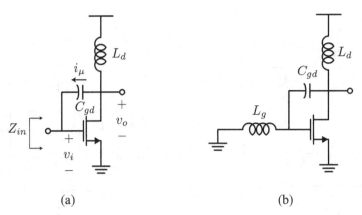

(a)   (b)

Figure 8.20 (a) The low-frequency feedback current is in phase with the voltage driving the gate, producing a negative input resistance. (b) If the gate is also biased with a choke, the net result is a low-frequency oscillator that can plague a high-frequency design.

Parasitic inductance

Earlier we remarked about the stability of a single-stage transconductance amplifier. Here we will study this issue in more depth. To begin with, consider the inductively loaded amplifier shown in Fig. 8.20a. The actual load may consist of a tank, but here we are concerned with low frequencies where the net impedance is inductive. We have explicitly drawn the Miller capacitance C_{gd} since it will play a crucial role. Let us first explore the problem by intuition. Notice that the phase of the gain (neglecting C_{gd}) is imaginary due to the load

$$A_v = -g_m Z_L = -g_m j\omega L_d \tag{8.47}$$

and the feedback current is given by

$$i_\mu = j\omega C_{gd}(v_o - v_i) \approx -j\omega C_{gd} v_i(g_m j\omega L_d + 1) \tag{8.48}$$

The total input current is thus

$$i_i = -i_\mu + j\omega C_{gs} v_i \approx j\omega C_{gs} v_i + j\omega C_{gd}(g_m j\omega L + 1)v_i \tag{8.49}$$

which leads to an input admittance with a negative conductance

$$Y_{in} = j\omega(C_{gs} + C_{gd}) - g_m \omega^2 C_{gd} L \tag{8.50}$$

Now it is not uncommon to use a choke to bias the gate of this transistor, which leads to a nice low-frequency oscillator shown in Fig. 8.20b. We did cheat a bit in our analysis, but a more rigorous calculation confirms the essence of our intuition. Low-frequency oscillations can often be traced to unforeseen problems like this.

8.2 Classical two-port amplifier design

As we have seen from a small sampling in this chapter, there are many ways to design amplifiers. We can generalize the various designs by observing that any amplifier can be

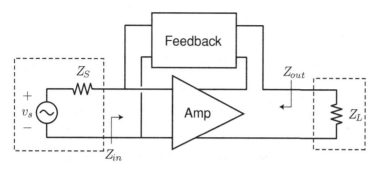

Figure 8.21 An amplifier decomposed into several functional blocks with a shunt-series feedback path.

decomposed into four functional blocks, as shown in Fig. 8.21. The core amplifier is a three terminal device with a common terminal to produce a two-port amplifier block.[4] We assume that the amplifier is biased appropriately for RF operation at DC and the AC circuit is presented with a load and source impedance Z_S and Z_L. Finally, optional feedback can be introduced and this is shown with the fourth two-port network. Four feedback topologies are possible, as the terminals of the feedback network can be connected either in series or parallel with the main amplifier.

For amplifiers without feedback, the design choices are to select Z_S and Z_L for a given transistor biased in a particular region of operation. If the load and source impedances are fixed, then matching networks can convert the load and source to the appropriate impedances seen by the amplifier. We shall shortly answer the important question: "What are the optimal values of Z_S and Z_L for a given amplifier?" To keep the discussion as general as possible, we shall represent the AC behavior of the transistor with two-port Y parameters.

The admittance parameters

Let us review some basic properties of the admittance or Y parameters. Recall that for any linear two-port we may write the following set of equations

$$i_1 = Y_{11}v_1 + Y_{12}v_2 \tag{8.51}$$
$$i_2 = Y_{21}v_1 + Y_{22}v_2 \tag{8.52}$$

where the parameters Y_{ij} are frequency and bias dependent. If you are a visual person, you may object to treating a transistor as four complex numbers! Hey, transistors don't like to be referred to by numbers, just as you don't. Thus you may prefer the equivalent circuit shown in Fig. 8.22a. You can show that at a fixed frequency the Y parameters are equivalent to this equivalent circuit. In fact, the circuit shown in Fig. 8.22b may be even more to your liking as it mirrors the actual topology of a transistor small-signal equivalent circuit. Note that this is not identical to the hybrid-π circuit as the Y parameters

---

[4] A triangle emphasizes the unilateral nature of the transistor-based amplifier.

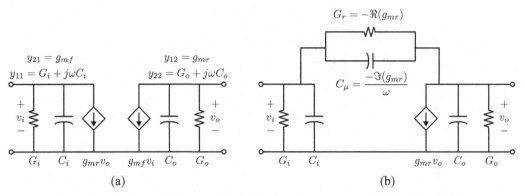

Figure 8.22 (a) An equivalent circuit for a generic two-port network described by admittance parameters. (b) A familiar hybrid-π equivalent circuit in terms of the admittance parameters.

vary as a function of frequency, correctly modeling high-frequency loss in the device. Whereas the hybrid-π model is derived based on physical grounds at low frequency, the Y parameters are a "black-box" representation that capture the frequency dependence of the device. In particular, the imaginary part of g_{mf} represents phase delay through the device, and G_i models the input resistance of a FET due to gate resistance and induced channel resistance.

We can now clearly see that Y_{11} and Y_{22} are the short-circuit input/output admittance. The parameter Y_{21} is the forward transconductance of the device, whereas Y_{12} represents the reverse transconductance or the intrinsic feedback in the amplifier. Note that the Y_{12} of a transistor is particularly "evil" as it accounts for the stability problems that plague many RF designs. For instance, for a common emitter transistor, $Y_{12} \approx j\omega C_\mu$, which is for all practical purposes zero at low frequency. But at higher frequencies, this term can feed back enough signal so that the loop gain approaches unity. We can thus see that for a "good" transistor, $Y_{21} \gg Y_{12}$, and we shall see that this condition greatly simplifies the amplifier design as it de-couples the input from the output. Such an amplifier is called *unilateral* for this reason.

You may wonder why we have chosen Y parameters to describe our transistor. Why not Z parameters or hybrid parameters? In fact all two-port parameters are equivalent and the following equations can be easily converted from one form to another. In fact, scattering or S parameters are perhaps the most appropriate parameters, but for the most part we omit discussion of amplifier design with S parameters in this book. Y parameters are the most appropriate choice when we employ shunt–shunt feedback, since the parallel connection of two two-ports is simply the sum of the individual Y parameters

$$Y_{\parallel} = Y_1 + Y_2 \tag{8.53}$$

This allows us to treat the shunt–shunt case with and without feedback in a unified manner as we can simply absorb the feedback network into our transistor two-port Y parameters. Likewise, series feedback is most easily handled by Z parameters, shunt–series feedback with G parameters, and series–shunt feedback with H parameters.

Properties of a two-port

Input/output admittance

The input and output impedance of a two-port will play an important role in our discussions. The stability and power gain of the two-port is determined by these quantities. In terms of Y parameters

$$Y_{in} = \frac{I_1}{V_1} = \frac{Y_{11}V_1 + Y_{12}V_2}{V_1} = Y_{11} + Y_{12}\frac{V_2}{V_1} \tag{8.54}$$

The voltage gain of the two-port is given by solving the following equations

$$-I_2 = V_2 Y_L = -(Y_{21}V_1 + V_2 Y_{22}) \tag{8.55}$$

$$\frac{V_2}{V_1} = \frac{-Y_{21}}{Y_L + Y_{22}} \tag{8.56}$$

Note that for a simple transistor $Y_{21} = g_m$ and so the above reduces to the familiar $-g_m R_o || R_L$. We can now solve for the input and output admittance

$$Y_{in} = Y_{11} - \frac{Y_{12}Y_{21}}{Y_L + Y_{22}} \tag{8.57}$$

And by symmetry (please don't redo the calculations!)

$$Y_{out} = Y_{22} - \frac{Y_{12}Y_{21}}{Y_S + Y_{11}} \tag{8.58}$$

Note that if $Y_{12} = 0$, then the input and output admittance are de-coupled

$$Y_{in} = Y_{11} \tag{8.59}$$

$$Y_{out} = Y_{22} \tag{8.60}$$

But in general they are coupled and changing the load will change the input admittance. A lot of us have been foolish enough to try to match the input and output impedance of a bilateral amplifier by tweaking the load and source matching networks one at a time. The problem is that as soon as you match one side, you de-tune the other side!

It is interesting to note the same formula derived above also works for the input/output impedance

$$Z_{in} = Z_{11} - \frac{Z_{12}Z_{21}}{Z_L + Z_{22}} \tag{8.61}$$

The same is true for the hybrid and inverse hybrid matrices.

Power gain

With reference to Fig. 8.23, the *power gain* G_p is defined as follows

$$G_p = \frac{P_L}{P_{in}} = f(Y_L, Y_{ij}) \neq f(Y_S) \tag{8.62}$$

We note that this power gain is a function of the load admittance Y_L and the two-port parameters Y_{ij}. We can define power gain in many different ways. The *available power gain*

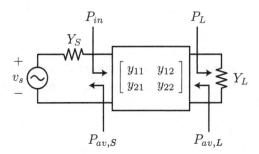

Figure 8.23 The various power definitions through a general two-port device.

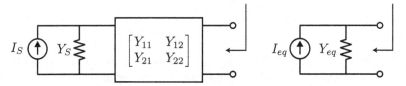

Figure 8.24 The Norton equivalent circuit of a two-port seen from the load.

is defined as follows

$$G_a = \frac{P_{av,L}}{P_{av,S}} = f(Y_S, Y_{ij}) \neq f(Y_L) \tag{8.63}$$

The available power from the two-port is denoted $P_{av,L}$, whereas the power available from the source is $P_{av,S}$. Finally, the *transducer gain* is defined by

$$G_T = \frac{P_L}{P_{av,S}} = f(Y_L, Y_S, Y_{ij}) \tag{8.64}$$

This is a measure of the efficacy of the two-port as it compares the power at the load to a simple conjugate match.

The power gain is readily calculated from the input admittance and voltage gain

$$P_{in} = \frac{|V_1|^2}{2} \Re(Y_{in}) \tag{8.65}$$

$$P_L = \frac{|V_2|^2}{2} \Re(Y_L) \tag{8.66}$$

$$G_p = \left|\frac{V_2}{V_1}\right|^2 \frac{\Re(Y_L)}{\Re(Y_{in})} \tag{8.67}$$

$$G_p = \frac{|Y_{21}|^2}{|Y_L + Y_{22}|^2} \frac{\Re(Y_L)}{\Re(Y_{in})} \tag{8.68}$$

To derive the available power gain, consider a Norton equivalent for the two-port shown in Fig. 8.24

$$I_{eq} = I_2 = Y_{21} V_1 = \frac{Y_{21}}{Y_{11} + Y_S} I_S \tag{8.69}$$

The Norton equivalent admittance is simply the output admittance of the two-port

$$Y_{eq} = Y_{22} - \frac{Y_{21} Y_{12}}{Y_{11} + Y_S} \tag{8.70}$$

The available power at the source and load are given by

$$P_{av,S} = \frac{|I_S|^2}{8\Re(Y_S)} \tag{8.71}$$

$$P_{av,L} = \frac{|I_{eq}|^2}{8\Re(Y_{eq})} \tag{8.72}$$

$$G_a = \frac{|Y_{21}|^2}{|Y_{11} + Y_S|^2} \frac{\Re(Y_S)}{\Re(Y_{eq})} \tag{8.73}$$

The transducer gain is given by

$$G_T = \frac{P_L}{P_{av,S}} = \frac{\frac{1}{2}\Re(Y_L)|V_2|^2}{\frac{|I_S|^2}{8\Re(Y_S)}} = 4\Re(Y_L)\Re(Y_S)\left|\frac{V_2}{I_S}\right|^2 \tag{8.74}$$

We need to find the output voltage in terms of the source current. Using the equations for voltage gain and input admittance we have

$$\left|\frac{V_2}{V_1}\right| = \left|\frac{Y_{21}}{Y_L + Y_{22}}\right| \tag{8.75}$$

$$I_S = V_1(Y_S + Y_{in}) \tag{8.76}$$

$$\left|\frac{V_2}{I_S}\right| = \left|\frac{Y_{21}}{Y_L + Y_{22}}\right| \frac{1}{|Y_S + Y_{in}|} \tag{8.77}$$

$$|Y_S + Y_{in}| = \left|Y_S + Y_{11} - \frac{Y_{12} Y_{21}}{Y_L + Y_{22}}\right| \tag{8.78}$$

We can now express the output voltage as a function of source current as

$$\left|\frac{V_2}{I_S}\right|^2 = \frac{|Y_{21}|^2}{|(Y_S + Y_{11})(Y_L + Y_{22}) - Y_{12} Y_{21}|^2} \tag{8.79}$$

And thus the transducer gain

$$G_T = \frac{4\Re(Y_L)\Re(Y_S)|Y_{21}|^2}{|(Y_S + Y_{11})(Y_L + Y_{22}) - Y_{12} Y_{21}|^2} \tag{8.80}$$

It is interesting to note that *all* of the gain expression, we have derived are in the exact same form for the impedance, hybrid, and inverse hybrid matrices.

Comparison of power gains

In general, $P_L \leq P_{av,L}$, with equality for a matched load. Thus we can say that

$$G_T \leq G_a \tag{8.81}$$

The maximum transducer gain as a function of the load impedance thus occurs when the load is conjugately matched to the two-port output impedance

$$G_{T,max,L} = \frac{P_L(Y_L = Y_{out}^*)}{P_{av,S}} = G_a \tag{8.82}$$

Likewise, since $P_{in} \leq P_{av,S}$, again with equality when the the two-port is conjugately matched to the source, we have

$$G_T \leq G_p \tag{8.83}$$

The transducer gain is maximized with respect to the source when

$$G_{T,max,S} = G_T(Y_{in} = Y_S^*) = G_p \tag{8.84}$$

When the input and output are simultaneously conjugately matched, or a *bi-conjugate match* has been established, we find that the transducer gain is maximized with respect to the source and load impedance

$$G_{T,max} = G_{p,max} = G_{a,max} \tag{8.85}$$

This is thus the recipe for calculating the optimal source and load impedance in order to maximize gain

$$Y_{in} = Y_{11} - \frac{Y_{12}Y_{21}}{Y_L + Y_{22}} = Y_S^* \tag{8.86}$$

$$Y_{out} = Y_{22} - \frac{Y_{12}Y_{21}}{Y_S + Y_{11}} = Y_L^* \tag{8.87}$$

Solution of the above four equations (real/imag) results in the optimal $Y_{S,opt}$ and $Y_{L,opt}$.

Another approach is to simply equate the partial derivatives of G_T with respect to the source/load admittance to find the maximum point

$$\frac{\partial G_T}{\partial G_S} = \frac{\partial G_T}{\partial B_S} = \frac{\partial G_T}{\partial G_L} = \frac{\partial G_T}{\partial B_L} = 0 \tag{8.88}$$

Again we have four equations. But we should be smarter about this and recall that the maximum gains are all equal. Since G_a and G_p are only a function of the source or load, we can get away with only solving two equations. Laziness is good when you are faced with four equations. For instance

$$\frac{\partial G_a}{\partial G_S} = \frac{\partial G_a}{\partial B_S} = 0 \tag{8.89}$$

This yields $Y_{S,opt}$ and by setting $Y_L = Y_{out}^*$ we can find the $Y_{L,opt}$. Likewise we can also solve

$$\frac{\partial G_p}{\partial G_L} = \frac{\partial G_p}{\partial B_L} = 0 \tag{8.90}$$

And now use $Y_{S,opt} = Y_{in}^*$. Let's outline the procedure for the optimal power gain. We'll use the power gain G_p and take partials with respect to the load. Let

$$Y_{jk} = m_{jk} + jn_{jk} \tag{8.91}$$

$$Y_L = G_L + jX_L \tag{8.92}$$

$$Y_{12}Y_{21} = P + jQ = Le^{j\phi} \tag{8.93}$$

$$G_p = \frac{|Y_{21}|^2}{D}G_L \tag{8.94}$$

$$\Re\left(Y_{11} - \frac{Y_{12}Y_{21}}{Y_L + Y_{22}}\right) = m_{11} - \frac{\Re(Y_{12}Y_{21}(Y_L + Y_{22})^*)}{|Y_L + Y_{22}|^2} \tag{8.95}$$

$$D = m_{11}|Y_L + Y_{22}|^2 - P(G_L + m_{22}) - Q(B_L + n_{22}) \tag{8.96}$$

$$\frac{\partial G_p}{\partial B_L} = 0 = -\frac{|Y_{21}|^2 G_L}{D^2}\frac{\partial D}{\partial B_L} \tag{8.97}$$

Solving the above equation we arrive at the following solution

$$B_{L,opt} = \frac{Q}{2m_{11}} - n_{22} \tag{8.98}$$

In a similar fashion, solving for the optimal load conductance

$$G_{L,opt} = \frac{1}{2m_{11}}\sqrt{(2m_{11}m_{22} - P)^2 - L^2} \tag{8.99}$$

If we substitute these values into the equation for G_p (lots of algebra ... fortunately someone else has done it for us), we arrive at

$$G_{p,max} = \frac{|Y_{21}|^2}{2m_{11}m_{22} - P + \sqrt{(2m_{11}m_{22} - P)^2 - L^2}} \tag{8.100}$$

Notice that for the solution to exist, G_L must be a real number. In other words

$$(2m_{11}m_{22} - P)^2 > L^2 \tag{8.101}$$

$$(2m_{11}m_{22} - P) > L \tag{8.102}$$

$$K = \frac{2m_{11}m_{22} - P}{L} > 1 \tag{8.103}$$

This factor K plays an important role as we shall show that it also corresponds to an unconditionally stable two-port. We can recast all of the work up to here in terms of K

$$Y_{S,opt} = \frac{Y_{12}Y_{21} + |Y_{12}Y_{21}|(K + \sqrt{K^2 - 1})}{2\Re(Y_{22})} \tag{8.104}$$

$$Y_{L,opt} = \frac{Y_{12}Y_{21} + |Y_{12}Y_{21}|(K + \sqrt{K^2 - 1})}{2\Re(Y_{11})} \tag{8.105}$$

$$G_{p,max} = G_{T,max} = G_{a,max} = \frac{Y_{21}}{Y_{12}}\frac{1}{K + \sqrt{K^2 - 1}} \tag{8.106}$$

Maximum gain

The maximum gain is usually written in the following insightful form

$$G_{max} = \frac{Y_{21}}{Y_{12}}(K - \sqrt{K^2 - 1}) \tag{8.107}$$

For a reciprocal network, such as a passive element, $Y_{12} = Y_{21}$ and thus the maximum gain is given by the second factor

$$G_{r,max} = K - \sqrt{K^2 - 1} \tag{8.108}$$

since $K > 1$, $|G_{r,max}| < 1$. The reciprocal gain factor is known as the efficiency of the reciprocal network. The first factor, on the other hand, is a measure of the non-reciprocity.

For a unilateral network, the design for maximum gain is trivial. For a bi-conjugate match

$$Y_S = Y_{11}^* \tag{8.109}$$

$$Y_L = Y_{22}^* \tag{8.110}$$

$$G_{T,max} = \frac{|Y_{21}|^2}{4m_{11}m_{22}} \tag{8.111}$$

Two-port stability and passivity

The stability and passivity of a two-port was the obsession of the circuit theory community for several years. Very smart men basically spent their good time playing with the four parameters of a two-port and generated hundreds of equations, equalities, and inequalities, all in the name of stability and passivity.

Crudely speaking, a stable two-port cannot oscillate. Furthermore, a passive network must always absorb power. For this reason, the input impedance looking into each port of a stable two-port is always positive. As we shall see, passivity and stability are related but distinct concepts.

Stability of a two-port

A two-port is unstable if the admittance of either port has a negative conductance for a passive termination on the second port. Under such a condition, the two-port can oscillate. Consider the input admittance

$$Y_{in} = G_{in} + jB_{in} = Y_{11} - \frac{Y_{12}Y_{21}}{Y_{22} + Y_L} \tag{8.112}$$

Using the following definitions

$$Y_{11} = g_{11} + jb_{11} \tag{8.113}$$

$$Y_{22} = g_{22} + jb_{22} \tag{8.114}$$

$$Y_{12}Y_{21} = P + jQ = L\angle\phi \tag{8.115}$$

$$Y_L = G_L + jB_L \tag{8.116}$$

Now substitute real/imaginary parts of the above quantities into Y_{in}

$$Y_{in} = g_{11} + jb_{11} - \frac{P + jQ}{g_{22} + jb_{22} + G_L + jB_L} \tag{8.117}$$

$$= g_{11} + jb_{11} - \frac{(P + jQ)(g_{22} + G_L - j(b_{22} + B_L))}{(g_{22} + G_L)^2 + (b_{22} + B_L)^2} \tag{8.118}$$

Taking the real part, we have the input conductance

$$\Re(Y_{in}) = G_{in} = g_{11} - \frac{P(g_{22} + G_L) + Q(b_{22} + B_L)}{(g_{22} + G_L)^2 + (b_{22} + B_L)^2} \tag{8.119}$$

$$= \frac{(g_{22} + G_L)^2 + (b_{22} + B_L)^2 - \frac{P}{g_{11}}(g_{22} + G_L) - \frac{Q}{g_{11}}(b_{22} + B_L)}{D} \tag{8.120}$$

Since $D > 0$ if $g_{11} > 0$, we can focus on the numerator. Note that $g_{11} > 0$ is a requirement since otherwise oscillations would occur for a short circuit at port 2. The numerator can be factored into several positive terms

$$N = (g_{22} + G_L)^2 + (b_{22} + B_L)^2 - \frac{P}{g_{11}}(g_{22} + G_L) - \frac{Q}{g_{11}}(b_{22} + B_L) \tag{8.121}$$

$$= \left(G_L + \left(g_{22} - \frac{P}{2g_{11}}\right)\right)^2 + \left(B_L + \left(b_{22} - \frac{Q}{2g_{11}}\right)\right)^2 - \frac{P^2 + Q^2}{4g_{11}^2} \tag{8.122}$$

Now note that the numerator can go negative only if the first two terms are smaller than the last term. To minimize the first two terms, choose $G_L = 0$ and $B_L = -\left(b_{22} - \frac{Q}{2g_{11}}\right)$ (reactive load)

$$N_{min} = \left(g_{22} - \frac{P}{2g_{11}}\right)^2 - \frac{P^2 + Q^2}{4g_{11}^2} \tag{8.123}$$

And thus the above must remain positive, $N_{min} > 0$, so

$$\left(g_{22} - \frac{P}{2g_{11}}\right)^2 - \frac{P^2 + Q^2}{4g_{11}^2} > 0 \tag{8.124}$$

$$g_{11}g_{22} > \frac{P + L}{2} = \frac{L}{2}(1 + \cos\phi) \tag{8.125}$$

Linvill/Llewellyn stability factors

Using the above equation, we define the Linvill stability factor

$$L < 2g_{11}g_{22} - P \tag{8.126}$$

$$C = \frac{L}{2g_{11}g_{22} - P} < 1 \tag{8.127}$$

The two-port is stable if $0 < C < 1$. It's more common to use the inverse of C as the stability measure

$$\frac{2g_{11}g_{22} - P}{L} > 1 \tag{8.128}$$

The above definition of stability is perhaps the most common

$$K = \frac{2\Re(Y_{11})\Re(Y_{22}) - \Re(Y_{12}Y_{21})}{|Y_{12}Y_{21}|} > 1 \qquad (8.129)$$

The above expression is identical if we interchange ports 1/2. Thus it's the general condition for stability. Note that $K > 1$ is the same condition for the maximum stable gain derived last section. The connection is now more obvious. If $K < 1$, then the maximum gain is infinity!

Stability from scattering parameters

We can also derive stability in terms of the input reflection coefficient. For a general two-port with load Γ_L we have

$$v_2^- = \Gamma_L^{-1}v_2^+ = S_{21}v_1^+ + S_{22}v_2^+ \qquad (8.130)$$

$$v_2^+ = \frac{S_{21}}{\Gamma_L^{-1} - S_{22}}v_1^- \qquad (8.131)$$

$$v_1^- = \left(S_{11} + \frac{S_{12}S_{21}\Gamma_L}{1 - \Gamma_L S_{22}}\right)v_1^+ \qquad (8.132)$$

$$\Gamma = S_{11} + \frac{S_{12}S_{21}\Gamma_L}{1 - \Gamma_L S_{22}} \qquad (8.133)$$

If $|\Gamma| < 1$ for all Γ_L, then the two-port is stable

$$\Gamma = \frac{S_{11}(1 - S_{22}\Gamma_L) + S_{12}S_{21}\Gamma_L}{1 - S_{22}\Gamma_L} = \frac{S_{11} + \Gamma_L(S_{21}S_{12} - S_{11}S_{22})}{1 - S_{22}\Gamma_L} \qquad (8.134)$$

$$= \frac{S_{11} - \Delta\Gamma_L}{1 - S_{22}\Gamma_L} \qquad (8.135)$$

To find the boundary between stability/instability, let's set $|\Gamma| = 1$

$$\left|\frac{S_{11} - \Delta\Gamma_L}{1 - S_{22}\Gamma_L}\right| = 1 \qquad (8.136)$$

$$|S_{11} - \Delta\Gamma_L| = |1 - S_{22}\Gamma_L| \qquad (8.137)$$

After some algebraic manipulations, we arrive at the following equation

$$\left|\Gamma - \frac{S_{22}^* - \Delta^*S_{11}}{|S_{22}|^2 - |\Delta|^2}\right| = \frac{|S_{12}S_{21}|}{|S_{22}|^2 - |\Delta|^2} \qquad (8.138)$$

This is of course the equation of a circle, $|\Gamma - C| = R$, in the complex plane with center at C and radius R. Thus a circle on the Smith Chart divides the region of instability from stability.

Consider the stability circle for a unilateral two-port

$$C_S = \frac{S_{11}^* - (S_{11}^*S_{22}^*)S_{22}}{|S_{11}|^2 - |S_{11}S_{22}|^2} = \frac{S_{11}^*}{|S_{11}|^2} \qquad (8.139)$$

$$R_S = 0 \qquad (8.140)$$

$$|C_S| = \frac{1}{|S_{11}|} \qquad (8.141)$$

The center of the circle lies outside of the unit circle if $|S_{11}| < 1$. The same is true of the load stability circle. Since the radius is zero, stability is only determined by the location of the center. If $S_{12} = 0$, then the two-port is unconditionally stable if $S_{11} < 1$ and $S_{22} < 1$. This result is trivial since

$$\Gamma_S\big|_{S_{12}=0} = S_{11} \tag{8.142}$$

The stability of the source depends only on the device and not on the load.

μ stability test

If we want to determine that a two-port is unconditionally stable, then we should use the μ-test

$$\mu = \frac{1 - |S_{11}|^2}{|S_{22} - \Delta S_{11}^*| + |S_{12} S_{21}|} > 1 \tag{8.143}$$

The μ-test not only is a test for unconditional stability, but the magnitude of μ is a measure of the stability. In other words, if one two-port has a larger μ, it is more stable.

The advantage of the μ-test is that only a single parameter needs to be evaluated. There are no auxiliary conditions like the K-test derivation earlier. The derivation of the μ-test can proceed as follows. First let $\Gamma_S = |\rho_s| e^{j\phi}$ and evaluate Γ_{out}

$$\Gamma_{out} = \frac{S_{22} - \Delta |\rho_s| e^{j\phi}}{1 - S_{11} |\rho_s| e^{j\phi}} \tag{8.144}$$

Next we can manipulate this equation into the equation for a circle $|\Gamma_{out} - C| = R$

$$\left|\Gamma_{out} + \frac{|\rho_s| S_{11}^* \Delta - S_{22}}{1 - |\rho_s| |S_{11}|^2}\right| = \frac{\sqrt{|\rho_s|} |S_{12} S_{21}|}{(1 - |\rho_s| |S_{11}|^2)} \tag{8.145}$$

For a two-port to be unconditionally stable, we'd like Γ_{out} to fall within the unit circle

$$||C| + R| < 1 \tag{8.146}$$

$$||\rho_s| S_{11}^* \Delta - S_{22}| + \sqrt{|\rho_s|} |S_{21} S_{12}| < 1 - |\rho_s| |S_{11}|^2 \tag{8.147}$$

$$||\rho_s| S_{11}^* \Delta - S_{22}| + \sqrt{|\rho_s|} |S_{21} S_{12}| + |\rho_s| |S_{11}|^2 < 1 \tag{8.148}$$

The worst case stability occurs when $|\rho_s| = 1$ since it maximizes the left-hand side of the equation. Therefore we have

$$\mu = \frac{1 - |S_{11}|^2}{|S_{11}^* \Delta - S_{22}| + |S_{12} S_{21}|} > 1 \tag{8.149}$$

K-Δ test

The K stability test has already been derived using Y parameters. We can also do a derivation based on S parameters. This form of the equation has been attributed to Rollett and Kurokawa. The idea is very simple and similar to the μ test. We simply require that all points in the instability region fall outside of the unit circle. The stability circle will intersect with the unit circle if

$$|C_L| - R_L > 1 \qquad (8.150)$$

or

$$\frac{|S_{22}^* - \Delta^* S_{11}| - |S_{12}S_{21}|}{|S_{22}|^2 - |\Delta|^2} > 1 \qquad (8.151)$$

This can be recast into the following form (assuming $|\Delta| < 1$)

$$K = \frac{1 - |S_{11}|^2 - |S_{22}|^2 + |\Delta|^2}{2|S_{12}||S_{21}|} > 1 \qquad (8.152)$$

N-port passivity

We would like to find if an N-port is active or passive. Passivity is different from stability, and plays an important role in determining the maximum frequency of operation for an "active" device. For instance, above a certain frequency every transistor will transition from an active device to a passive device, setting an upper limit for amplification or oscillation with a given device. By definition, an N-port is passive if it can only absorb net power. The total net complex power flowing into or out of an N-port is given by

$$P = (V_1^* I_1 + V_2^* I_2 + \cdots) = (I_1^* V_1 + I_2^* V_2 + \cdots) \qquad (8.153)$$

If we sum the above two terms, we have

$$P = \frac{1}{2}(v^*)^T i + \frac{1}{2}(i^*)^T v \qquad (8.154)$$

for vectors of current and voltage i and v. Using the admittance matrix $i = Yv$, this can be recast as

$$P = \frac{1}{2}(v^*)^T Y v + \frac{1}{2}(Y^* v^*)^T v = \frac{1}{2}(v^*)^T Y v + \frac{1}{2}(v^*)^T (Y^*)^T v \qquad (8.155)$$

$$P = (v^*)^T \frac{1}{2}(Y + (Y^*)^T)v = (v^*)^T Y_H v \qquad (8.156)$$

Thus for a network to be passive, the Hermitian part of the matrix Y_H should be positive semi-definite.

For a two-port, the condition for passivity can be simplified as follows. Let the general hybrid admittance matrix for the two-port be given by

$$H(s) = \begin{pmatrix} k_{11} & k_{12} \\ k_{21} & k_{22} \end{pmatrix} = \begin{pmatrix} m_{11} & m_{12} \\ m_{21} & m_{22} \end{pmatrix} + j \begin{pmatrix} n_{11} & n_{12} \\ n_{21} & n_{22} \end{pmatrix} \qquad (8.157)$$

$$H_H(s) = \frac{1}{2}(H(s) + H^*(s)) \qquad (8.158)$$

$$= \begin{pmatrix} m_{11} & \frac{1}{2}((m_{12} + m_{21}) + j(n_{12} - n_{21})) \\ ((m_{12} + m_{21}) + j(n_{21} - n_{12})) & m_{22} \end{pmatrix} \qquad (8.159)$$

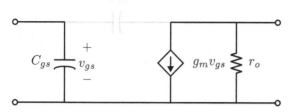

Figure 8.25 A simplified hybrid-π equivalent circuit.

This matrix is positive semi-definite if

$$m_{11} > 0 \tag{8.160}$$
$$m_{22} > 0 \tag{8.161}$$
$$\det H_n(s) \geq 0 \tag{8.162}$$

or

$$4m_{11}m_{22} - |k_{12}|^2 - |k_{21}|^2 - 2\Re(k_{12}k_{21}) \geq 0 \tag{8.163}$$
$$4m_{11}m_{22} \geq |k_{12} + k_{21}^*|^2 \tag{8.164}$$

Example 13
A simple equivalent circuit for a FET without any feedback, shown in Fig. 8.25, is of course absolutely stable if the resistors of the model are positive. The Z matrix for the circuit is given by

$$Z = \begin{bmatrix} \frac{1}{j\omega C_{gs}} & 0 \\ \frac{-g_m r_o}{j\omega C_{gs}} & r_o \end{bmatrix} \tag{8.165}$$

Since $Z_{12} = 0$, the stability factor $K = \infty$

$$K = \frac{2\Re(Z_{11})\Re(Z_{22}) - \Re(Z_{12}Z_{21})}{|Z_{12}Z_{21}|} \tag{8.166}$$

Example 14
The hybrid-pi model for a transistor is shown in Fig. 8.26. Under what conditions is this two-port active? The hybrid matrix is given by

$$H(s) = \frac{1}{G_\pi + s(C_\pi + C_\mu)} \left(\begin{matrix} 1 & sC_\mu \\ g_m - sC_\mu & q(s) \end{matrix} \right) \tag{8.167}$$
$$q(s) = (G_\pi + sC_\pi)(G_0 + sC_\mu) + sC_\mu(G_\pi + g_m) \tag{8.168}$$

Applying the condition for passivity we arrive at

$$4G_\pi G_0 \geq g_m^2 \tag{8.169}$$

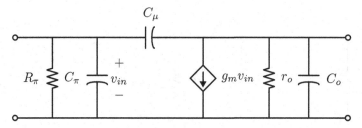

Figure 8.26 The simple hybrid-pi model for a transistor.

The above equation is either satisfied for the two-port or not, regardless of frequency. Thus our analysis shows that the hybrid-pi model is not physical. We know from experience that real two-ports are active up to some frequency f_{max}.

Mason's invariant U function

In 1954, Samuel Mason discovered the function U given by [39]

$$U = \frac{|k_{21} - k_{12}|^2}{4(\Re(k_{11})\Re(k_{22}) - \Re(k_{12})\Re(k_{21}))} \tag{8.170}$$

For the hybrid matrix formulation (H or G), the U function is given by

$$U = \frac{|k_{21} + k_{12}|^2}{4(\Re(k_{11})\Re(k_{22}) + \Re(k_{12})\Re(k_{21}))} \tag{8.171}$$

where k_{ij} are the two-port Y, Z, H, or G parameters.

This function is invariant under lossless reciprocal embeddings. Stated differently, any two-port can be *embedded* into a lossless and reciprocal circuit and the resulting two-port will have the same U function. This is a very important property, because this invariant property does not depend on any lossless matching circuitry that we employ before or after the two-port, or any lossless feedback.

Okay, U is pretty cool, but what does U signify?

Properties of U

The invariant property is shown in Fig. 8.27. The U of the original two-port is the same as U_a of the overall two-port when a four-port lossless reciprocal four-port is added.

The U function has several important properties:

1 If $U > 1$, the two-port is active. Otherwise, if $U \leq 1$, the two-port is passive.
2 U is the maximum unilateral power gain of a device under a lossless reciprocal embedding.
3 U is the maximum gain of a three-terminal device regardless of the common terminal.

With regards to the previous diagram, any lossless reciprocal embedding can be seen as an interconnection of the original two-port to a four-port, with the following block admittance

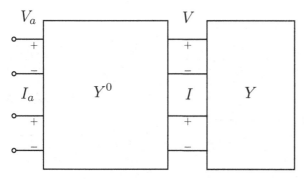

Figure 8.27 A general two-port described by Y is embedded into a lossless, reciprocal four-port device, described by the matrix Y^0.

matrix [5]

$$\begin{pmatrix} I_a \\ -I \end{pmatrix} = \begin{pmatrix} Y_{11}^0 & Y_{12}^0 \\ Y_{21}^0 & Y_{22}^0 \end{pmatrix} \begin{pmatrix} V_a \\ V \end{pmatrix} \tag{8.172}$$

Note that Y_{ij} is a 2×2 imaginary symmetric sub-matrix

$$Y_{jk}^0 = j B_{jk} \tag{8.173}$$

$$B_{jk} = B_{kj}^T \tag{8.174}$$

Since $I = YV$, we can solve for V from the second equation

$$-I = Y_{21}^0 V_a + Y_{22}^0 V = -YV \tag{8.175}$$

$$V = -\left(Y + Y_{22}^0\right)^{-1} Y_{21}^0 V_a \tag{8.176}$$

From the first equation we have the composite two-port matrix

$$I_a = \left(Y_{11}^0 - Y_{12}^0 \left(Y + Y_{22}^0\right)^{-1} Y_{21}^0\right) V_a = Y_a V_a \tag{8.177}$$

By definition, the U function is given by

$$U = \frac{\det\left(Y_a - Y_a^T\right)}{\det(Y_a + Y_a^*)} \tag{8.178}$$

Note that Y_a can be written as

$$Y_a = j B_{11} - j B_{12}(Y + j B_{22})^{-1} j B_{12}^T \tag{8.179}$$

$$Y_a = j B_{11} + B_{12}(Y + j B_{22})^{-1} B_{12}^T \tag{8.180}$$

Focus on the denominator of U

$$Y_a + Y_a^* = B_{12}(W^{-1} + (W^*)^{-1})B_{12}^T \tag{8.181}$$

where $W = Y + Y_{22}^0 = Y + j B_{22}$.
Factoring W^{-1} from the left and $(W^*)^{-1}$ from the right, we have

$$= B_{12} W^{-1}(W^* + W)(W^*)^{-1} B_{12}^T \tag{8.182}$$

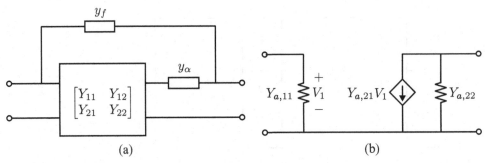

Figure 8.28 (a) A general two-port can be *unilaterized* by adding lossless feedback elements y_f and y_α. (b) The equivalent circuit for the unilaterized two-port.

But $W + W^* = Y + Y^*$ resulting in

$$Y_a + Y_a^* = B_{12} W^{-1}(Y + Y^*)(W^*)^{-1} B_{12}^T \tag{8.183}$$

In a like manner, we can show that

$$Y_a - Y_a^T = B_{12} W^{-1}(Y^T - Y)(W^*)^{-1} B_{12}^T \tag{8.184}$$

Taking the determinants and ratios

$$\det\left(Y_a + Y_a^*\right) = \frac{(\det B_{12})^2 \det(Y + Y^*)}{(\det W)^2} \tag{8.185}$$

$$\det\left(Y_a - Y_a^T\right) = \frac{(\det B_{12})^2 \det(Y^T - Y)}{(\det W)^2} \tag{8.186}$$

$$U = \frac{\det\left(Y_a - Y_a^T\right)}{\det\left(Y_a + Y_a^*\right)} = \frac{\det(Y - Y^T)}{\det(Y + Y^*)} \tag{8.187}$$

Maximum unilateral gain

Consider Fig. 8.28a, a feedback structure where y_f and y_α are lossless reactances. We can derive the overall two-port equations by a cascade connection followed by a shunt connection of two-ports

$$Y_a = \frac{y_\alpha}{y_\alpha + y_{22}} \begin{bmatrix} y_{11} + \Delta_y/y_\alpha & y_{12} \\ y_{21} & y_{22} \end{bmatrix} + \begin{bmatrix} y_f & -y_f \\ -y_f & y_f \end{bmatrix} \tag{8.188}$$

To unilaterize the device, we select

$$y_f = \frac{y_{12} y_\alpha}{y_{22} + y_\alpha} \tag{8.189}$$

We can solve for b_α and b_f

$$b_f = \Im(y_{12}) - \frac{\Re(y_{12})}{\Re(y_{22})} \Im(y_{22}) \tag{8.190}$$

$$b_\alpha = b_f \frac{\Re(y_{22})}{\Re(y_{12})} \tag{8.191}$$

It can be shown that the overall Y_a matrix is given by

$$Y_a = \frac{j\Im(y_{22}^* y_{12})}{y_{12}\Re(y_{22})} \begin{bmatrix} y_{11} + y_{12} - j\frac{\Delta_y\Re(y_{12})}{\Im(y_{22}^* y_{12})} & 0 \\ y_{21} - y_{12} & y_{22} + y_{12} \end{bmatrix} \quad (8.192)$$

Unilaterized two-port

The two-port equivalent circuit under unilaterization is shown in Fig. 8.28b. Notice now that the maximum power gain of this circuit is given by

$$G_{U,max} = \frac{|Y_{a_{21}}|^2}{4\Re(Y_{a_{11}})\Re(Y_{a_{22}})} = U_a \quad (8.193)$$

We can now attribute physical significance to U_a as the maximum unilateral gain. Furthermore, due to the invariance of U, $U_a = U$ for the original two-port network. It is important to note that *any* unilaterization scheme will yield the same maximum power! Thus U is a good metric for the device.

Neutralization

The dominant feedback path in most transistor amplifiers is C_μ, due to overlap capacitance in a MOS transistor and due to the intrinsic reverse-bias junction capacitance of a base collector in a BJT transistor. In a bipolar device, increasing the reverse bias reduces the capacitance somewhat, but in a MOS device the capacitance is fixed in strong inversion. There are, however, circuit techniques to eliminate or *neutralize* this capacitance. Thus to unilaterize the device, we only need to employ a lossless feedback inductor. Don't confuse neutralization with unilaterization. If the feedback Y_{12} is imaginary, they are the same thing. This is true for transistors at moderate frequencies.

The cascode topology is a popular amplifier because the common-source device provides good power gain and low noise and the second common-gate stage provides isolation. With good layout, there is essentially no feedback and $Y_{12} \approx 0$. Any parasitic coupling from the output to the input is due to substrate, magnetic, or package coupling.

Some classic approaches to neutralization dating back to the pre-IC era are shown in Figures 8.29. In the first approach, we simply resonate out the capacitance at the desired operating frequency. A coupling capacitor is needed if the output bias is different from the input. The problem, of course, is that complete neutralization only occurs at a single frequency. The frequency range of the neutralization can be improved if we could simply connect a negative capacitor across the input-to-output terminals. This is effectively the approach of Fig. 8.29b, where the output voltage is inverted through a tapped inductor and a feedback current of opposite phase is generated through the capacitance C_n. If a linear capacitor is used to implement C_n for a bipolar amplifier, this technique is only effective for small output swings. This is because the junction capacitance is bias dependent. A diode junction capacitor can be used in the place of C_n to improve the large-swing performance of the neutralization. A transformer is also handy for generating the output phase inversion.

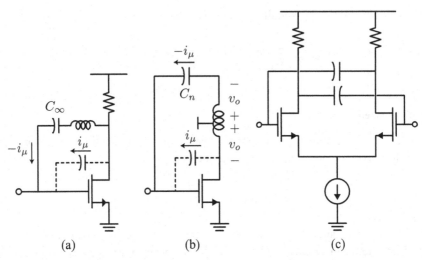

Figure 8.29 (a) A simple neutralizing scheme employs an inductor to resonate the intrinsic capacitive feedback C_μ. (b) A second neutralization scheme employing a center-tapped inductor is used to invert the phase of the output voltage, mimicking a negative capacitor. (c) A differential amplifier with cross-coupled capacitor neutralization.

An IC solution is to use more transistors in place of passive devices. We have already seen that a cascode device solves this problem. In a differential circuit, an elegant solution, shown in Fig. 8.29c, is to cross-couple the differential outputs using capacitors C_n to generate the correct phase feedback current.

Single-stage feedback revisited

With new tools at hand, let's revisit the problem of inductive and shunt feedback amplifiers.

Inductive degeneration

Although $Z_{12} \approx 0$ for a FET at low frequency, the input impedance is purely capacitive. To introduce a real component, we found that inductive degeneration can be employed, shown schematically in Fig. 8.30. The Z matrix for the inductor is simply

$$Z_L = j\omega L_s \begin{bmatrix} 1 & 1 \\ 1 & 1 \end{bmatrix} \tag{8.194}$$

Adding the Z matrix (due to series connection) to the Z matrix of the FET

$$Z = \begin{bmatrix} j\omega L_s + \frac{1}{j\omega C_{gs}} & j\omega L_s \\ j\omega L_s - \frac{g_m r_o}{j\omega C_{gs}} & r_o + j\omega L_s \end{bmatrix} \tag{8.195}$$

This feedback introduces a Z_{12} and thus the stability must be carefully examined

$$K = \frac{2 \cdot 0 \cdot r_o - \left(-\omega^2 L_s^2 - \frac{g_m L_s r_o}{C_{gs}}\right)}{\omega^2 L_s^2 + \frac{g_m r_o L_s}{C_{gs}}} = 1 \tag{8.196}$$

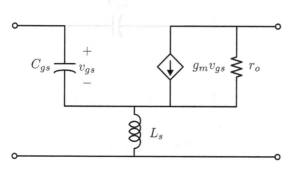

Figure 8.30 A FET with inductive degeneration.

We see that this circuit is unconditionally stable. More importantly, the stability factor is frequency independent. In reality parasitics can destabilize the transistor.

The maximum gain is thus given by

$$G_{max} = \left| \frac{Z_{21}}{Z_{12}} \right| \left(K - \sqrt{K^2 - 1} \right) = \left| \frac{Z_{21}}{Z_{12}} \right|$$

(8.197)

$$= \frac{\omega L_s + \frac{g_m r_o}{\omega C_{gs}}}{\omega L_s} = 1 + \frac{g_m r_o}{\omega^2 L_s C_{gs}}$$

(8.198)

$$= 1 + \left(\frac{\omega_T}{\omega_0} \right)^2 \left(\frac{r_o}{\omega_T L_s} \right)$$

(8.199)

We saw that the synthesized real input resistance is given by $\omega_T L_s$, and so the last term is the ratio of r_o / R_S under matched conditions.

Capacitive degeneration

Using the same approach, the Z matrix for capacitive degeneration is given by

$$Z = \begin{bmatrix} \frac{1}{j\omega C_s} + \frac{1}{j\omega C_{gs}} & \frac{1}{j\omega C_s} \\ \frac{1}{j\omega C_s} - \frac{g_m r_o}{j\omega C_{gs}} & r_o + \frac{1}{j\omega C_s} \end{bmatrix}$$

(8.200)

The stability factor is given by

$$K = \frac{2 \cdot 0 \cdot r_o - \left(\frac{g_m r_o}{\omega^2 C_s C_{gs}} - \frac{1}{\omega^2 C_s^2} \right)}{\left| \frac{g_m r_o}{\omega^2 C_s C_{gs}} - \frac{1}{\omega^2 C_s^2} \right|}$$

(8.201)

Note this is simply

$$K = \frac{-a + b}{|a - b|} = \begin{cases} \frac{b-a}{a-b} < 0 & a > b \\ \frac{b-a}{b-a} = 1 & b < a \end{cases}$$

(8.202)

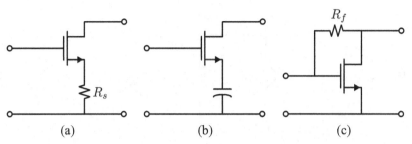

Figure 8.31 Common-source amplifier with (a) capacitive degeneration, (b) resistive degeneration, (c) shunt feedback.

The condition for stability is therefore

$$\frac{g_m r_o}{C_{gs}} > \frac{1}{C_s} \tag{8.203}$$

So far we have dealt with $K > 0$. Suppose that $|\Delta| > 1$. We know that for $0 < K < 1$ the two-port is conditionally stable. In other words, the stability circle intersects with the unit circle with the overlap (usually) corresponding to the unstable region. Instability can also occur if $K > 1$ and $|\Delta| > 1$, but this is less common (occurs with feedback).

On the other hand, if $-1 < K < 0$, we can show graphically that the entire unit circle on the Smith Chart is unstable. In other words, the stability circle does not intersect with the unit circle or the instability circle contains the entire circle.

Unintentional capacitive degeneration is very common. For instance a common drain (source follower) driving a capacitive load may have stability problems. Likewise, as shown in Fig. 8.19, a cascode amplifier may become unstable at high frequencies since the g_m input stage presents capacitive degeneration to the cascode device at high frequency.

Resistive degeneration

Resistive degeneration is commonly employed to stabilize the bias point of a transistor. The overall Z matrix is given by

$$Z = \begin{bmatrix} R_s + \frac{1}{j\omega C_{gs}} & R_s \\ R_s - \frac{g_m r_o}{j\omega C_{gs}} & r_o + R_s \end{bmatrix} \tag{8.204}$$

The K factor is computed as before

$$K = \frac{2R_s(r_o + R_s) - R_s^2}{R_s \sqrt{R_s^2 + \frac{g_m^2 r_o^2}{\omega^2 C_{gs}^2}}} \tag{8.205}$$

At low frequencies, we have

$$K = \frac{2r_o + R_s}{\frac{g_m r_o}{\omega C_{gs}}} \approx \frac{2\omega C_{gs}}{g_m} = \frac{2\omega}{\omega_T} < 1 \tag{8.206}$$

Shunt feedback

We have seen that shunt feedback is a common broadband matching approach. Now working with the Y matrix of the transistor (simplified as before)

$$Y_{fet} = \begin{bmatrix} j\omega C_{gs} & 0 \\ g_m & G_o + j\omega C_{ds} \end{bmatrix} \tag{8.207}$$

The feedback element has a Y matrix

$$Y_f = G_f \begin{bmatrix} +1 & -1 \\ -1 & +1 \end{bmatrix} \tag{8.208}$$

And thus the overall amplifier Y matrix is given by

$$Y = \begin{bmatrix} G_f + j\omega C_{gs} & -G_f \\ g_m - G_f & G_f + G_o + j\omega C_{ds} \end{bmatrix} \tag{8.209}$$

The stability factor for the shunt feedback amplifier is given by

$$K = \frac{2G_f(G_o + G_f) - G_f(G_f - g_m)}{G_f|g_m - G_f|} \tag{8.210}$$

Suppose that $g_m R_f > 1$

$$= \frac{g_m + G_f}{g_m - G_f} = \frac{g_m R_f + 1}{g_m R_f - 1} > 1 \tag{8.211}$$

The choice of R_f and g_m is governed by the current consumption, power gain, and impedance matching. For a bi-conjugate match

$$G_{max} = \left|\frac{Y_{21}}{Y_{12}}\right| (K - \sqrt{K^2 - 1}) \tag{8.212}$$

$$= \frac{g_m - G_f}{G_f} \left(\left(\frac{g_m R_f + 1}{g_m R_f - 1}\right) - \sqrt{\left(\frac{g_m R_f + 1}{g_m R_f - 1}\right)^2 - 1} \right) = \left(1 - \sqrt{g_m R_F}\right)^2 \tag{8.213}$$

The input admittance is calculated as follows

$$Y_{in} = Y_{11} - \frac{Y_{12} Y_{21}}{Y_{22} + Y_L} \tag{8.214}$$

$$= j\omega C_{gs} + G_f - \frac{-G_f(g_m - G_f)}{G_o + G_f + G_L + j\omega C_{ds}} \tag{8.215}$$

$$= j\omega C_{gs} + G_f + \frac{G_f(g_m - G_f)(G_o + G_f + G_L - j\omega C_{ds})}{(G_o + G_f + G_L)^2 + \omega^2 C_{ds}^2} \tag{8.216}$$

At lower frequencies, $\omega < \frac{1}{C_{ds}R_f||R_L}$ we have (neglecting G_o)

$$\Re(Y_{in}) = G_f + \frac{G_f(g_m - G_f)}{G_f + G_L} \tag{8.217}$$

$$= \frac{1 + g_m R_L}{R_F + R_L} \tag{8.218}$$

$$\Im(Y_{in}) = \omega \left(C_{gs} - \frac{C_{ds}}{1 + \frac{R_f}{R_L}} \right) \tag{8.219}$$

8.3   Transistor figures of merit

A common figure of merit to characterize transistors is the device unity gain frequency, f_T, which we have seen to be connected to the fundamental device physics (see Eq. (1.2) and Eq. (1.5)). But RF device characterization is based upon f_{max}, or the maximum frequency, where we can extract power gain from the device. Essentially, beyond the f_{max} frequency, the device is passive and it cannot be used to build an amplifier with power gain. Likewise, beyond f_{max} we cannot build an oscillator from an amplifier since oscillators need nearly infinite power gain, usually realized through feedback.[5] If a device does not have power gain, it certainly cannot have infinite power gain with feedback, and so the f_{max} frequency also corresponds to the maximum frequency of oscillation.

By definition, therefore, the frequency point where G_{max} crosses unity is the f_{max} of a two-port. Recall that G_{max} is only defined when the transistor is unconditionally stable, or $K > 1$. If $K < 1$, G_{max} is undefined and we usually speak of the maximum stable gain G_{MSG}, which corresponds to the maximum gain when the transistor is stabilized by adding positive conductance at the input and/or output ports so that $K = 1$.

In practice, the device f_{max} is usually estimated by plotting the device maximum unilateral power gain, or Mason's Gain U, and either observing or extrapolating the unity gain frequency point. This procedure should be performed with care and extrapolations should be avoided for maximal accuracy. If data are not available (e.g. above 100 GHz), it is better to model the device with an equivalent circuit up to the limits of measurements and then to use the f_{max} from directly evaluating the model up to the point when the power gain crosses unity.

In Fig. 8.32, the device G_{MSG} is plotted for low frequencies where $K < 1$. At the breakpoint, $K = 1$ and the device is unconditionally stable and thus G_{max} is plotted. Note that the U curve is always larger than G_{max} but both curves cross 0 dB together. At this point, the f_{max} of the device, the two-port becomes passive. f_{max} is a good metric for characterizing a three terminal device with a common-terminal, such as a transistor. Since U is invariant to the common terminal, a common-gate amplifier has the same U as a common-source amplifier.

---

[5]   Nearly infinite because in any real circuit there is noise and thus the oscillator power gain is extremely large, but not infinite.

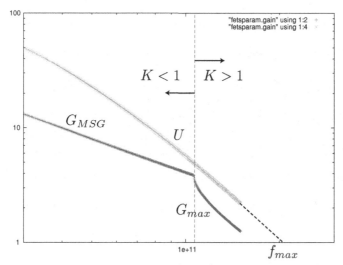

Figure 8.32 The various gain curves for a two-port device. The device is unstable at low frequency, $K < 1$, and thus we plot the G_{MSG} in this region. At the breakpoint, the device is stable. At high frequency the device is stable and we plot the G_{max} curve. The maximum unilateral gain U is also shown.

Using the unilateral gain U, the f_{max} of a BJT transistor can be estimated by

$$f_{max} \approx \sqrt{\frac{f_T}{8\pi r_b C_\mu}} \tag{8.220}$$

where the base resistance r_b and feedback capacitance C_μ are seen to set the ultimate frequency of operation for a device. It is interesting to observe that in most low frequency designs, both of these effects are ignored with negligible error. But design close to the limits of a device f_{max} requires careful modeling of all parasitic feedback and loss mechanisms. In particular, the distributed nature of the feedback C_μ into r_b requires a sectional model.

The cross section of a MOSFET device is shown in Fig. 8.33. The f_{max} of a modern FET transistor can be estimated by [42]

$$f_{max} \approx \frac{f_T}{2\sqrt{R_g(g_m C_{gd}/C_{gg}) + (R_g + r_{ch} + R_S)g_{ds}}} \tag{8.221}$$

In contrast to the device f_T, the f_{max} is a strong function of the losses in the device. As MOS technology scaling continues, f_T improves almost proportionally to channel length due to velocity saturation. But shorter channel devices may have higher gate, source, and drain losses. It is interesting to note that the effect of gate resistance on f_{max} can be reduced by scaling the width of the transistor W through a multi-finger layout, as shown in Fig. 8.34. The drain and source resistances, though, do not scale and pose a challenge for next-generation technologies. This is in stark contrast to MESFET devices where a low-resistance metal gate is employed. In deeply scaled MOS technology, metal gate work-function engineering

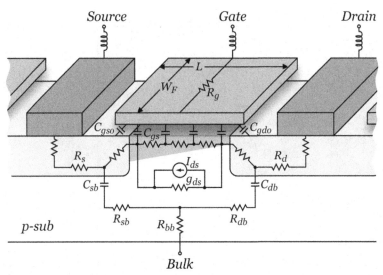

Figure 8.33 The cross section of a FET device showing the important high-frequency parasitics. *Courtesy of Chinh Doan.*

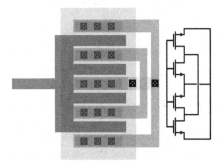

Figure 8.34 A high-frequency multi-finger FET layout minimizes the poly gate resistance.

may replace doping as a means to set the threshold voltage of a device [51], leading to enhanced RF performance.

8.4 References

Circuit design is a broad field with a long history. Along the way two separate schools of thought have developed. The more prevalent approach is widely taught in classical books such as Gray and Meyer [50] with specific applications to analog circuit design. In the approach you work directly with the small-signal model of the transistor and the small-signal parameters are derived from physical considerations. To analyze feedback amplifiers, two-port analysis is often employed, but otherwise the two-port theory is not used to design amplifiers. On the other hand, the microwave community uses two-port design extensively, employing scattering parameters. Two-port parameters are more abstract and are often derived from measurement (as opposed to a physical model). Several classic references, such as Vendelin [63] and Gonzalez [20], discuss this approach extensively. Since the goal of this chapter is to build a bridge between traditional analog circuit design and RF and microwave design, a holistic approach was taken. I found that the classic y-parameter design approach to be very insightful, and Carson's book [4] is particularly good in this regard. Interestingly, I found that even though Mason's Unilateral gain is widely used in the microwave and device community to characterize devices, little is known about this classic piece of work. Some good sources include Mason's own paper [39] and a recently published tutorial paper on the subject [23]. To get a good sense for the mathematical basis of network theory, I highly recommend Wei-Kai Chen's textbook [5].

9 Transmission lines

Transmission line behavior represents a true departure from lumped circuit theory. This is most poignant when we consider the input impedance of a quarter-wavelength shorted transmission line. Circuit theory cannot account for the fact that the input impedance is actually infinite, or an open, rather than a short. But circuit theory can be used as a foundation to understand these effects. This comes about when we expand our circuit theory to account for the *distributed* nature of the circuit elements.

In circuit theory we implicitly assume that all signals travel throughout the circuit infinitely fast. In reality, signals cannot travel faster than the speed of light. Thus there is always a delay from one point in a circuit to another point. In fact circuit theory is strictly valid in the limit of a truly *lumped* circuit, or a circuit with zero physical dimension.

Thus distributed effects become important when circuits become electrically large. In time-harmonic problems, we can relate the speed of light to the wavelength, hence the distributed effects become important when physical circuit dimensions approach the wavelength of electromagnetic propagation in the medium λ. For example, a circuit with dimensions of 3 cm in free space is electrically large when the operating frequency approaches $f = c/(.1\lambda)$ or about 10 GHz. We have arbitrarily used $\lambda/10$ to denote this boundary, whereas in reality the cutoff is application dependent. If the waveforms are non-sinusoidal, then the high-frequency spectral content of the waveform is important. For instance, in digital applications the clocking frequency may be low, but, due to the fast edges of the waveform (risetime and falltime), there is appreciable energy at higher harmonics.

9.1 Distributed properties of a cable

Consider a long cable, say a transatlantic cable running from London to New York (5585 km) shown in Fig. 9.1. The cable has a uniform cross-section, and if we ignore any non-uniformity caused by the bends, we can consider it a uniform two-line wire. How do we model such a cable? At very low frequencies, we can think of this wire as a series inductor (with loss). As shown in Fig. 9.2a, we can calculate the equivalent series impedance by shorting the end of the wire and measuring the input impedance. But also, as shown in Fig. 9.2b, if we leave the other end of the cable open, certainly there is a substantial shunt capacitance associated with the cable. If the cables are unshielded and uninsulated, then due to the conductivity

Figure 9.1 A long transatlantic cable from New York to London.

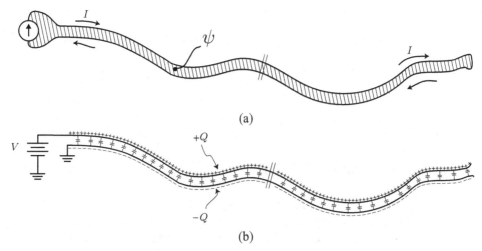

Figure 9.2 (a) The series impedance (inductance and resistance) associated with the transatlantic cable can be measured at very low frequencies by measuring the short-circuit input impedance of the line. (b) Likewise, the shunt admittance (capacitance and conductance) can be measured using the open-circuit admittance at very low frequency.

of sea-water, there is also a shunt conductance associated with the cable, even when left open.

Since the cable is very long and the delay associated with signal propagation is significant ($t_d \approx 20\,\mu\mathrm{s}$ in air over the same distance), we know that a lumped circuit model will not properly account for the behavior of the line. The large series inductance would render the wire an open circuit, even at very low frequencies, whereas the large shunt capacitance would make the wire a short circuit. So how do signals propagate through this cable? Before tossing lumped circuit theory into the sea, note that any small section of the wire will behave like a lumped circuit. For instance, if we are concerned with signal propagation up to say 1 GHz, we can break up our long cable into short sections $\ell_{sec} \ll \lambda = 30\,\mathrm{cm}$, and use circuit theory on the resulting large number of cascaded lumped circuits.

We can now take the total series impedance and shunt admittance and break it up into $N = \ell/\ell_{sec}$ section, as shown in Fig. 9.3. For simplicity, we have ignored loss in the circuit. For instance with $\ell = 1\,\mathrm{cm}$, we would need about 1 billion inductors and capacitors. Even with today's powerful computers, running a SPICE simulation with that many elements would tax our computers. So how did people solve this problem when the first Transatlantic Cable was laid out in 1857?

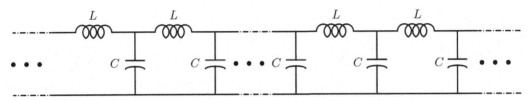

Figure 9.3 An LC ladder network representation of the transatlantic cable (neglecting cable loss). For a given frequency the required number of sections can be computed to obtain accurate results.

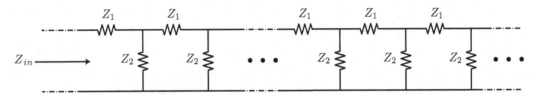

Figure 9.4 An infinite ladder network with regular structure of series impedance Z_1 shunt impedance Z_2.

9.2 An infinite ladder network

We begin our study of transmission lines by first studying an infinite lumped ladder network shown in Fig. 9.4. It is interesting that we can find the input impedance of such a network (often a freshman physics problem). Simply observe that since the ladder network is infinite, addition of a single section to the *front* should not alter the impedance. With this observation we can show that

$$Z_{in} = Z_1 + Z_2 || Z_{in} \tag{9.1}$$

or

$$Z_{in}^2 - Z_1 Z_{in} - Z_2 Z_2 = 0 \tag{9.2}$$

As shown in Fig. 9.3, suppose now that $Z_1 = j\omega L$ and $Y_2 = j\omega C$. Then the input impedance of such a line is

$$Z_{in} = \frac{j\omega L}{2} \pm \sqrt{-\frac{(\omega L)^2}{4} + \frac{L}{C}} \tag{9.3}$$

We would now like to make the leap from lumped to distributed. As such, we will assume that each inductor and capacitor in the ladder is very very small, in fact, infinitesimal in size. Therefore, for any finite frequency, the input impedance degenerates to

$$Z_{in} \approx \sqrt{\frac{L}{C}} \tag{9.4}$$

This is a very important and profound result. The input impedance is positive and real! Notice that we started out with strictly reactive elements and managed to construct a circuit with positive real input impedance. This final result is puzzling because the power dissipated

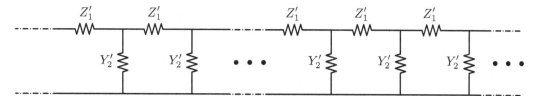

Figure 9.5 A distributed ladder network with a series impedance per unit length Z'_1 and a shunt admittance per unit length of Y_2.

by such a network is proportional to $\Re(Z_{in})$. But since each section of the ladder is incapable of dissipating power, where does the energy go?

We will answer this question in due course but for now we will hint at the solution. The energy is absorbed by the network because it is infinite in extent. The energy pumped into the network keeps vacillating from inductive to capacitive energy as it travels through the ladder network. Since there are an infinite number of capacitors and inductors to charge and discharge, the process goes on indefinitely.

Observe that if we choose to terminate the ladder network at any point with an impedance of Z_{in}, then the above results remain valid. The behavior of a finite ladder section terminated with Z_{in} is indistinguishable from that of the infinite network. In the case of a distributed line, the termination resistance is real and given by $Z_0 = \sqrt{L/C}$. The energy injected into the network, therefore, is absorbed by this resistor.

9.3 Transmission lines as distributed ladder networks

We can analyze the two-wire transmission line shown in Fig. 9.1 using the concept of distributed circuits. We note that this two-wire line stores both magnetic and electrical energy everywhere along the line in a distributed fashion. In other words, we cannot say that in some region the magnetic energy storage dominates and thus the line behaves inductively, whereas in another region the electrical energy storage dominates and thus the line behaves capacitively. Thus we now use a purely distributive circuit perspective shown in Fig. 9.5, a distributed ladder network with series impedance Z'_1 and shunt admittance Y'_2. The prime denotes that the element in question is distributed, in other words there is Z'_1 impedance per unit length.

To approximately find Z'_1, we ignore the effects of the shunt admittance by focusing on the magnetic energy stored in the line. To do this, short the line at a distance ℓ from the input, as shown in Fig. 9.2a, and the line behaves like an inductor with impedance Z_1 accounting for its magnetic energy storage and loss. Since the line is uniform, we may define the impedance per unit length

$$Z'_1 = \frac{Z_1}{\ell} \tag{9.5}$$

Similarly, if we break the two-wire line at a distance ℓ and keep the end open, as shown in Fig. 9.2b, and only consider the electrical energy storage in the line, the line behaves

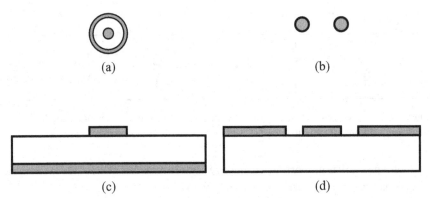

(a)   (b)

(c)   (d)

Figure 9.6 Several examples of common transmission line structures. (a) A familiar coaxial transmission line is ubiquitous in video applications. (b) A pair of wires are employed in some antenna
feedlines and more importantly in *twisted* form as telephone and data cables. (c) A microstrip line
and (d) a co-planar line are common structures for building transmission lines over a printed-circuit
board (PCB) or in a Microwave Integrated Circuit. In the co-planar structure the middle conductor
carries the signal, whereas the outer conductors form the ground return path.

capacitively. And so we define the admittance per unit length

$$Y_2' = \frac{Y_2}{\ell} \tag{9.6}$$

In general, the electrical and magnetic energy storage are intermingled and thus we must
distribute the series and shunt impedances uniformly along the line.

So far we have used the two-wire transmission line as an example. But our discussion
applies to any structure with a uniform two-dimensional cross section. Common examples
are shown in Fig. 9.6. For now we restrict our discussion to structures with two conductors,
where one conductor is chosen as the reference plane. If more than one conductor is present,
it is assumed that all but one conductor is at the same reference or ground potential. This
is true, for instance, in the co-planar line shown in Fig. 9.6d, where the conducting planes
on the left and right of the conductor are the ground return path for the signal conductor
situated in the middle.

Telegrapher's time harmonic equations

Now we are in position to derive the famous Telegrapher's equations. To do this, consider the infinitesimal section of length δz of the line at an arbitrary distance z from the
input. As shown in Fig. 9.7, the voltage at the output of the section is related to the input
by

$$v(z + \delta z) = v(z) - i(z)Z_1' \delta z \tag{9.7}$$

Similarly the currents are related by

$$i(z) = i(z + \delta z) + \delta z \, Y_2' v(z + \delta z) \tag{9.8}$$

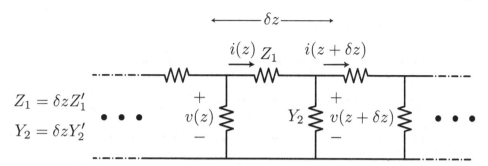

Figure 9.7 An infinitesimal section of the distributed transmission line.

These equations can be converted into differential form if we take the limit of $\delta z \to 0$

$$\frac{dv}{dz} = -Z_1' i(z) \tag{9.9}$$

$$\frac{di}{dz} = -Y_2' v(z) \tag{9.10}$$

These equations can be combined by taking the derivation of Eq. 9.9 and substituting from Eq. 9.10

$$\frac{d^2v}{dz^2} = -Z_1' \frac{di}{dz} = Z_1' Y_2' v(z) \tag{9.11}$$

The same equation applies to the current. Let the constant $Z_1' Y_2'$ be denoted as γ^2, resulting in the following set of boundary value differential equations

$$\frac{d^2v}{dz^2} = \gamma^2 v(z) \tag{9.12}$$

$$\frac{d^2i}{dz^2} = \gamma^2 i(z) \tag{9.13}$$

The general solution to the above equation is a complex exponential function

$$g(x) = G^+ e^{-\gamma z} + G^- e^{\gamma z} \tag{9.14}$$

The G^+ term is known as the "forward" traveling wave and the G^- term is known as the "reverse" or "backward" traveling wave. This is because when the time-harmonic solution is written explicitly as a function of time and position

$$g(x, t) \propto e^{j(\omega t \pm \gamma z)} \tag{9.15}$$

which represents a propagating sinusoidal wave function in the forward and reverse z-direction. The propagation constant γ is in general complex

$$\gamma = \alpha + j\beta \tag{9.16}$$

We can thus factor the wave propagation function into a constant amplitude wave propagation multiplied by an envelope decay or growth

$$g(x, t) \propto e^{\pm \alpha z} e^{j(\omega t \pm \beta z)} \tag{9.17}$$

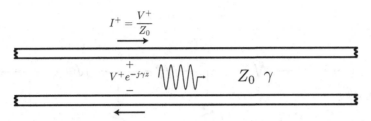

Figure 9.8 An electrical representation for a transmission line of characteristic impedance Z_0 and propagation constant γ.

The phase of the complex exponential $\theta = \omega t \pm \beta z$ changes as a function of time

$$\frac{d\theta}{dt} = \omega \pm \beta \frac{dz}{dt} \tag{9.18}$$

To find the speed of the wave, we can follow a point on the wavefront. Such a point has constant phase θ, we set the above derivative equal to zero

$$\frac{d\theta}{dt} = 0 \tag{9.19}$$

and calculate the phase velocity v_p of the wave by noting how the position z must change in order to satisfy this condition. We find that

$$v_p = \frac{dz}{dt} = \pm \frac{\omega}{\beta} \tag{9.20}$$

Returning to the Telegrapher's equation, we may write the general solution in the following form

$$v(z) = V^+ e^{-\gamma z} + V^- e^{\gamma z} \tag{9.21}$$

and

$$i(z) = I^+ e^{-\gamma z} + I^- e^{\gamma z} \tag{9.22}$$

An arbitrary transmission line will be denoted as shown in Fig. 9.8, an electrical representation for an arbitrary two conductor structure. At a given frequency, the line can be characterized by two complex numbers, the characteristic impedance Z_0 and the propagation constant γ. For the lossless line, there are only two relevant numbers, Z_0 and β, for all frequencies of propagation.

Transmission line properties

The coefficients V^+, V^-, I^+, and I^- are related in a simple manner. By taking the derivative of the above equations and substituting the original relations from Eqs. 9.9 and 9.10

$$-V^+ \gamma e^{-\gamma z} + V^- \gamma e^{\gamma z} = -Z_1'(I^+ e^{-\gamma z} + I^- e^{\gamma z}) \tag{9.23}$$

$$-I^+ \gamma e^{-\gamma z} + I^- \gamma e^{\gamma z} = -Y_2'(V^+ e^{-\gamma z} + V^- e^{\gamma z}) \tag{9.24}$$

These equations are satisfied for all values of z, and in particular for $z = 0$

$$-V^+\gamma + V^-\gamma = -Z_1'(I^+ + I^-) \tag{9.25}$$
$$-I^+\gamma + I^-\gamma = -Y_2'(V^+ + V^-) \tag{9.26}$$

which shows that only two of the four coefficients can possibly be independent. Let us define the impedance $Z_0 = \sqrt{Z_1'/Y_2'}$ and recall that $\gamma = \sqrt{Z_1'Y_2'}$. Then we can solve for I^+ and I^- in terms of V^+ and V^- to arrive at the following classic equations

$$v(z) = V^+e^{-\gamma z} + V^-e^{\gamma z} \tag{9.27}$$
$$i(z) = \frac{V^+}{Z_0}e^{-\gamma z} - \frac{V^-}{Z_0}e^{\gamma z} \tag{9.28}$$

In these equations the individual forward and backward waves begin to take on a life of their own. Notice that the current waves are related to the voltage waves by the characteristic impedance of the line Z_0. The backward wave, though, has a minus sign associated with it

$$Z_0^- = -\frac{V^-}{I^-} \tag{9.29}$$

A common case is the lossless line with $Z_1 = j\omega L$ and $Y_2 = j\omega C$. The characteristic impedance of the line is thus

$$Z_0 = \sqrt{\frac{j\omega L}{j\omega C}} = \sqrt{\frac{L}{C}} \tag{9.30}$$

and the propagation constant is

$$\gamma = j\sqrt{LC}\omega \tag{9.31}$$

which is purely imaginary. The propagation velocity is given by

$$v = \pm\frac{\omega}{\beta} = \pm\sqrt{\frac{1}{LC}} \tag{9.32}$$

In Section 9.7, we shall show that this velocity is in fact the speed of light in the medium. For a two-wire transmission line suspended in air, this velocity is well known at approximately 3×10^8 m/s.

9.4 Transmission line termination

Up to now we have only considered an infinitely long uniform transmission in the z-direction. We found that such a structure supports voltages and currents which we have labeled as "forward" waves and "backward" waves (Eqs. 9.27 and 9.28)

$$v(z) = V^+e^{-\gamma z} + V^-e^{\gamma z}$$
$$i(z) = \frac{V^+}{Z_0}e^{-\gamma z} - \frac{V^-}{Z_0}e^{\gamma z}$$

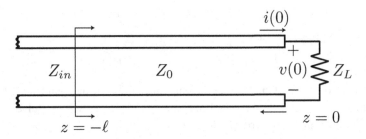

Figure 9.9 A transmission line terminated in a load impedance Z_L.

We would now like to focus on a terminated transmission line as shown in Fig. 9.9. In the figure we show a coaxial line terminated in a load impedance Z_L at $z = 0$. Therefore, at the load, the following relation must hold

$$Z_L = \frac{v(0)}{i(0)} \tag{9.33}$$

or substituting $z = 0$ into the above equations

$$\frac{V^+ + V^-}{V^+ - V^-} = \frac{Z_L}{Z_0} \tag{9.34}$$

Henceforth we shall denote normalized impedances such as $z_L = Z_L/Z_0$ with lowercase letters. Since there are two free variables V^+ and V^- and the constraint imposed by the load (Eq. 9.33), these parameters are now related. The variable $\rho_L = V^-/V^+$ parameterizes this relationship. For obvious reasons, we call ρ_L the load reflection coefficient since it represents the fraction of the wave "reflected" from the load relative to the forward wave. Thus rewriting the previous equation in our new notation, we have

$$z_L = \frac{1 + \rho_L}{1 - \rho_L} \tag{9.35}$$

The above equation is easily inverted

$$\rho_L = \frac{z_L - 1}{z_L + 1} \tag{9.36}$$

We see that when a transmission line is terminated, the voltage and current on the line take on more specific forms due to the constraint of Eq. 9.33, which results in a definite relationship between the "forward" and "backward" waves. Note in particular that $\rho_L = 0$ if $z_L = 1$, or $Z_L = Z_0$. This is the so-called "matched line" load impedance that results in zero reflections. Its importance will be doubly highlighted when we consider transients on the transmission line. For a short circuit termination, $z_L = 0$ implies that $\rho_L = -1$, or, in other words, the reflected backward wave has equal magnitude and opposite phase from the forward wave. This makes sense intuitively since the short circuit requires that $v(z = 0) = 0$, or the reflected wave must interfere destructively with the forward wave.

The current reflection coefficient is very simply related to the voltage reflected coefficient

$$\rho_{L,i} = \frac{I^-}{I^+} = \frac{-V^-/Z_0}{V^+/Z_0} = -\frac{V^-}{V^+} = -\rho_L \tag{9.37}$$

Thus for an an open termination $z_L = \infty$, zero current at the load implies that $\rho_{L,i} = -1$, or equivalently $\rho = +1$. Thus the reflected voltage signal has equal magnitude and phase from an open circuit.

9.5 Lossless transmission lines

The lossless transmission line is an important idealization for many everyday computations. Consider the arbitrary transmission line shown in Fig. 9.9 terminated in a load impedance Z_L, at some point $z = -\ell$ from the load. Since $\gamma = j\beta$ and $\alpha = 0$ for a lossless line, we may write Eqs. 9.27 and 9.28 as

$$v(z) = V^+ e^{-j\beta z} + V^- e^{j\beta z} \tag{9.38}$$

$$i(z) = \frac{V^+}{Z_0} e^{-j\beta z} - \frac{V^-}{Z_0} e^{j\beta z} \tag{9.39}$$

Since the line is lossless, the propagation constant $\gamma = j\beta$ is imaginary and the characteristic impedance Z_0 is real. The constraint of the load impedance creates a reflected wave as before

$$\rho_L = \frac{Z_L - Z_0}{Z_L + Z_0} \tag{9.40}$$

and in terms of ρ_L

$$v(z) = V^+ (e^{-j\beta z} + \rho_L e^{j\beta z}) \tag{9.41}$$

$$i(z) = \frac{V^+}{Z_0} (e^{-j\beta z} - \rho_L e^{j\beta z}) \tag{9.42}$$

The average power flow into the transmission line at an arbitrary point on the line is calculated by

$$P_{av} = \frac{1}{2} \Re[v(z) i(z)^*] \tag{9.43}$$

which can be written as

$$P_{av} = \frac{1}{2} \Re \left[(e^{-j\beta z} + \rho_L e^{j\beta z})(e^{j\beta z} - \rho_L^* e^{-j\beta z}) \right] \times \frac{|V^+|^2}{Z_0} \tag{9.44}$$

and expanded

$$P_{av} = \frac{1}{2} \Re \left[1 + \rho_L e^{2j\beta z} - \rho_L^* e^{-2j\beta z} - |\rho_L|^2 \right] \times \frac{|V^+|^2}{Z_0} \tag{9.45}$$

Notice that the middle terms in the brackets sum to a purely imaginary number, [1] simplifying the calculations to

$$P_{av} = \frac{|V^+|^2}{2Z_0} (1 - |\rho_L|^2) \tag{9.46}$$

---

[1] Since for any complex number a, $a - a^* = 2j\Im(a)$

which is a constant independent of position. In particular, it is equal to the power delivered to the load. This of course follows from the lossless property of the line. Even though the power flow is constant on the line, the amplitude of the voltage and current waveforms are *not* constant unless $\rho_L = 0$.

Voltage standing wave ratio (VSWR)

When the termination is matched to the line impedance $Z_L = Z_0$, $\rho_L = 0$ and thus the voltage along the line $|v(z)| = |V^+|$ is constant. Otherwise

$$|v(z)| = |V^+||1 + \rho_L e^{2j\beta z}| = |V^+||1 + \rho_L e^{-2j\beta \ell}| \qquad (9.47)$$

where, as before, ℓ is the distance away from the load. The reflection coefficient is in general a complex number

$$\rho_L = |\rho_L|e^{j\theta} \qquad (9.48)$$

So the magnitude of voltage along the line can be written as

$$|v(-\ell)| = |V^+||1 + |\rho_L|e^{j(\theta - 2\beta\ell)}| \qquad (9.49)$$

The voltage is maximum when the $2\beta\ell$ is equal to $\theta + 2k\pi$, for any integer k; in other words, the reflection coefficient phase modulo 2π

$$v_{max} = |V^+|(1 + |\rho_L|) \qquad (9.50)$$

and, similarly, minimum when $\theta + k\pi$, where k is an integer $k \neq 0$

$$v_{min} = |V^+|(1 - |\rho_L|) \qquad (9.51)$$

The ratio of the maximum voltage to minimum voltage is an important metric and commonly known as the voltage standing wave ratio, VSWR[2]

$$VSWR = \frac{v_{max}}{v_{min}} = \frac{1 + |\rho_L|}{1 - |\rho_L|} \qquad (9.52)$$

It is easy to show that the current standing wave ratio is the same as for the voltage case, so it is also common to call the ratio the standing wave ratio, or SWR. It follows that for a shorted or open transmission line the VSWR is infinite, since $|\rho_L| = 1$.

Physically the maxima occur when the reflected wave adds in phase with the incoming wave, and minima occur when destructive interference takes place. The distance between maxima and minima is $180°$ in phase, or $2\beta\Delta z = \pi$, or

$$\Delta z = \frac{\pi}{2\beta} = \frac{\lambda}{4} \qquad (9.53)$$

VSWR is an important concept because it can be deduced with a relative measurement. Absolute measurements of impedance are difficult and impractical at microwave frequencies

---

[2]   Sometimes pronounced viswar.

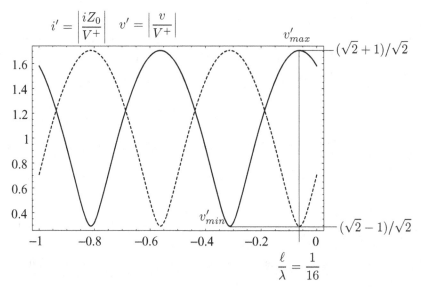

Figure 9.10 The magnitude of the steady-state time-harmonic current (dotted) and voltage (solid) waveforms along a transmission line with $z_L = (1 + 2j)$.

and a VSWR measurement allows us to deduce the impedance of an arbitrary load terminating a transmission line. By measuring VSWR, we can readily calculate $|\rho_L|$ by Eq. (9.52). By measuring the location of the voltage minima from an unknown load, we can solve for the load reflection coefficient phase θ

$$\psi_{min} = \theta - 2\beta\ell_{min} = \pi \tag{9.54}$$

Thus an unknown impedance can be characterized at microwave frequencies by measuring VSWR and ℓ_{min} and computing the load reflection coefficient. This important measurement technique has been largely supplanted by a modern network analyzer with built-in digital calibration and correction.

Example 15

Consider a transmission line terminated in a load impedance $Z_L = (1 + 2j)Z_0$. The reflection coefficient at the load is given by

$$\rho_L = \frac{z_L - 1}{z_L + 1} = \frac{1 + 2j - 1}{1 + 2j + 1} = \frac{j}{1 + j} = \sqrt{\frac{1}{2}} e^{j\frac{\pi}{4}}$$

Since $1 + |\rho_L| = (\sqrt{2} + 1)/\sqrt{2}$ and $1 - |\rho_L| = (\sqrt{2} - 1)/\sqrt{2}$, the VSWR is equal to $\frac{\sqrt{2}+1}{\sqrt{2}-1}$. A plot of the voltage and current along a transmission line is shown in Fig. 9.10. Notice the location of the voltage maxima occurs when the reflection coefficient $|\rho_L| e^{j\frac{\pi}{4}} e^{-j2\beta\ell}$ is a real number, which occurs when $2\beta\ell = \frac{\pi}{4}$, or $\ell/\lambda = \frac{1}{16}$.

Transmission line input impedance

We have seen that the voltage and current along a transmission line are altered by the presence of a load termination. At an arbitrary point z shown in Fig. 9.9, we wish to calculate the input impedance, or the ratio of the voltage to current as a function of Z_L

$$Z_{in}(-\ell) = \frac{v(-\ell)}{i(-\ell)} \tag{9.55}$$

It is convenient to define an analogous reflection coefficient at an arbitrary position along the line

$$\rho(-\ell) = \frac{V^- e^{-j\beta\ell}}{V^+ e^{j\beta\ell}} = \rho_L e^{-2j\beta\ell} \tag{9.56}$$

which has a constant magnitude equal to the reflection coefficient at the load but a periodic phase. From this we may infer that the input impedance of a transmission line is also periodic, since the relation between the reflection coefficient and impedance is one-to-one

$$Z_{in}(-\ell) = \frac{V^+(1 + \rho_L e^{-2j\beta\ell})}{V^+(1 - \rho_L e^{-2j\beta\ell})} Z_0 \tag{9.57}$$

or eliminating the common voltage

$$Z_{in}(-\ell) = Z_0 \frac{1 + \rho_L e^{-2j\beta\ell}}{1 - \rho_L e^{-2j\beta\ell}} \tag{9.58}$$

The above equation is of paramount importance as it expresses the input impedance of a transmission line as a function of position ℓ away from the termination. This equation can be transformed into another more useful form by substituting the value of ρ_L

$$\rho_L = \frac{Z_L - Z_0}{Z_L + Z_0} \tag{9.59}$$

and collecting terms

$$Z_{in}(-\ell) = Z_0 \frac{Z_L(1 + e^{-2j\beta\ell}) + Z_0(1 - e^{-2j\beta\ell})}{Z_0(1 + e^{-2j\beta\ell}) + Z_L(1 - e^{-2j\beta\ell})} \tag{9.60}$$

Using the common complex expansions for sine and cosine

$$\tan(x) = \frac{\sin(x)}{\cos(x)} = \frac{(e^{jx} - e^{-jx})/2j}{(e^{jx} + e^{-jx})/2} \tag{9.61}$$

which allows Eq. (9.60) to be replaced by the most important equation of the chapter, known sometimes as the "transmission line equation"

$$Z_{in}(-\ell) = Z_0 \frac{Z_L + jZ_0 \tan(\beta\ell)}{Z_0 + jZ_L \tan(\beta\ell)} \tag{9.62}$$

Several special cases warrant individual attention.

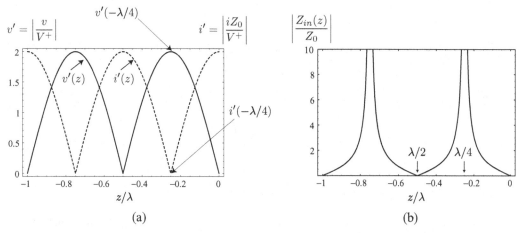

Figure 9.11 (a) The normalized magnitude of the current and voltage along a short-circuited transmission line as a function of position. (b) The magnitude of the impedance of a short-circuited transmission line as a function of position.

Shorted transmission line

As already noted, the shorted transmission line has infinite VSWR and $\rho_L = -1$. Thus the minimum voltage $v_{min} = |V^+|(1 - |\rho_L|) = 0$, as expected. At any given point along the transmission line

$$v(z) = V^+(e^{-j\beta z} - e^{j\beta z}) = -2jV^+ \sin(\beta z) \qquad (9.63)$$

whereas the current is given by

$$i(z) = \frac{V^+}{Z_0}(e^{-j\beta z} + e^{j\beta z}) \qquad (9.64)$$

or

$$i(z) = \frac{2V^+}{Z_0} \cos(\beta z) \qquad (9.65)$$

and so the impedance at any point along the line takes on a simple form

$$Z_{in}(-\ell) = \frac{v(-\ell)}{i(-\ell)} = jZ_0 \tan(\beta \ell) \qquad (9.66)$$

which is just a special case of the more general transmission line equation with $Z_L = 0$. In particular note that the impedance is purely imaginary since a short and a lossless transmission line cannot dissipate any power. The line, though, stores reactive power in a distributed fashion. A plot of the normalized voltage and current, and input impedance as a function of z is shown in Fig. 9.11a,b.

Since the tangent function takes on infinite values when $\beta \ell$ approaches $\pi/2$ modulo 2π, the shorted transmission line can have infinite input reactance! This is particularly surprising since the load is in effect transformed from a short of $Z_L = 0$ to an infinite impedance.

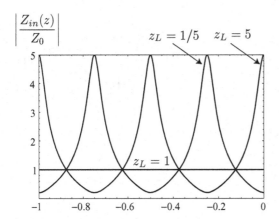

Figure 9.12 The magnitude of the impedance as a function of position on the transmission line for $z_L = 1, z_L = 5$ and $z_L = 1/5$.

Example 16

A transmission line with $Z_0 = 100\Omega$ is terminated in a load impedance of either $Z_L = 500\Omega$ or $Z_L = 20\Omega$. Can a VSWR measurement be used to determine the load?

Observe that the VSWR can be written as

$$VSWR = \frac{1 + |\rho_L|}{1 - |\rho_L|} = \frac{|1 + z_L| + |1 - z_L|}{|1 + z_L| - |1 - z_L|}$$

which can be simplified to

$$VSWR = \begin{cases} \frac{1}{z_L} & \text{if } z_L < 1 \\ z_L & \text{if } z_L > 1 \end{cases}$$

Thus the two cases above both result in a $VSWR = 5$. A plot of the normalized voltage is shown in Fig. 9.12. It is clear that the maxima occurs first on the transmission line with $z_L = 5$. This is easily remembered using a Smith Chart (see Section 9.10).

9.6 Lossy transmission lines

We can account for loss in a transmission line by working directly with Eqs. (9.27) and (9.28). In particular, all the calculation of the previous section can be redone to yield similar expressions.

Transmission line input impedance

Taking the ratio of voltage to current on the transmission line, we have the general expression

$$Z_{in}(z) = \frac{v(z)}{i(Z)} = Z_0 \frac{V^+ e^{-j\gamma z} + V^- e^{j\gamma z}}{V^+ e^{-j\gamma z} - V^- e^{j\gamma z}} \tag{9.67}$$

Using the definition of the reflection position, $\rho(z)$, we have

$$Z_{in}(z) = \frac{1 + \rho(z)}{1 - \rho(z)} \tag{9.68}$$

which can be written explicitly in terms of the load impedance

$$Z_{in}(-\ell) = Z_0 \frac{Z_L + Z_0 \tanh(\gamma \ell)}{Z_0 + Z_L \tanh(\gamma \ell)} \tag{9.69}$$

It's easy to show that the above equation degenerates to Eq. (9.5) under the lossless case.

Example 17
Calculate the input impedance of a short-circuited transmission line of length ℓ at low frequency. Assume the line is very short, e.g. $\gamma \ell \approx 0$.
From Eq. (9.6), substituting $Z_L = 0$, we have

$$Z_{in}(-\ell) = Z_0 \tanh(\gamma \ell) \approx Z_0 \gamma \ell$$

Recalling the definition of γ and Z_0, we have

$$Z_{in}(-\ell) \approx \sqrt{Z_1' Y_2'} \sqrt{Z_1' Y_2'} \ell = Z_1' \ell = R_T + j\omega L_T$$

where R_T is the resistance of the line and L_T is the inductance of the line. Thus a shorted short transmission line behaves like a lumped inductor.

Dispersionless line

In general, a lossy transmission line introduces distortion due to dispersion. Dispersion occurs when the propagation speed and attenuation is frequency dependent. If a group of frequencies are excited along the line, they travel along the line with different velocity and experience different attenuation. Thus, if an arbitrary waveform (say a pulse) is excited on the line, after significant propagation it will arrive with a completely distorted waveform since its different frequency components will be time shifted and attenuated unevenly.

For a dispersionless line, the output should be a linearly scaled delayed version of the input $v_{out}(t) = K v_{in}(t - \tau)$, or in the frequency domain

$$V_{out}(j\omega) = K V_{in}(j\omega) e^{-j\omega \tau} \tag{9.70}$$

The transfer function has constant magnitude $|H(j\omega)|$ and linear phase $\angle H(j\omega) = -\omega \tau$. The propagation constant $j\beta$ should therefore be a linear function of frequency and α should be a constant. We can find the conditions for the transmission line to be dispersionless in terms of the R, L, C, G.

$$\gamma = \sqrt{(j\omega L' + R')(j\omega C' + G')} \tag{9.71}$$

$$= \sqrt{(j\omega)^2 LC \left(1 + \frac{R}{j\omega L} + \frac{G}{j\omega C} + \frac{RG}{(j\omega)^2 LC}\right)} \tag{9.72}$$

$$= \sqrt{(j\omega)^2 LC} \sqrt{\square} \tag{9.73}$$

Suppose that $R/L = G/C$ and simplify the $\square$ term

$$\square = 1 + \frac{2R}{j\omega L} + \frac{R^2}{(j\omega)^2 L^2} \tag{9.74}$$

For $R/L = G/C$, the propagation constant simplifies to

$$\square = \left(1 + \frac{R}{j\omega L}\right)^2 \tag{9.75}$$

$$\gamma = -j\omega\sqrt{LC}\left(1 + \frac{R}{j\omega L}\right) \tag{9.76}$$

Breaking γ into real and imaginary components

$$\gamma = R\sqrt{\frac{C}{L}} - j\omega\sqrt{LC} = \alpha + j\beta \tag{9.77}$$

The attenuation constant α is independent of frequency. For low loss lines, $\alpha \approx -\frac{R}{Z_0}$. The propagation constant β is a linear function of frequency.

Therefore, to design a dispersionless lossy line, we must strive to equalize R/L and G/C. One way to achieve this is to periodically load the line with a lumped capacitor or inductor to force equality. For instance, if series lumped inductors of value L_x are used, this forms an *artificial* transmission line with inductance per unit length given by $L'' = L' + L_x/d$, if the distance d between elements is much smaller than the wavelength of propagation. Then the line behaves very much like an ideal transmission line for frequencies up to a cutoff frequency.

Power flow on a lossy line

It is interesting to calculate the steady-state power delivered to a load termination of a transmission line

$$P_L = \frac{1}{2}\Re\left[V_L I_L^*\right] \tag{9.78}$$

Expanding the voltage and current at the load in terms of the forward and backward waves, we can rewrite the above as

$$P_L = \frac{1}{2}\Re\left[(V^+ + V^-)(V^{+*} - V^{-*})\frac{1}{Z_0^*}\right] \tag{9.79}$$

which can be rewritten in the following form

$$P_L = \frac{1}{2}|V^+|^2 \Re(Y_0)(1 - |\rho_L|^2) \tag{9.80}$$

The first term can be interpreted to represent the incident power of the forward wave, which implies that the second term represents the power reflected from the load. Clearly, maximum power delivery to the load occurs when $\rho_L = 0$, or when the load is matched to the transmission line impedance Z_0.

For a lossy transmission line, the power delivered into the line at a point z is non-constant and decaying exponentially

$$P_{av}(z) = \frac{1}{2}\Re(v(z)i(z)^*) = \frac{|v^+|^2}{2|Z_0|^2}e^{-2\alpha z}\Re(Z_0) \tag{9.81}$$

Compare the above equation to Eq. (9.5), where all the power injected into a lossless line flows to the load. In the lossy case, more power is required due to the attenuation constant α.

For instance, if $\alpha = .01\text{m}^{-1}$, then a transmission line of length $\ell = 10\text{m}$ will attenuate the signal by $10 \log(e^{2\alpha\ell})$ or 2 dB. At $\ell = 100\text{m}$, the line attenuates the signal by $10 \log(e^{2\alpha\ell})$ or 20 dB. The attenuation constant α plays a very important role since it essentially determines the maximum length of a transmission line before requiring signal amplification. If the signal is attenuated too much, it will be buried in the natural noise of the system.

If the load is mismatched, then we must consider the reflected waves

$$P_{in} = \frac{1}{2}\Re(v(-\ell)i(-\ell)^*) \tag{9.82}$$

$$= \frac{1}{2}\frac{|V^+|^2}{Z_0}\Re[(e^{\gamma\ell} + \rho e^{-\gamma\ell})(e^{\gamma^*\ell} - \rho^* e^{-\gamma^*\ell})] \tag{9.83}$$

where we assumed a low loss line so that Z_0 is essentially a real quantity. Expanding the expression we have

$$= \frac{1}{2}\frac{|V^+|^2}{Z_0}\Re\left[e^{2\alpha\ell} + \underbrace{\rho e^{-2j\beta\ell} - \rho^* e^{2j\beta\ell}}_{\text{imaginary}} - |\rho|^2 e^{-2\alpha\ell}\right] \tag{9.84}$$

which simplifies to

$$P_{in} = \frac{|V^+|^2}{2Z_0}\Re[e^{2\alpha\ell} - |\rho|^2 e^{-2\alpha\ell}] \tag{9.85}$$

At the load we have

$$P_L = P_{in}(0) = \frac{|V^+|^2}{2Z_0}\Re(1 - |\rho|^2) \tag{9.86}$$

and so subtracting the above from Eq. (9.85) gives the power dissipated by the transmission line

$$P_{diss} = P_{in} - P_L = \frac{|V^+|^2}{2Z_0}[(e^{2\alpha\ell} - 1) + |\rho|^2(1 - e^{-2\alpha\ell})] \tag{9.87}$$

We can interpret the first term as the power dissipated by the incoming wave and the second term as the power dissipated by the reflected wave.

9.7   Field theory of transmission lines

In more general terms, a transmission line is a structure that supports transverse electromagnetic (TEM) wave propagation. By "transverse" we mean that the fields are always perpendicular to the direction of propagation. For a transmission line aligned with the z-axis, the fields cannot have a z-component. There are several important and simplifying consequences to this assumption. First, since currents flow into and out of the conductors of a transmission line, $J_z \neq 0$ but $E_z \equiv 0$. This implies that the conductors must be infinitely conductive, or lossless. Furthermore, since the axial magnetic field is zero, it follows that the E-field behaves statically in the transverse plane

$$\int_{t\text{-plane}} \mathbf{E} \cdot \mathbf{dl} = -\frac{d}{dt} \int_{t\text{-plane}} \mathbf{B} \cdot \mathbf{dS} \equiv 0 \tag{9.88}$$

and we can thus define a unique voltage in the transverse plane

$$v(z, t) = -\int_1^2 \mathbf{E} \cdot \mathbf{dl} \tag{9.89}$$

Similarly, since the axial electric field is zero, the magnetic field is solely dependent on the current

$$\int_{t\text{-plane}} \mathbf{H} \cdot \mathbf{dl} = -\frac{d}{dt} \int_{t\text{-plane}} \mathbf{E} \cdot \mathbf{dS} + \mu I \tag{9.90}$$

Since the displacement current is identically zero, we define a unique current

$$\int_{t\text{-plane}} \mathbf{H} \cdot \mathbf{dl} = \mu I \tag{9.91}$$

and the magnetic field also behaves statically.

In summary, for TEM propagation, the fields have a zero axial component, $E_z \equiv H_z \equiv 0$, and the transverse components of the fields behave like static fields. Thus, for a uniform structure, the laws of electromagnetics reduce to statics regardless of the frequency of excitation! While only lossless conductors can truly support TEM waves, in practice, low-loss conductors are often employed and behave essentially like TEM guides. Finally, it can be proved that TEM waves can only be supported by *two* or more conductors. Any single conductor cannot support TEM waves because a single ideal conductor cannot support a static field solution. For instance, consider the structures shown in Fig. 9.13. The solution to Poisson's equation, for example, is found by noting that the constant potential is a permissible solution, e.g. $\mathbf{E} \equiv 0$ everywhere including on the surface of the conductor. But since the solution to Poisson's equation is unique, this is the *only* solution. Thus more than one conductor is required for TEM modes. While such a conductor cannot support TEM waves, it can support transverse electric (TE) or transverse magnetic (TM) waves, or any combination thereof.

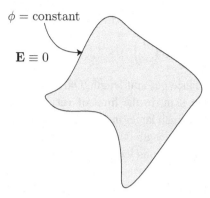

$\phi = \text{constant}$

$\mathbf{E} \equiv 0$

Figure 9.13 A single conductor cannot support TEM waves since the electrostatic solution has only a constant potential solution.

It is important to observe that an ideal transmission line has a constant characteristic impedance Z_0 versus frequency. Since the fields are the solution of the static fields in the transverse plane, there is no frequency dependence in the fields and thus no variation in the inductance and capacitance. This can be seen in another way. Since all conductors are perfectly conducting, no fields can penetrate the conductors and therefore only external inductance contributes to magnetic energy storage in the line. Thus the inductance is constant as a function of frequency. In practice, of course, all transmission lines have loss. We have already modeled the loss in our distributed circuits with an equivalent series and shunt resistor in the ladder network. So, in practice, the transmission line properties do depend on frequency. While it is possible to account for the loss in our field equations, it's much better to treat the loss as a perturbation to an ideal line rather than manipulate the full-blown Maxwell's equations from the outset.

9.8 T-line structures

The coaxial line

The coaxial line shown in Fig. 9.6a is perhaps the most commonly encountered transmission line. In many residential areas, cable TV and Internet data services are delivered to homes via 75Ω transmission lines. Due to circular symmetry, the inductance and capacitance per unit length are readily calculated. An important observation comes from Eq. (9.32), which shows that the inductance and capacitance are in fact related. Thus, only one needs to be calculated and the other is easily inferred.

In Fig. 9.6a, the coaxial transmission line is shown with an inner conductor of radius a and outer conductor of radius b. We have already calculated the inductance and capacitance per unit length for such a line

$$C' = \frac{2\pi\epsilon}{\ln(b/a)} \tag{9.92}$$

since the propagation velocity of a TEM line is constant $v_p = 1/\sqrt{\epsilon\mu} = 1/\sqrt{L'C'}$

$$L' = \frac{\epsilon\mu}{C'} = \frac{\mu}{2\pi}\ln\left(\frac{b}{a}\right) \tag{9.93}$$

Keep in mind that this is the external inductance per unit length. Only a perfect conductor will carry all of its current on the outer skin. In the limit of very high frequency, though, or when the radius of the conductor is much larger than the skin depth, $a \gg \delta$, the above formula applies. The conductance per unit length due to a lossy dielectric is easily derived since $C'/G' = \epsilon/\sigma$. The series resistance per unit length, though, is more difficult to calculate.

To minimize the loss of a coaxial transmission line, we should minimize the conductive and resistive losses. The conductive losses can be minimized by using a material with the lowest possible dielectric loss. Air is a pretty good insulator with virtually zero loss, vacuum is even better. But in practice a solid dielectric is preferred since it provides mechanical support.

Coaxial lines have several advantages. Since the outer conductor completely surrounds the inner conductor, no fields are to be found outside the structure that originate from charge and currents inside. This follows by application of Gauss' Theorem and Ampere's Law. From Gauss' Theorem, consider the volume of a cylinder completely surrounding the coaxial line. The surface integral of the electric field is proportional to the charge inside the volume

$$\int_{cylinder} \mathbf{E}\cdot\mathbf{dS} = \epsilon Q_{inside} \tag{9.94}$$

The outer conductor is grounded and thus charges flow onto the inner surface of the outer conductor and effectively shield the conductor. For a properly grounded coaxial transmission line, the ground charge on the outer conductor is equal and opposite to the charge on the inner conductor. Therefore the surface integral is identically zero. But, by symmetry, the field is everywhere radial and of equal magnitude on the surface of the cylinder

$$\int_{cylinder} \mathbf{E}\cdot\mathbf{dS} = 2\pi R E_r \ell \tag{9.95}$$

where ℓ is the length of the cylinder. Thus the field is identically zero, $E_r \equiv 0$, everywhere outside of the coaxial transmission line. A similar argument can be made for the magnetic field implying that $H_t \equiv 0$ everywhere outside of the conductors.

Balanced two-wire line

Consider the two-wire transmission line shown in Fig. 9.14. The wires are separated by a distance d and each conductor has a radius of a. Each conductor carries an equal but opposite current. To find the magnetic field at a given point in space, notice that Ampère's Law does not help much since the second conductor ruins the symmetry of the problem. But an approximate formula for the inductance per unit length is easy to derive if we assume that $d \gg a$. In that case the currents in the wires do not interact and the total field is simply

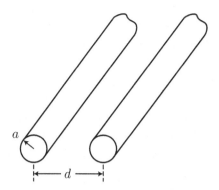

Figure 9.14 A balanced two-wire transmission line.

the sum of the magnetic fields of each wire

$$B(r) = \frac{\mu I}{2\pi r_1} + \frac{\mu I}{2\pi r_2} \tag{9.96}$$

where r_1 and r_2 are the distance to the center of each conductor. Let's integrate the magnetic field along the plane containing the conductors to obtain the magnetic flux per unit length

$$\psi' = \int_a^{a+d} \frac{\mu I}{2\pi} \left(\frac{1}{x} + \frac{1}{d + 2a - x} \right) dx \tag{9.97}$$

$$= \frac{\mu I}{\pi} \ln \frac{d+a}{a} \tag{9.98}$$

So that the inductance per unit length is simply

$$L' \approx \frac{\mu}{\pi} \ln \frac{d+a}{a} \tag{9.99}$$

An exact derivation will take some more work but the solution is considerably simplified if we simply consider the contours of the constant magnetic field for two filamental currents with opposite polarity, as shown in Fig. 9.15. This field was plotted in Chapter 5, and is repeated here. Notice that the contours are circles, allowing us to conveniently slip in two conductors with finite radius without changing the field distribution! Note that the filaments are *not* at the centers of the conductors. If you like to do algebra, then you can show that this is exactly true by plotting the contours and noting that, on any given contour, the ratio of the distances to one filament must be fixed, or $r_2/r_1 = k$.

In the exact derivation, we find that

$$L' = \frac{\mu}{\pi} \cosh^{-1} \frac{d}{2a} \tag{9.100}$$

which agrees with our previous derivation for $d \gg a$.

The two-wire transmission line is used as a low-cost feedline for differential antennas. For instance, a UHF loop antenna of 300Ω is common, and a two-wire transmission line can be used to feed the structure. Note that most television sets use 75Ω transmission lines, so an impedance matching circuit is needed (otherwise you would see a ghost on the TV

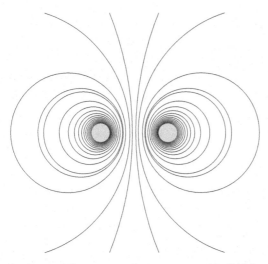

Figure 9.15 The contours of constant magnetic field for two filaments carrying opposite currents.

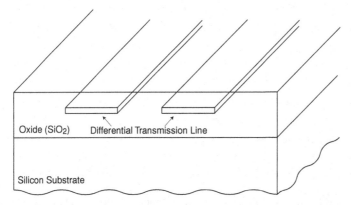

Figure 9.16 A balanced two-conductor transmission line implemented in an IC process.

screen). But since the matching network must work for all UHF channels, a narrowband matching network will not work. A broadband matching network using transmission line transformers will be discussed in Chapter 11.

On-chip differential transmission line

Since differential circuits are commonly used on-chip, a two-wire rectangular transmission line is very commonly used. As shown in Fig. 9.16, a differential pair can excite equal and opposite currents on a differential two-wire transmission line. In reality a pseudo-TEM wave propagates since the dielectric constant of the Si substrate differs from the oxide SiO_2, and thus any fields leaking into the substrate will travel at a slower velocity than the oxide. Such a structure cannot support a TEM wave, but the actual waves are a good approximation to TEM waves if the spacing between the conductors is not too large.

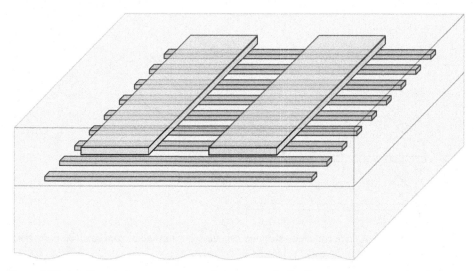

Figure 9.17 A balanced slow-wave two-conductor transmission line.

One way to avoid leakage into the substrate is to place a solid shield under the conductors. This lowers the characteristic impedance of the line since the inductance per unit length decreases, while the capacitance per unit length increases. The wave velocity is practically unchanged, though, since the product LC is constant. We can lower the wave velocity with the slow-wave structure shown in Fig. 9.17. Here, conductive strips are placed beneath the two wires to increase the capacitance per unit length without altering the inductance per unit length. The inductance is not lowered since current cannot flow perpendicular to the strips, and thus the transmission line current flow is confined to the differential conductors. Since L is unchanged while C increases, the product LC is increased, corresponding to a lower phase velocity. Does this contradict our earlier claim that LC is a constant? No, because this structure is not a uniform transmission line. In fact, you can view this structure as an artificial dielectric with $\epsilon' > \epsilon$ due to the metal strips.

Stripline and microstrip line

The stripline, shown in Fig. 9.18, is a planar version of the coaxial line, with a ground plane above and below the rectangular conductor. Such a structure can be constructed on a PCB or even on-chip. In practice the grounds are not infinite in extent, but wide enough to absorb all the fringing fields. The microstrip is a simpler structure, shown in Fig. 9.19, as one ground plane is eliminated. Since the fields fringe into the air on the top side, though, pure TEM waves cannot be supported. In practice most of the fields remain in the dielectric since $\epsilon > \epsilon_0$ and a TEM approximation is a good one.

One advantage of a stripline over a microstrip is the loss per unit length. In a stripline the ground current flows over two conductors, thus halving the return current losses. More importantly, the conductor current flows on the top and bottom faces, thus reducing the resistance per unit length. For the microstrip, the current flows primarily on the bottom face of the conductor.

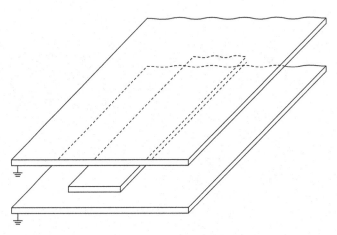

Figure 9.18 A three conductor stripline structure. The top and bottom conductors are both grounded.

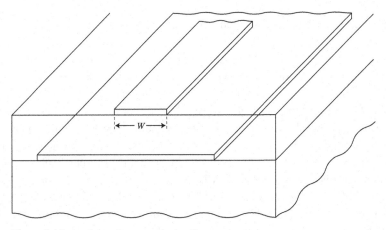

Figure 9.19 A microstrip transmission line.

Co-planar lines

Microstrip lines have some disadvantages when implemented on-chip. Since the return current flows on the bottom metal, the spacing between the conductor and ground is fixed. Thus, the only way to control the line impedance is to change the conductor width W. For a high impedance line, we require small W to minimize C and maximize L. But this introduces more loss into the line. A co-planar line, shown in Fig. 9.20, uses ground return currents on the same metal layer. The line impedance is now a strong function of the spacing S. A high impedance line can be realized by increasing S. But a co-planar line is less compact since the ground return must be sufficiently wide (say 3–4 times the skin depth).

A grounded co-planar waveguide (GCPW) structure is shown in Fig. 9.21. At first glance, the GCPW may seem to have superior characteristics over a normal co-planar line since it does not consume any extra area while the extra ground shields the transmission line from the lossy Si substrate. To make the comparison more fair though, we should first note that the ground shield can lower the characteristic impedance, especially if the spacing between the

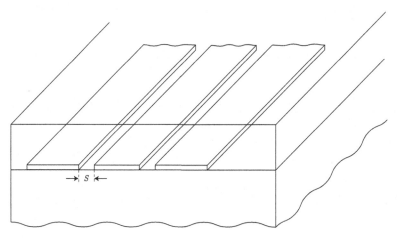

Figure 9.20 A co-planar transmission line.

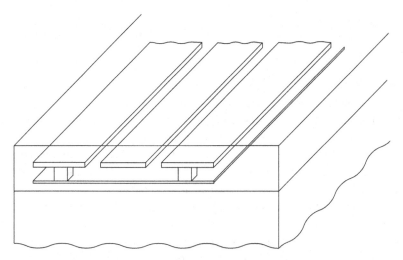

Figure 9.21 A grounded co-planar waveguide (GCPW) structure implemented in a typical IC process utilizes an array of vias to connect the co-planar ground with the lower ground plane layer.

signal and ground plane is smaller than the gap spacing. For instance, as a resonator the GCPW has a lower α and consequently a higher-quality factor. At resonance, a shorted quarter wave line appears as a resistor of value $R_{eq} = Z_0/\alpha\ell$. For low power operation, we would like to maximize this resistance and so the ratio Z_0/α is important. In practice we can optimize the GCPW dimensions to realize a higher overall quality factor and a higher equivalent resistance.

The other important consideration is that we should carefully distinguish between the line loss per unit length, α, and the loss that actually matters for a particular application. For example, in many situations, we employ transmission in matching networks as open or short stubs of length $\ell < \lambda/4$. In such a situation, imagine using the transmission line as an equivalent inductor. Then we really care about the series resistance of the structure. Will a

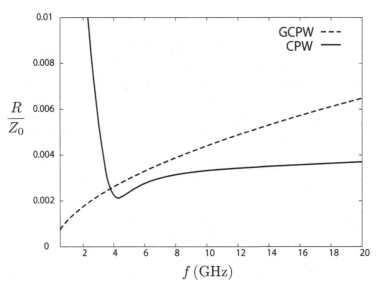

Figure 9.22 A comparison of the resistance of a co-planar waveguide (CPW) structure and a grounded co-planar waveguide (GCPW) structure.

grounded co-planar have lower resistance? Certainly this is true at low frequency, as adding resistors in parallel should lower the overall resistance. What about high frequency?

Consider a typical CMOS process with a thin metal 1 of thickness $0.2\,\mu$ and a thicker top metal of thickness $1\,\mu$ situated about $6\,\mu$ above from the substrate. Let's compare a short $100\,\mu$ CPW line with gap spacing of $10\,\mu$ to a GCPW line, with the extra ground plane realized on the first metal layer. Since current always flows along the path of least impedance, at low frequency we would expect the ground current to flow on both the top and bottom layer, resulting in a lower overall resistance. But at high frequency, we note that the ground plane path has lower inductance since the spacing between the top and bottom layer is only about $5\,\mu$, as opposed to the gap spacing of $10\,\mu$. In fact, at high frequencies $\omega L' \gg R'$ and so the current flow path is determined by inductive considerations, even if the ground plane path is more lossy. This is certainly conceivable in the case at hand, since the ground plane is thinner and has higher sheet resistance. A full-wave simulation confirms the above analysis and the results are summarized in Fig. 9.22. As evident from the figure, at lower frequencies the GCPW has substantially lower normalized resistance R/Z_0, but at even moderately high frequencies the situation reverses and the CPW line has much lower normalized resistance. At 20 GHz, for instance, the GCPW line has nearly twice the resistance.

9.9   Transmission line circuits

Thus far we have explored the properties of a generic transmission line. We shall now exploit the transmission line as a circuit element, realizing any desired reactance, using it for impedance matching, or as a resonator. The Smith Chart will be introduced as a powerful graphical aid in the design of transmission line circuits. The Smith Chart, a graphical representation of the complex bilinear transform, is also an excellent visualization tool that can give us intuition and insight into transmission line behavior.

Open and short transmission lines

Open transmission line

The open line is the dual of the shorted line analyzed in the previous section. The open transmission line has infinite VSWR and $\rho_L = 1$. At any given point along the transmission line

$$v(z) = V^+(e^{-j\beta z} + e^{j\beta z}) = 2V^+ \cos(\beta z) \tag{9.101}$$

whereas the current is given by

$$i(z) = \frac{V^+}{Z_0}(e^{-j\beta z} - e^{j\beta z}) \tag{9.102}$$

or

$$i(z) = \frac{-2jV^+}{Z_0} \sin(\beta z) \tag{9.103}$$

The impedance at any point along the line takes on a simple form

$$Z_{in}(-\ell) = \frac{v(-\ell)}{i(-\ell)} = -jZ_0 \cot(\beta \ell) \tag{9.104}$$

This is a special case of the more general transmission line equation with $Z_L = \infty$. Note that the impedance is purely imaginary since an open lossless transmission line cannot dissipate any power. We have learned, though, that the line stores reactive energy in a distributed fashion.

A plot of the input impedance as a function of z is shown in Fig. 9.23a. As evident from the plot of the current and voltage, shown in Fig. 9.23b, the current must be zero at the end of the line due to the open boundary condition, but it reaches a maximum value at every odd multiple of $\lambda/4$.

The cotangent function takes on zero values when $\beta \ell$ approaches $\pi/2$ modulo 2π. The open transmission line can have zero input impedance! This is particularly surprising since the open load is in effect transformed to a short. A plot of the voltage/current as a function of z is shown below.

Notice that if $\ell \ll \lambda/4$, the open line behaves like a lumped capacitor. This is true for any open line at very low frequencies. But even as the frequency increases, and the line approaches a quarter wavelength, $\ell < \lambda/4$, the open line appears as a distributed capacitive reactance. At exactly a quarter wavelength, though, $\ell = \lambda/4$, the line becomes a short circuit. At wavelengths slightly displaced from quarter wavelength, we shall see that the line behaves like a series resonant LC circuit. For $\ell > \lambda/4$ but $\ell < \lambda/2$, the line has an inductive reactance. And since the impedance is periodic in $\lambda/2$, the process repeats. This is illustrated in Fig. 9.24.

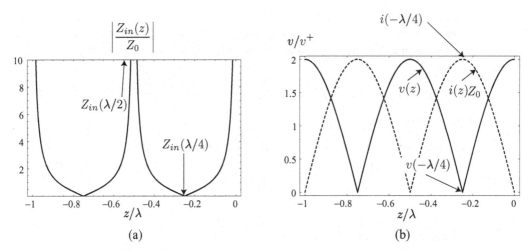

Figure 9.23 (a) The input impedance of an open transmission line as a function of the line length. (b) The voltage and current waveforms on an open line as a function of position.

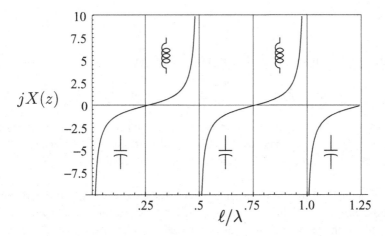

Figure 9.24 The reactance of an open transmission line alternates between capacitive and inductive behavior.

Shorted transmission line

In a previous section we showed that the behavior of the shorted transmission line varies with length, in a completely dual fashion to the open line. If $\ell \ll \lambda/4$, then the line behaves as a lumped inductor. As long as $\ell < \lambda/4$, the reactance is purely inductive. At a quarter wavelength, $\ell = \lambda/4$, the line looks like an open circuit. We shall see that in fact it acts like a resonant parallel LC circuit about small deviations from this point. Beyond quarter wavelength, $\ell > \lambda/4$ but $\ell < \lambda/2$, the line becomes capacitive. As before, the process repeats periodically with period $\lambda/2$. This is summarized in Fig. 9.25.

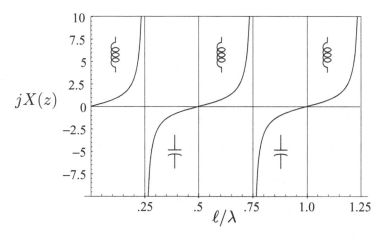

Figure 9.25 The reactance of a shorted transmission line alternates between inductive and capacitive behavior.

Half-wave line

As we have seen, the impedance on a transmission line is periodic about $\lambda/2$. Thus, a section of transmission line of length $\lambda/2$ has an interesting property. Plug into the general T-line equation for any multiple of $\lambda/2$

$$Z_{in}(-m\lambda/2) = Z_0 \frac{Z_L + jZ_0 \tan(-\beta\lambda/2)}{Z_0 + jZ_L \tan(-\beta\lambda/2)} \qquad (9.105)$$

Since $\beta\lambda m/2 = \frac{2\pi}{\lambda} \frac{\lambda m}{2} = \pi m$, then $\tan m\pi = 0$ if $m \in \mathcal{Z}$. Or $Z_{in}(-\lambda m/2) = Z_0 \frac{Z_L}{Z_0} = Z_L$. Therefore, as expected, the load impedance does not change as a result of the transmission line. It's as if there were no transmission line connecting the load to the source.

Quarter-wave line

Unlike a half-wave line, the quarter-wave line has the most dramatic impact on the load. Starting from the general T-line equation, we see that $\beta\lambda m/4 = \frac{2\pi}{\lambda} \frac{\lambda m}{4} = \frac{\pi}{2}m$, and $\tan m\frac{\pi}{2} = \infty$ if m is an odd integer. We finally have that

$$Z_{in}(-\lambda m/4) = \frac{Z_0^2}{Z_L} \qquad (9.106)$$

The $\lambda/4$ line transforms or "inverts" the impedance of the load. This is precisely the observed behavior of the open and short lines. For instance, the shorted line acts like an open when we drive it from a transmission line of length $\lambda/4$, since the zero voltage at the load is transformed to a maximum value at the source. Likewise, the finite current at the load is transformed into zero current at the source.

The property of the $\lambda/4$ line can be exploited to perform impedance matching. As shown in Fig. 9.26, a load resistor R_L is matched to the source resistance R_S by a judicious choice of the line characteristic impedance. In this case, therefore, we equate this to the desired source impedance $Z_{in} = \frac{Z_0^2}{R_L} = R_s$. The quarter-wave line should therefore have a

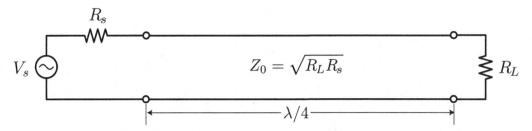

Figure 9.26 The quarter-wave transmission line can transform the load resistance to the source resistance by choosing $Z_0 = \sqrt{R_L R_S}$.

characteristic impedance that is the geometric mean $Z_0 = \sqrt{R_s R_L}$. This only works if the source and load are real resistors.

Since $Z_0 \neq R_L$, the line has a non-zero reflection coefficient

$$\rho = \frac{R_L - \sqrt{R_L R_s}}{R_L + \sqrt{R_L R_s}} \tag{9.107}$$

It also therefore has standing waves on the T-line. The non-unity SWR is given by $\frac{1+|\rho_L|}{1-|\rho_L|}$.

There is a simple explanation for how the $\lambda/4$ line does its magic. Consider a generic lossless transformer ($R_L > R_s$). Thus to make the load look smaller to match to the source, the voltage of the source should be increased in magnitude. But since the transformer is lossless, the current will likewise decrease in magnitude by the same factor. With the $\lambda/4$ transformer, the location of the voltage minimum to maximum is $\lambda/4$ from load (since the load is real). The voltage/current is thus increased/decreased by a factor of $1 + |\rho_L|$ at the load. Hence the impedance decreased by a factor of $(1 + |\rho_L|)^2$.

We can evaluate the insertion loss IL of a lossy quarter wave transformer by noting that the power injected into a low-loss transmission line is given by Eq. (9.85)

$$P_{in} = \frac{|V^+|^2}{2Z_0}(e^{2\alpha\ell} - |\rho(\lambda/4)|^2 e^{-2\alpha\ell}) \tag{9.108}$$

where

$$|\rho(\lambda/4)| = |\rho_L|e^{-2\alpha\lambda/4} \tag{9.109}$$

The power delivered to the load is simply given by

$$P_L = \frac{|V^+|^2}{2Z_0}(1 - |\rho_L|^2) \tag{9.110}$$

So the insertion loss is given by

$$IL = \frac{P_L}{P_{in}} = \frac{1 - |\rho_L|^2}{e^{2\alpha\lambda/4} - |\rho(\lambda/4)|e^{-2\alpha\lambda/4}} \tag{9.111}$$

The above expression can be simplified. Let the matching ratio be defined as the ratio of the higher to lower resistance

$$m = \frac{R_{hi}}{R_{lo}} \geq 1 \tag{9.112}$$

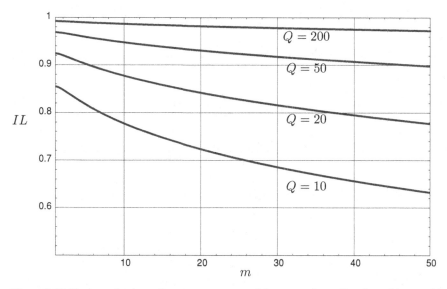

Figure 9.27 The insertion loss of a quarter-wave matching network as a function of the matching ratio m and the transmission line quality factor Q.

For instance, if $R_S > R_L$, then $m = R_S/R_L$. Then we can write IL as

$$IL = \frac{1}{\cosh(2\alpha\lambda/4) + \frac{1+m}{2\sqrt{m}} \sinh(2\alpha\lambda/4)} \tag{9.113}$$

In terms of the transmission line $Q = \frac{\beta}{2\alpha}$

$$2\alpha\lambda/4 = \frac{\alpha\lambda}{2} = \frac{\beta}{2Q}\lambda 2 = \frac{\pi}{2Q} \tag{9.114}$$

and thus we can parameterize the insertion loss in terms of the unitless parameters Q and m

$$IL(Q, m) = \frac{1}{\cosh\left(\frac{\pi}{2Q}\right) + \frac{1+m}{2\sqrt{m}} \sinh\left(\frac{\pi}{2Q}\right)} \tag{9.115}$$

For a low loss line $2\alpha\ell \ll 1$ and the above reduces to

$$IL_{ll}(Q, m) = 1 - \frac{(1+m)\pi}{4\sqrt{m}Q} \tag{9.116}$$

A plot of the insertion loss as a function of the matching ratio is shown in Fig. 9.27. Similar to the lumped matching networks, the loss is a function of the matching ratio m, quickly dropping for high ratios.

As already noted, since the reflection coefficient is zero only at a single frequency when the structure is electrically exactly $\lambda/4$ long, the quarter-wave transmission line is a narrowband matching network. A plot of the reflection coefficient ρ versus frequency is shown in Fig. 9.28. Note that the higher the matching ratio m, the smaller the bandwidth. In practice, we can usually tolerate a reflection coefficient no larger than a certain amount, $\rho < \rho_m$. For instance, we may tolerate a reflection as large as -10 dB. Thus we can calculate

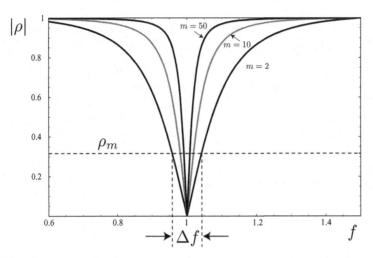

Figure 9.28 The reflection coefficient of a quarter-wave matching network as a function of frequency and matching ratio m.

the effective bandwidth of the match. For large matching ratio m, the bandwidth can be extended by employing a cascade of quarter-wave sections [9].

Transmission line resonance

We would like to now demonstrate that transmission lines act like resonant circuits around frequencies where the reactance changes sign. Intuitively it seems like a shorted transmission acts like a parallel LC resonant circuit about odd multiples of $\lambda/4$ and like a series LC resonant circuit about even multiples of $\lambda/4$.

The impedance of a series resonator near resonance $Z(\omega) = j\omega L + \frac{1}{j\omega C} + R$ can be written in terms of the Q factor, $Q = \omega_0 L/R$.

For a small frequency shift from resonance $\delta\omega \ll \omega_0$

$$Z(\omega_0 + \delta\omega) = j\omega_0 L + j\delta\omega L + \frac{1}{j\omega_0 C}\left(\frac{1}{1 + \frac{\delta\omega}{\omega_0}}\right) + R \tag{9.117}$$

which can be simplified using the fact that $\omega_0 L = \frac{1}{\omega_0 C}$

$$Z(\omega_0 + \delta\omega) = j2\delta\omega L + R \tag{9.118}$$

Using the definition of Q

$$Z(\omega_0 + \delta\omega) = R\left(1 + j2Q\frac{\delta\omega}{\omega_0}\right) \tag{9.119}$$

For a parallel line, the same formula applies to the admittance

$$Y(\omega_0 + \delta\omega) = G\left(1 + j2Q\frac{\delta\omega}{\omega_0}\right) \tag{9.120}$$

where $Q = \omega_0 C/G$.

Shorted half-wave line resonance

A shorted transmission line of length ℓ has input impedance of $Z_{in} = Z_0 \tanh(\gamma \ell)$. For a low-loss line, Z_0 is almost real. Expanding the tanh term into real and imaginary parts

$$\tanh(\alpha\ell + j\beta\ell) = \frac{\sinh(2\alpha\ell)}{\cos(2\beta\ell) + \cosh(2\alpha\ell)} + \frac{j\sin(2\beta\ell)}{\cos(2\beta\ell) + \cosh(2\alpha\ell)} \quad (9.121)$$

Since $\lambda_0 f_0 = c$ and $\ell = \lambda_0/2$ (near the resonant frequency), we have $\beta\ell = 2\pi\ell/\lambda = 2\pi\ell f/c = \pi + 2\pi\delta f\ell/c = \pi + \pi\delta\omega/\omega_0$. If the lines are low loss, then $\alpha\ell \ll 1$.

Simplifying the above relation we come to

$$Z_{in} = Z_0 \left(\alpha\ell + j\frac{\pi\delta\omega}{\omega_0} \right) \quad (9.122)$$

The above form for the input impedance of the series resonant T-line has the same form as that of the series LRC circuit. We can define equivalent elements

$$R_{eq} = Z_0\alpha\ell = Z_0\alpha\lambda/2 \quad (9.123)$$

$$L_{eq} = \frac{\pi Z_0}{2\omega_0} \quad (9.124)$$

$$C_{eq} = \frac{2}{Z_0\pi\omega_0} \quad (9.125)$$

The equivalent Q factor is given by

$$Q = \frac{1}{\omega_0 R_{eq} C_{eq}} = \frac{\pi}{\alpha\lambda_0} = \frac{\beta_0}{2\alpha} \quad (9.126)$$

For a low-loss line, this Q factor can be made very large. A good T-line might have a Q of 1000 or 10 000 or more. It's difficult to build a lumped circuit resonator with such a high Q factor.

Shorted quarter-wave line resonance

The shorted quarter-wave line will likewise behave like a parallel resonant circuit. For a short-circuited $\lambda/4$ line

$$Z_{in} = Z_0 \tanh(\alpha + j\beta)\ell = Z_0\frac{\tanh\alpha\ell + j\tan\beta\ell}{1 + j\tan\beta\ell\tanh\alpha\ell} \quad (9.127)$$

Multiply numerator and denominator by $-j\cot\beta\ell$

$$Z_{in} = Z_0\frac{1 - j\tanh\alpha\ell\cot\beta\ell}{\tanh\alpha\ell - j\cot\beta\ell} \quad (9.128)$$

For $\ell = \lambda/4$ at $\omega = \omega_0$ and $\omega = \omega_0 + \delta\omega$

$$\beta\ell = \frac{\omega_0\ell}{v} + \frac{\delta\omega\ell}{v} = \frac{\pi}{2} + \frac{\pi\delta\omega}{2\omega_0} \quad (9.129)$$

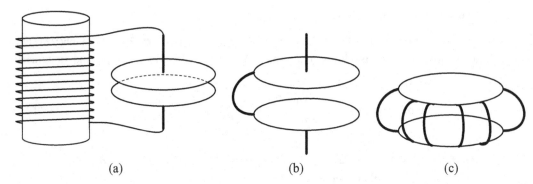

Figure 9.29 (a) A lumped LC resonator is formed by a parallel plate capacitor and a coil inductor. (b) A higher-frequency lumped LC resonator employing a single short wire as L. (c) Employing more short wires in parallel further reduces the overall inductance.

So $\cot \beta \ell = -\tan \frac{\pi \delta \omega}{2\omega_0} \approx \frac{-\pi \delta \omega}{2\omega_0}$ and $\tanh \alpha \ell \approx \alpha \ell$, which leads to

$$Z_{in} = Z_0 \frac{1 + j\alpha \ell \pi \delta \omega / 2\omega_0}{\alpha \ell + j\pi \delta \omega / 2\omega_0} \approx \frac{Z_0}{\alpha \ell + j\pi \delta \omega / 2\omega_0} \tag{9.130}$$

This has the same form for a parallel resonant RLC circuit

$$Z_{in} = \frac{1}{1/R + 2j\delta \omega C} \tag{9.131}$$

The equivalent circuit elements are

$$R_{eq} = \frac{Z_0}{\alpha \ell} \tag{9.132}$$

$$C_{eq} = \frac{\pi}{4\omega_0 Z_0} \tag{9.133}$$

$$L_{eq} = \frac{1}{\omega_0^2 C_{eq}} \tag{9.134}$$

The quality factor is thus

$$Q = \omega_0 RC = \frac{\pi}{4\alpha \ell} = \frac{\beta}{2\alpha} \tag{9.135}$$

the same as before.

Feynman's can

In Richard Feynman's class lecture series, he provokes a thought experiment that transforms a simple LC tank into a cylindrical resonator. The idea is as follows. Let's say we want to design a high-frequency resonator with an inductor and a capacitor. Let's start with a simple coil and capacitor, shown in Fig. 9.29a. How do we increase the resonant frequency? Simply decrease C and L as much as possible. We can make a small capacitor from two parallel plates by moving the plates far apart. Likewise, we can create a low inductance in shunt with the plates by simply connecting a straight wire from the top plate to the bottom plate. This arrangement is shown in Fig. 9.29b. There is a tradeoff between the distance between the

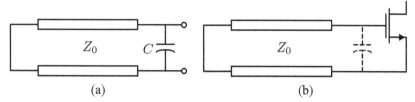

(a)   (b)

Figure 9.30 (a) A lumped/distributed resonant circuit. The capacitor is in resonance with the transmission line. (b) A common example is the use of a shorted transmission line stub to resonate the input capacitance of a transistor.

plates and the inductance, since moving the plates further apart will decrease C but increase L. But we can decrease L by putting multiple inductors in parallel, as shown in Fig. 9.29c. The best we can do, though, is to fill the entire surface between the top and bottom plate with wires, forming a cylinder. Thus we see that a "can" is a high-frequency resonator!

Lumped/distributed resonant networks

Often transmission lines are used as resonant elements along with lumped elements. A good example, shown in Fig. 9.30, is a short section of transmission line resonating with the input capacitance of a transistor. For simplicity assume that the lumped input capacitance is lossless. What is the Q factor of the resulting resonant circuit?

It's important to note that $Q \neq \frac{1}{2}\beta/\alpha$, since this only applies to the transmission line in resonance when the magnetic and electric energy are equal on the transmission line. In our case, we would like to use the transmission line as an inductor, so we will be concerned with the net magnetic energy on the line. The Q factor is therefore given by

$$Q = 2\omega_0 \frac{\text{net energy stored}}{\text{avg. power loss}} = \frac{2\omega_0(W_m - W_e)}{P_R + P_G} \qquad (9.136)$$

where W_m and W_e are the average magnetic and electric energy stored, and P_R represents the "series" resistive losses and P_G the "shunt" conductive losses. Defining the series inductive and shunt capacitive Q we have [14]

$$Q_L = 2\omega_0 \frac{W_m}{P_R} \qquad (9.137)$$

$$Q_C = 2\omega_0 \frac{W_e}{P_G} \qquad (9.138)$$

We can express the overall Q as

$$\frac{1}{Q} = \frac{1}{\eta_L Q_L} + \frac{1}{\eta_C Q_C} \qquad (9.139)$$

where

$$\eta_L = 1 - \frac{W_e}{W_m} \qquad (9.140)$$

and

$$\eta_C = \frac{W_m}{W_e} - 1 \tag{9.141}$$

For a shorted transmission line, under the assumption of low loss, we can show that

$$W_m \approx \frac{1}{2} \frac{L V^{+2} \ell}{Z_0^2} \left(1 + \mathrm{sinc}\left(\frac{4\pi \ell}{\lambda} \right) \right) \tag{9.142}$$

and

$$W_e \approx \frac{1}{2} C V^{+2} \ell \left(1 - \mathrm{sinc}\left(\frac{4\pi \ell}{\lambda} \right) \right) \tag{9.143}$$

Thus we have

$$\frac{1}{\eta_L} = \frac{1}{2 \, \mathrm{sinc}\left(\frac{4\pi \ell}{\lambda} \right)} + \frac{1}{2} \tag{9.144}$$

and

$$\frac{1}{\eta_C} = \frac{1}{2 \, \mathrm{sinc}\left(\frac{4\pi \ell}{\lambda} \right)} - \frac{1}{2} \tag{9.145}$$

For a shorted line, say $\ell \ll \lambda$, then $\eta_C \gg \eta_L$. For instance, if $\ell < 0.1\lambda$, then $\eta_C > 7\eta_L$. The net Q of such a resonant circuit is therefore $Q \approx \eta_L Q_L$.

9.10   The Smith Chart

The Smith Chart, shown in Fig. 9.31, is simply a graphical calculator for computing impedance as a function of reflection coefficient $z = f(\rho)$. More importantly, many problems can be easily visualized with the Smith Chart. This visualization leads to an insight about the behavior of transmission lines. All the knowledge is coherently and compactly represented by the Smith Chart. Why else study the Smith Chart? It's beautiful and extremely insightful! There are deep mathematical connections in the Smith Chart. The study of the complex bilinear transform sheds much insight into the geometry of the Smith Chart.

Smith Chart construction

Let's begin with the voltage on the line

$$v(z) = v^+(z) + v^-(z) = V^+(e^{-\gamma z} + \rho_L e^{\gamma z})$$

Recall that we can define the reflection coefficient anywhere by taking the ratio of the reflected wave to the forward wave

$$\rho(z) = \frac{v^-(z)}{v^+(z)} = \frac{\rho_L e^{\gamma z}}{e^{-\gamma z}} = \rho_L e^{2\gamma z} \tag{9.146}$$

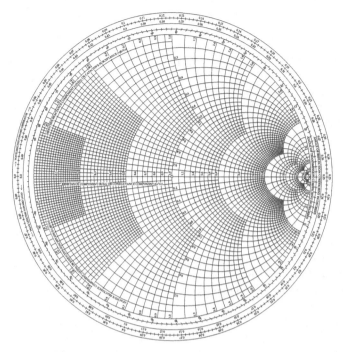

Figure 9.31 The Smith Chart.

Therefore the impedance on the line

$$Z(z) = \frac{v^+ e^{-\gamma z}(1 + \rho_L e^{2\gamma z})}{\frac{v^+}{Z_0} e^{-\gamma z}(1 - \rho_L e^{2\gamma z})} \tag{9.147}$$

can be expressed in terms of $\rho(z)$

$$Z(z) = Z_0 \frac{1 + \rho(z)}{1 - \rho(z)} \tag{9.148}$$

It is extremely fruitful to work with normalized impedance values $z = Z/Z_0$

$$z(z) = \frac{Z(z)}{Z_0} = \frac{1 + \rho(z)}{1 - \rho(z)} \tag{9.149}$$

Let the normalized impedance be written as $z = r + jx$ (note small case). The reflection coefficient is "normalized" by default since for passive loads $|\rho| \leq 1$. Let $\rho = u + jv$. Now simply equate the $\Re$ and $\Im$ components in the above equation

$$r + jx = \frac{(1 + u) + jv}{(1 - u) - jv} = \frac{(1 + u + jv)(1 - u + jv)}{(1 - u)^2 + v^2} \tag{9.150}$$

To obtain the relationship between the (r, x) plane and the (u, v) plane

$$r = \frac{1 - u^2 - v^2}{(1 - u)^2 + v^2} \tag{9.151}$$

$$x = \frac{v(1 - u) + v(1 + u)}{(1 - u)^2 + v^2} \tag{9.152}$$

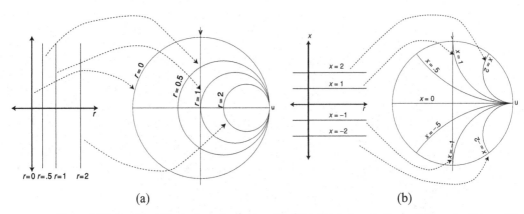

Figure 9.32 Graphical illustration of mappings from the (r, x) plane to the complex reflection coefficient unit circuit (u, v).

The above equations can be simplified and put into a nice form. If you remember your grade school algebra, you can derive the following equivalent equations

$$\left(u - \frac{r}{1+r}\right)^2 + v^2 = \frac{1}{(1+r)^2} \tag{9.153}$$

$$(u - 1)^2 + \left(v - \frac{1}{x}\right)^2 = \frac{1}{x^2} \tag{9.154}$$

These are circles in the (u, v) plane! Circles are good! We see that vertical and horizontal lines in the (r, x) plane (complex impedance plane) are transformed into circles in the (u, v) plane (complex reflection coefficient).

Some special mappings, shown in Fig. 9.32a,b, are worth noting

- $r = 0$ maps to $u^2 + v^2 = 1$ (unit circle)
- $r = 1$ maps to $(u - 1/2)^2 + v^2 = (1/2)^2$ (matched real part)
- $r = .5$ maps to $(u - 1/3)^2 + v^2 = (2/3)^2$ (load R less than Z_0)
- $x = \pm 1$ maps to $(u - 1)^2 + (v \mp 1)^2 = 1$
- $x = \pm 2$ maps to $(u - 1)^2 + (v \mp 1/2)^2 = (1/2)^2$
- $x = \pm 1/2$ maps to $(u - 1)^2 + (v \mp 2)^2 = 2^2$

Inductive reactance maps to the upper half of the unit circle. Capacitive reactance maps to the lower half of the unit circle. A simpler Smith Chart, shown in Fig. 9.33, can be constructed from the above mappings.

Load on Smith Chart

As shown in Fig. 9.34, simply plot z_L on the Smith Chart. We can read off ρ_L as a polar complex number. To read off the impedance on the T-line at any point on a lossless line, simply move on a circle of constant radius since $\rho(z) = \rho_L e^{2j\beta z}$. Moving towards the generator means $\rho(-\ell) = \rho_L e^{-2j\beta\ell}$, or a clockwise motion. For a lossy line, this corresponds

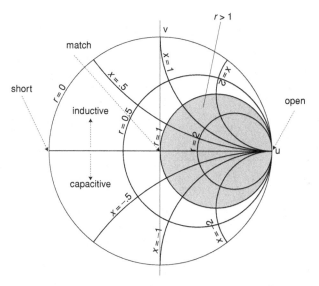

Figure 9.33 A simple Smith Chart illustrating some key points.

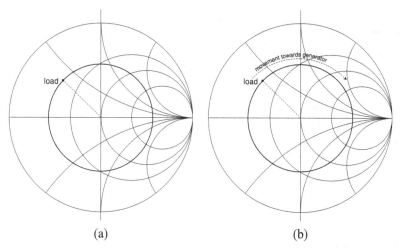

Figure 9.34 (a) A load impedance is plotted on the Smith Chart. (b) By moving clockwise on a circle of constant radius, we can read off the impedance at any point on the transmission line.

to a spiral motion. We're back to where we started when $2\beta\ell = 2\pi$, or $\ell = \lambda/2$. Thus the impedance is periodic (as we know).

Since SWR is a function of $|\rho|$, a circle at the origin in the (u, v) plane is called an SWR circle. Recall the voltage max occurs when the reflected wave is in phase with the forward wave, so $\rho(z_{min}) = |\rho_L|$. This corresponds to the intersection of the SWR circle with the positive real axis, shown in Fig. 9.35. Likewise, the intersection with the negative real axis is the location of the voltage min.

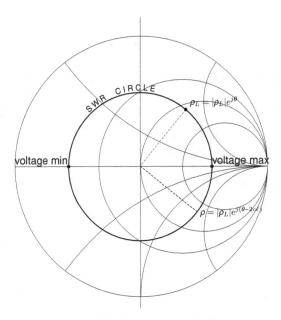

Figure 9.35 An SWR circle on the Smith Chart.

Example 18

This example will illustrate the utility of the Smith Chart for visualization. Let's prove that if Z_L has an inductance reactance, then the position of the first voltage maximum occurs before the voltage minimum as we move towards the generator. A visual proof is easy using the Smith Chart. On the Smith Chart start at any point in the upper half of the unit circle. Moving towards the generator corresponds to a clockwise motion on a circle. Therefore, we will always cross the positive real axis first and then the negative real axis.

The Admittance Chart

Since $y = 1/z = \frac{1-\rho}{1+\rho}$, you can imagine that an Admittance Smith Chart looks very similar. In fact everything is switched around a bit and you can buy or construct a combined Admittance/Impedance Smith Chart. You can also use an impedance chart for admittance if you simply map $x \to b$ and $r \to g$. Be careful as the capacitors are now on the top of the chart and the inductors on the bottom. The short and open likewise swap positions.

Sometimes you may need to work with both impedances and admittances. This is easy on the Smith Chart due to the impedance inversion property of a $\lambda/4$ line

$$Z' = \frac{Z_0^2}{Z} \tag{9.155}$$

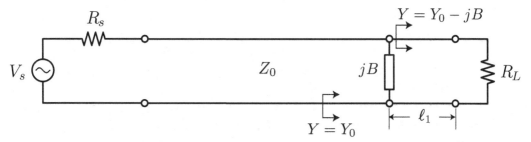

Figure 9.36 A section of transmission line and a shunt reactance can be used to match a load to a desired impedance.

If we normalize Z' we get y

$$\frac{Z'}{Z_0} = \frac{Z_0}{Z} = \frac{1}{z} = y \tag{9.156}$$

Thus if we simply rotate π degrees on the Smith Chart and read off the impedance, we are actually reading off the admittance! Rotating π degrees is easy. Simply draw a line through the origin and z_L and read off the second point of intersection on the SWR circle.

9.11 Transmission line-matching networks

Matching with lumped elements

Recall the input impedance looking into a T-line varies periodically

$$Z_{in}(-\ell) = Z_0 \frac{Z_L + jZ_0 \tan(\beta\ell)}{Z_0 + jZ_L \tan(\beta\ell)} \tag{9.157}$$

As shown in Fig. 9.36, move a distance ℓ_1 away from the load such that the real part of Z_{in} has the desired value. Then place a shunt or series impedance on the T-line to obtain the desired reactive part of the input impedance (e.g. zero reactance for a real match). For instance, for a shunt match, the input admittance looking into the line is

$$y(z) = Y(z)/Y_0 = \frac{1 - \rho_L e^{j2\beta z}}{1 + \rho_L e^{j2\beta z}} \tag{9.158}$$

At a distance ℓ_1, we desire the normalized admittance to be $y_1 = 1 - jb$. Substitute $\rho_L = \rho e^{j\theta}$ and solve for ℓ_1 and let $\psi = 2\beta z + \theta$

$$\frac{1 - \rho e^{j\psi}}{1 + \rho e^{j\psi}} = \frac{1 - \rho^2 - j2\rho \sin\psi}{1 + 2\rho \cos\psi + \rho^2} \tag{9.159}$$

Solve for ψ (and then ℓ_1) from $\Re(y) = 1$

$$\psi = \theta - 2\beta\ell = \cos^{-1}(-\rho) \tag{9.160}$$

$$\ell_1 = \frac{\theta - \psi}{2\beta} = \frac{\lambda}{4\pi}(\theta - \cos^{-1}(-\rho)) \tag{9.161}$$

At ℓ_1, the imaginary part of the input admittance is

$$b = \Im(y_1) = \pm\frac{2\rho}{\sqrt{1 - \rho^2}} \tag{9.162}$$

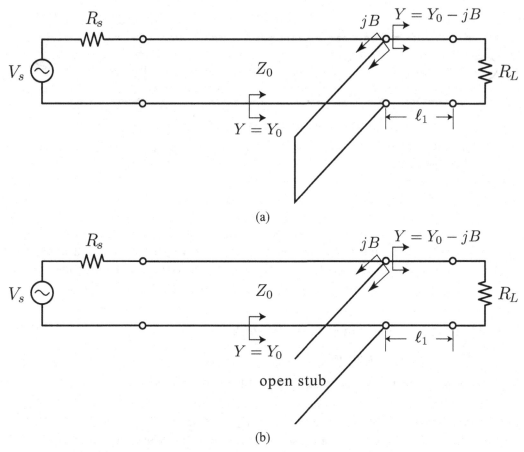

Figure 9.37 A (a) short or open (b) shunt stub can be used in place of the lumped element to obtain the desired imaginary component.

Placing a reactance of value $-b$ in shunt provides impedance match at this particular frequency. If the location of ℓ_1 is not convenient, we can achieve the same result by moving back a multiple of $\lambda/2$.

Matching with T-line stubs

At high frequencies the matching technique discussed above is difficult due to the lack of lumped passive elements (inductors and capacitors). But short/open pieces of transmission lines simulate fixed reactance over a narrow band. A shorted stub with $\ell < \lambda/4$ looks like an inductor. An open stub with $\ell < \lambda/4$ looks like a capacitor. The procedure is identical to the case with lumped elements, but instead of using a capacitor or inductor we use shorted or open transmission lines. For most transmission line structures, shunt stubs, shown in Fig. 9.37, are easier to fabricate than series stubs. But in theory, either shunt or series stubs can be used to obtain the match.

Matching with the aid of the Smith Chart

Single stub impedance matching is easy to do with the Smith Chart. Simply find the intersection of the SWR circle with the $r = 1$ circle. The match is at the center of the circle. Grab a reactance in series or shunt to move you there. To do a shunt stub, though, we need to use the admittance chart. To solve the same matching problem with a shunt stub, find the shunt stub value, simply convert the value of $z = 1 + jx$ to $y = 1 + jb$ and place a reactance of $-jb$ in shunt.

Example 19

Consider matching a load impedance $Z_L = 150 - j80$ to Z_0 (usually 50 Ω at a frequency of 1 GHz).

The normalized impedance $z_L = Z_L/Z_0$ is plotted in Fig. 9.38 and labeled point 1. Since we are connecting a shunt stub, we convert to $y_l = 1/z_L$ by drawing a constant

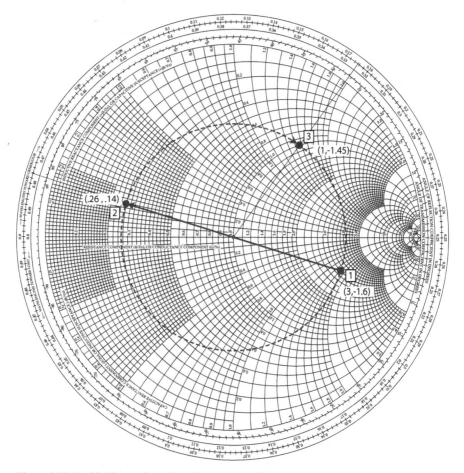

Figure 9.38 Smith Chart calculations for stub-matching example.

VSWR circle and project through the origin. This point labeled 2 corresponds to an admittance of $0.26 + j0.14$, which we can read directly from the chart. Now, interpreting the chart as an admittance chart, we move to the $g = 1$ circle along the same constant VSWR circle and arrive at point 3. Reading off the distance traveled on the Smith Chart, we have $\ell = \lambda(0.175 - 0.024) = \lambda 0.151$. This distance corresponds to an electrical length of about $55°$.

At point 3, the normalized admittance is about $1 + 1.45j$. Now, all we need to do is connect a shunt stub with susceptance $-1.45j$ to complete the design. This can be realized as an open or short circuit stub. Since the desired susceptance is negative, the shorter stub is a short circuited "inductive" line of appropriate length. The length of the stub can be calculated directly from

$$y_{stub} = -j \cot \beta \ell$$

or read from the Smith Chart. From the Smith Chart, we start at the short circuit point and travel along the $r = 0$ circle and note that $x = 1/1.45 = .69$ requires a stub length of about 0.096λ, or about $35°$. This completes the design of the matching network.

Another great application of the Smith Chart is matching with lumped elements. In Chapter 7, we described lumped matching networks configurations such as the L, Π, and T matching circuits. Instead of computing element values explicitly, we can use the Smith Chart as a graphical calculator to design matching networks.

Example 20

Consider again matching a load impedance $Z_L = 150 - j80$ to Z_0 (usually 50 Ω at a frequency of 1 GHz using lumped elements). From our previous discussion, we know that an L-matching network connected in shunt with the output is needed in order to step down from $|Z_L| > Z_0$.

The normalized impedance $z_L = Z_L/Z_0$ is plotted in Fig. 9.39 and labeled point 1. Since we are connecting a reactance in shunt, we convert to $y = 1/z_L$ by drawing a constant SWR circle and project through the origin. This point labeled 2 corresponds to an admittance of $0.26 + j0.14$, which we can read directly from the chart. Next we draw the admittance circle $1 + jb$ on the chart. This is a reflection of the $1 + jx$ circle. To move on to this circle, we observe that adding a susceptance of about $0.3j$ moves us to a point labeled 3 on the chart. This corresponds to adding a shunt capacitor of value

$$B_C = 0.3/50$$
$$C = \frac{B_C}{\omega} = 955 \text{ fF}$$

Next we convert back to impedance by once again projecting through the origin to arrive at point 4, which corresponds to the normalized impedance of $1 - 1.65j$. It's important to note that this is a consistency check. If we had not ended up on a point

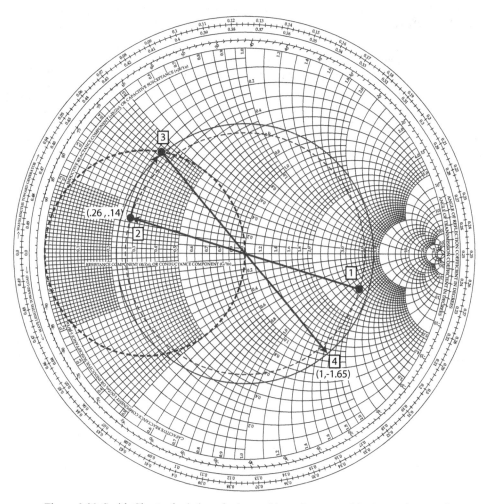

Figure 9.39 Smith Chart calculations for lumped impedance matching network example.

$1 + jx$, then there would be an error in our analysis. From here forward it's easy to see that a series reactance of value $+1.65$ will move us to the origin. This corresponds to a series inductor of value

$$X_L = 50 \times 1.65$$
$$L = \frac{X_L}{\omega} = 13 \, \text{nH}$$

This completes the design of the matching network. It's interesting to note that we could have traveled to the origin by subtracting reactance from point 2. This would have resulted in another perfectly valid solution to the problem. Generally we would check both solutions and choose the network that best meets our needs (such as bandwidth).

9.12   References

I have taught this material as part of an undergraduate electromagnetics course and as part of a graduate microwave circuits book, and much of the material is adapted from my lecture notes. In preparing my notes, I have relied on several sources, including Collin's great books [10] [9], Pozar's *Microwave Engineering* book [47], and undergraduate electromagnetics books such as Cheng [6] and by Inan and Inan [62].

10 Transformers

Transformers find wide and important applications in RF circuits. Historically, transformers were used in power systems for voltage step-up and step-down. Since power transmission is more efficient at high voltages, transformers are used to boost signals to tens of thousands of volts for long-range transmission. For safety, though, we prefer to work with much lower voltage levels, and thus a transformer is used to step down the voltage to hundreds of volts before delivery into homes and factories. Electronic components, though, even operate at lower voltages. Before the widespread use of switching power supplies, transformers were ubiquitous in performing this task. For ultra low-noise applications, such as sensitive measurements, transformer-based designs still reign supreme as the method of choice.

The name "transformer," in fact, derives from this function. As such, it's an ideal element for transforming voltages. Consequently, due to conservation of energy, transformers are equally good at stepping down/up currents. Thus, transformers are doubly good at impedance transformation. A desirable quality of transformers is their broadband operation. Whereas passive LC circuits can easily double as impedance transformation circuits, they do so only over a narrow bandwidth.

An ideal transformer is broadband, faithfully duplicating, inverting, and scaling voltages and currents independent of frequency with one important exception. Since its behavior derives from magnetic induction, it cannot perform it's function for static or DC signals. This simple fact is probably responsible for the worldwide deployment of an AC electrical network. Early proponents of AC power successfully argued that high voltage AC power delivery was more efficient and flexible due to the existence of the transformer. If an efficient DC transformer existed at the dawn of power, we would all no doubt be much less accustomed to AC voltages.

10.1 Ideal transformers

An ideal two-winding transformer is shown in Fig. 10.1a. Each winding defines a port, or a terminal pair with voltage $V_{1,2}$ and current $I_{1,2}$. An ideal transformer is characterized by the turns ratio N, which defines the voltage step up/down from one port to another

$$V_2 = NV_1 \tag{10.1}$$

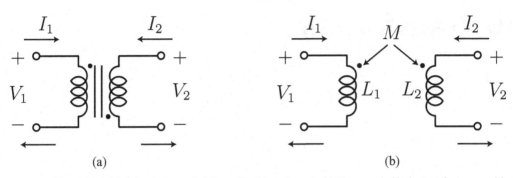

Figure 10.1 (a) Schematic symbol for an ideal transformer. (b) Two coupled inductors form a non-ideal transformer.

As already noted, this relationship is valid for any AC frequency $\omega \neq 0$. Since the transformer is a passive element, it conserves power and thus

$$V_1 I_1 + V_2 I_2 = 0 \tag{10.2}$$

which implies

$$I_2 = -\frac{1}{N} I_1 \tag{10.3}$$

The impedance transformation property is therefore a function of N^2

$$Z_1 = \frac{V_1}{I_1} = \frac{\frac{1}{N} V_2}{-N I_2} = \left(\frac{1}{N^2}\right) \frac{V_2}{-I_2} = \frac{Z_2}{N^2} \tag{10.4}$$

Since the ports of an ideal transformer are isolated, we can easily redefine a port of opposite polarity $V_2' = -V_2$ and thus perform voltage inversion with no additional effort.

10.2 Dot convention

A common point of confusion with transformers is the dot convention. As shown in Fig. 10.1, it's standard practice to place dots on one side of the terminals of transformers or coupled inductors. The dot convention can be stated in the following way [12]: If current is sent *into* the dotted terminals of the coupled inductors, the magnetic fluxes linking the coils will reinforce each other. With this convention, the mutual inductance M is always positive.

When confusion over notation arises, it's often fruitful to go back to the fundamental physics. In general, when a time-varying magnetic field impinges on a conductor, the induced currents will always flow in a direction so as to oppose the magnetic field. This is generally valid, as the current generates its own magnetic field to oppose the incoming field so as to lower the overall energy of the system. This is known as Lenz's Law. The same is true for electric fields. When an electric field impinges on a conductor, the charges redistribute so as to cancel the incoming field. This is why the magnetic and electric fields in a perfect conductor are driven to zero.

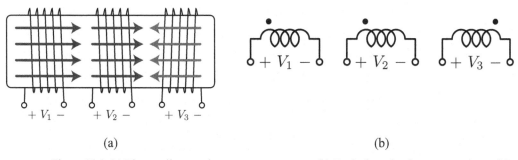

(a) (b)

Figure 10.2 (a) Three coils wound on to a common core. (b) Equivalent circuit representations of the coils using the dot convention.

An example will illustrate the dot convention more clearly. Three coils are wound on to a common core as shown in Fig. 10.2a. The terminals have been defined as positive, as indicated in the figure. Since coils 1 and 2 have a similar clockwise rotation, their magnetic fields will always add (when current flows into the positive terminals). So they should have dots placed on the same terminal. We place the dots at the positive terminal as shown in Fig.10.2b. Equivalently we could have chosen both of the negative terminals. The third coil, though, is wound counterclockwise. Thus, the magnetic field of coil 3 with positive current will oppose the fields of the first two coils. As such, we must place the dot at the negative terminal to indicate that a negative current in the third coil is necessary in order to reinforce the fields of coils 1 and 2.

10.3 Coupled inductors as transformers

Ideal transformers are an abstraction as all real transformers deviate from ideal behavior in one form or another. As shown in Fig. 10.1b, real transformers are implemented as coupled inductors. As we have learned already, if two inductors are coupled magnetically, then

$$V_1 = j\omega L_1 I_1 + j\omega M I_2 \tag{10.5}$$
$$V_2 = j\omega M I_1 + j\omega L_2 I_2 \tag{10.6}$$

Solving for I_2 from the second equation

$$I_2 = \frac{1}{j\omega L_2}(V_2 - j\omega M I_1) \tag{10.7}$$

and substituting for I_2 from the second equation into the first

$$V_1 = j\omega L_1 I_1 + \frac{j\omega M}{j\omega L_2}(V_2 - j\omega M I_1) \tag{10.8}$$

and collecting terms we have

$$V_1 = j\omega \left(L_1 - \frac{M^2}{L_2}\right) I_1 + \frac{M}{L_2} V_2 \tag{10.9}$$

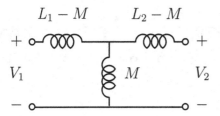

Figure 10.3 T model equivalent circuit for two coupled inductors.

It is interesting to observe that the transformer behavior is contained in the second term, which is frequency independent. The first term, on the other hand, is undesirable since it represents an inductive component. In fact, Eq. (10.8) explicitly emphasizes that an alternating voltage at port 2 generates a voltage through port 1 by first generating an alternating current in the second inductor, which in turn generates a time-varying magnetic field, in turn inducing a solenoidal electric field, which impresses a voltage at port 1.

The mutual inductance is usually parameterized in terms of the magnetic coupling coefficient $M = k\sqrt{L_1 L_2}$. Substituting this into the above equation we arrive at

$$V_1 = j\omega L_1(1 - k^2)I_1 + k\sqrt{\frac{L_1}{L_2}}V_2 \tag{10.10}$$

The transformer turns ratio is defined by coupling factor k and by the square root of the ratio of the inductances of each winding

$$\frac{1}{N} = k\sqrt{\frac{L_1}{L_2}} \tag{10.11}$$

The above equation is very important as it reveals that an ideal transformer can be constructed by two tightly wound coils such that $k \to 1$.

10.4 Coupled inductor equivalent circuits

It is sometimes more convenient to view two coupled inductors as a T network, as shown in Fig. 10.3. It is easy to show that this model is identical to two coupled inductors. To see this, consider the Z matrix of two coupled inductors

$$\mathbf{v} = \mathbf{Zi} = j\omega \begin{pmatrix} L_1 & M \\ M & L_2 \end{pmatrix} \mathbf{i} \tag{10.12}$$

Now consider the coupled inductors as a black box. Since a two-port is completely characterized by its Z matrix, any other circuit that has the same two-port behavior, i.e. Z matrix, must be identical and indistinguishable from the external world. The voltage at port 1 of the T network is

$$V_1 = j\omega(L_1 - M)I_1 + j\omega M(I_1 + I_2) = j\omega L_1 I_1 + j\omega M I_2 \tag{10.13}$$

and so it's clear that the circuits are equivalent.

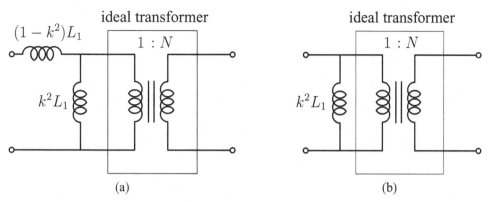

Figure 10.4 (a) Equivalent circuit model for two coupled inductors. (b) Approximation of (a) valid for two tightly coupled inductors ($k \to 1$).

When two inductors are tightly coupled ($k \approx 1$), then it's sometimes useful to view the coupled inductors as a perfect transformer marred by parasitics. In fact, it's not difficult to show that the equivalent circuit shown in Fig. 10.4a is such an exact representation. At port 1 the current into the shunt inductor $k^2 L_1$ is the sum of the current into port 1 and N times the transformer secondary current

$$V_1 = j\omega(1 - k^2)L_1 I_1 + j\omega k^2 L_1 I_1 + j\omega k^2 N L_1 I_2 \tag{10.14}$$

and recalling the definition of the magnetic coupling coefficient k, we have $k^2 N L_1 = M$. Similarly, the voltage at port two is just N times the voltage V_p at the transformer primary

$$V_2 = N V_p = j\omega k^2 N L_1 (I_1 + N I_2) \tag{10.15}$$

Simplifying the above and noting that $N^2 k^2 L_1 = L_2$ we have the desired result.

In the limit that $k \to 1$, we arrive at the circuit shown in Fig. 10.4b. It's interesting to note that even if $k = 1$, all real transformers have *leakage inductance*.

The leakage inductance reminds us that a real transformer cannot operate at DC and that it always stores magnetic energy. At DC this shunt inductance prevents any current from flowing into the ideal transformer. The leakage inductance $L_{leak} = k^2 L_1$ will form a high-pass network with the source resistance driving port 1 with cutoff frequency R_s/L_{leak}. A decade or so past the cutoff frequency, the leakage inductance has little effect on the performance of the transformer. For a sinusoidal excitation, the magnetic energy stored in this inductor, though, is proportional to the current through the inductor

$$E_L = \frac{1}{2} L_1 I_1^2 \approx \frac{1}{2} L_1 \left(\frac{V_1}{\omega L_1}\right)^2 \tag{10.16}$$

so the reactive power stored in the inductor is inversely proportional to frequency

$$P_L = \omega E_L \approx \frac{1}{2} \frac{V_1^2}{\omega L} \tag{10.17}$$

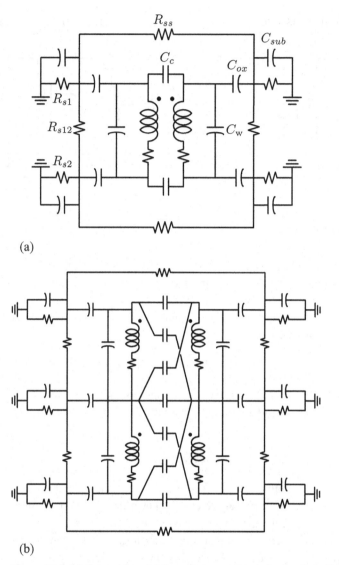

(a)

(b)

Figure 10.5 (a) Compact model for an integrated transformer including the parasitic electrical coupling. (b) A more distributed model for wideband applications.

In narrow band applications, we may tune out this inductance with a shunt capacitor in order to avoid delivering reactive power to the transformer.

At a given frequency $\omega \neq 0$, we may reduce the effects of the leakage inductance by winding more turns into the coils. Eventually, we are limited by higher-order parasitics, such as winding capacitance, that will limit the frequency response of the transformer in the frequency range of interest.

A more complete model, more useful for computer simulation, is shown in Fig. 10.5a. Here the magnetically coupled inductors are at the core of a more complicated model, which includes the oxide capacitance to the substrate, C_{ox}, substrate resistors, R_{s1} and R_{s2},

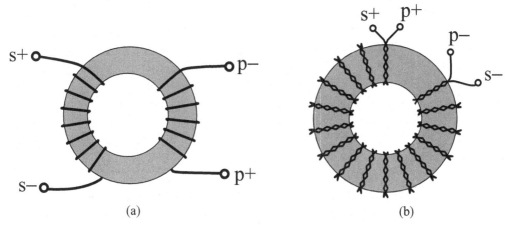

Figure 10.6 (a) A transformer wound on a toroid magnetic core. (b) A 1:1 high-frequency transformer with primary and secondary windings twisted together.

and winding capacitance C_w. All of these parasitic effects were also seen in the inductor compact model. The model includes additional coupling mechanisms, such as the inter-winding capacitance C_c, and the coil-to-coil coupling resistance R_{ss}. To capture an even larger frequency range, the model can be broken up into multiple sections in order to model the distributed effects. Such a two-section model is shown in Fig. 10.5b. Such a model can fit measured or 3D EM simulation results over a wide bandwidth. This is important because transformers are often employed for their wideband matching capability, or to add signals at disparate frequencies (say the LO and RF signals), and thus a single compact model should cover a wide frequency range.

10.5 Transformer design and layout

At low frequencies, inductor coils wound on a common magnetic core produce good trans-formers. The core confines and magnifies the magnetic fields producing large coupling fac-tors k. A typical transformer using a toroid as a core is shown in Fig. 10.6a. Unequal turns ratios are simply produced by winding more turns on side. Another technique prominent in high-frequency applications, shown in Fig. 10.6b, first twists the primary and secondary wires together, which are then wound around the core to produce a 1:1 transformer. To create other turns ratios using the same technique, the primary or secondary can be con-nected in parallel multiple times. These bifilar structures are generalized to build multifilar transformers having three or more ports.

When confined to planar structures, turns wind inward to produce spirals. Some common transformer layouts are shown in Figs. 10.7–10.10. Two spirals can be wound side by side to form a bifilar transformer as shown in Fig. 10.7a. To create non-unity turns ratio, turns can be avoided in one of the windings. A variation of this bifilar transformer is shown Fig. 10.7b where the primary and secondary windings are wound in a symmetric fashion. This structure preserves symmetry between the primary and secondary windings but each

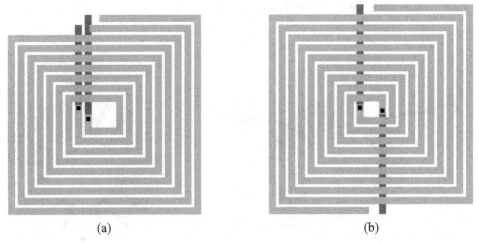

Figure 10.7 (a) Planar bifilar transformer layout. (b) Symmetric bifilar transformer layout.

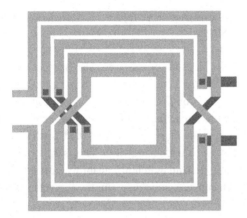

Figure 10.8 Fully symmetric transformer layout. Transformer windings can be tapped at center.

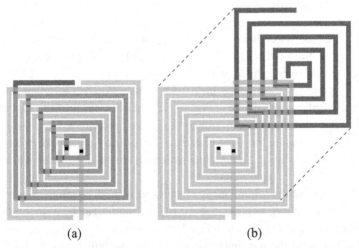

Figure 10.9 (a) Multi-layer transformer layout. The primary is a series-connected structure on two layers. (b) Bottom metal layer is displaced to show structure more clearly.

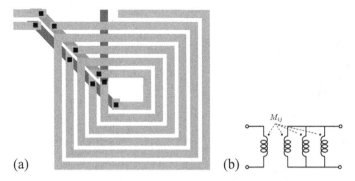

(a)   (b)

Figure 10.10 Planar spiral transformer with non-unity turns ratio. The secondary turns are connected in shunt with bridge connections.

winding itself is not symmetric. In other words, if a given port is excited by a single-ended signal by grounding the inside port first versus driving it from the inside and grounding the outside port, the input impedance would differ in each case. This is due to the capacitive asymmetries, due the bridge connection and the fringing capacitance (the outer turn sees more fringing capacitance due to the open nature of its surroundings). At higher frequencies, due to current constriction, the inner turns also experience higher resistive losses due to diminishing effective conductor area.

A fully symmetric layout as shown in Fig. 10.8 solves these problems by proper inter-connection of concentric rectangles or circles. Due to the bridge connections, the layout is not perfectly symmetric but the geometric center of each winding is nearly the electrical center of the structure and can be tapped to produce more interesting layout structures.

In a modern IC process, multiple metal layers can be utilized to create even more com-plicated structures. For instance, as shown in Fig. 10.9, one winding can be made of two inductors in series, whereas another winding can be constructed with an ordinary spiral or multiple spirals in parallel. Thus the inductance of the primary goes up (faster than linearly due to magnetic coupling k), whereas the inductance of the secondary windings remains roughly constant (again due to non-zero magnetic coupling k). This can be exploited to create compact transformers of high effective turns ratio.

To create $N : 1$ turns ratios, many different approaches are possible. As already alluded to, the most obvious approach is to simply leave out inner turns in one winding. Another approach that utilizes area more effectively is to include the inner turns but employs shunt bridge connections to place turns in parallel as opposed to in series. This layout technique is shown in Fig. 10.10a and schematically in Fig. 10.10b.

10.6 Baluns

Since transformers can invert signals for free, a common application is to convert signals from differential to single-ended form. Differential signals are immune to common-mode noise and even-order distortion provided that the process of conversion from differential to single-ended form is done with amplitude and phase balance.

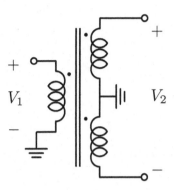

Figure 10.11 A balun converts single-ended signal to a differential or *balanced* signal. The reverse is also easily accomplished.

The balun shown in Fig. 10.11 converts *balanced* signals to *unbalanced* form by employing two coils wound on a common core. The primary coil is grounded and driven by a single-ended signal. The secondary coil, though, is grounded (or tapped) at the center. The voltage pickup on the top half of the secondary coil is therefore in phase with the primary, and the bottom half is 180° out of phase.

The layout of Fig. 10.12a is a common way to layout a balun in an integrated circuit environment, imposing planar geometry. Since it is desirable for ports 2 and 3 to be symmetric, a center tapped spiral is split in half to form two secondary inductors and a third inductor is inter-wound to form the primary. An alternative layout is shown in Fig. 10.12b. Here two metal layers are employed, one layer forming the primary side and a symmetric transformer forming the secondary coil. Laying out a balun with three metal layers, though, would not be symmetric with respect to the primaries due to capacitive and magnetic coupling to the substrate.

10.7 Hybrid transformer

Consider the ideal transformer shown in Fig. 10.13. The turns ratio from primary to secondary is $2N : 1$, but the primary is tapped at the center to produce two transformers of turns ratio $N : 1$. In any practical implementation, the primary windings would also couple to each other, but for now we assume ideal behavior so that $V_1 = V_2 = NV_3$.

Writing loop equations for the transformer, we arrive at [58]

$$V_{s1} - V_{s4} = I_1(Z_1 + Z_4) - I_2 Z_4 + NV_3 \tag{10.18}$$
$$-V_{s2} + V_{s4} = -I_1 Z_4 + I_2(Z_2 + Z_4) + NV_3 \tag{10.19}$$
$$V_{s3} = I_3 Z_3 + V_3 \tag{10.20}$$

Since the transformer is ideal, no power is dissipated in the windings so that

$$V_1 I_1 + I_2 V_2 + I_3 V_3 = 0 \tag{10.21}$$

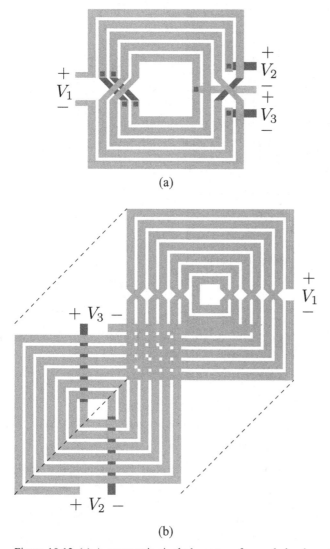

(a)

(b)

Figure 10.12 (a) A symmetric single-layer transformer balun layout with common center tap. (b) A two-layer balun transformer layout. The primary and secondaries are offset to show geometry more clearly.

substituting for $V_2 = V_1 = NV_3$

$$NV_3(I_1 + I_2) = -I_3V_3 \tag{10.22}$$

or more simply

$$I_3 = -N(I_1 + I_2) \tag{10.23}$$

Equations (10.18), (10.19), and (10.23) can be put into matrix form involving the transformer currents and source voltages. The voltage at the secondary winding can be eliminated by

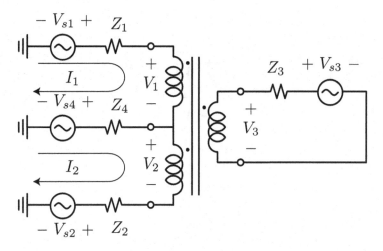

Figure 10.13 A general three-winding transformer.

(10.20).

$$\begin{pmatrix} V_{s1} - V_{s4} - N V_{s3} \\ -V_{s2} + V_{s4} - N V_{s3} \\ 0 \end{pmatrix} = A \begin{pmatrix} I_1 \\ I_2 \\ I_3 \end{pmatrix} \tag{10.24}$$

where

$$A = \begin{pmatrix} Z_1 + Z_4 & -Z_4 & -N Z_3 \\ -Z_4 & Z_2 + Z_4 & -N Z_3 \\ N & N & 1 \end{pmatrix} \tag{10.25}$$

In many applications, it is desirable to isolate certain ports. For instance, in a diode or single transistor mixer, we desire to sum the LO and RF signals to drive a non-linear diode or transistor to perform the mixing function. But if the LO port also couples to the RF port, undesirable LO leakage and radiation will occur. Thus, ideally, a passive device capable of power summation and isolation is desired.

In the three-winding transformer, we can isolate ports 3 and 4 by proper selection of impedances. Without doing the math, it should be fairly obvious that this condition is satisfied if $Z_1 = Z_2$. This occurs since the center tap is a virtual ground and thus isolated. It is not as obvious, though, that ports 1 and 2 can be similarly isolated. Consider the response of I_2 to the source voltage V_{s1}. Using Cramer's rule, we find

$$I_2 = \frac{-V_{s1}(N^2 Z_3 - Z_4)}{|A|} \tag{10.26}$$

The numerator is zero if

$$\frac{Z_4}{Z_3} = N^2 \tag{10.27}$$

By symmetry, of course, port 1 is similarly isolated from port 2.

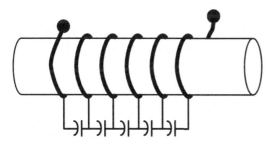

Figure 10.14 The distributed turn-to-turn capacitance in a coil.

Finally, the three-winding transformer can also be designed to provide an impedance match at each port as well as isolation. This occurs when

$$Z_1 = Z_2 = Z \tag{10.28}$$

and

$$Z_3 = \frac{Z_4}{N^2} = \frac{Z^*}{2N^2} \tag{10.29}$$

when these conditions are satisfied, the three-winding structure is commonly known as a hybrid transformer. We discuss hybrid *four-port* devices in more detail in Chapter 15.

10.8 Transformer parasitics

At high frequencies many problems begin to plague transformers. Recall that parasitic inter-turn electric coupling occurs in inductors and ultimately limits the performance or the resonant frequency of inductors. Even air wound inductors have turn-to-turn coupling, as shown in Fig. 10.14, limiting the high-frequency performance. For inductors, though, this problem is mild since these inter-turn capacitors are in effect shorted by the turn-to-turn inductance. As long as the turn-to-turn inductance impedance is small compared with the capacitive impedance, the structure will behave as an inductor.

But for multifilar transformers, the turn-to-turn coupling is not shorted by an inductor since one turn is on the primary and one is on the secondary. Thus the capacitive coupling directly competes with the magnetic coupling. In fact, if we model the system as a distributed circuit, the transformer is essentially a non-uniform transmission line. This idea is exploited by turning the transformer 90° to form a transmission line transformer. See Section 11.2 for details. The Z parameters of the distributed transformer deviate from ideal behavior by creating a real component in Z_{21}. At low frequencies, the coupling is close to the ideal value of $j\omega M$ but at frequencies approaching the resonant frequency of the structure, a large deviation is observed.

10.9 Transformer figures of merit

Transformers are very convenient elements for a host of applications. Their downside is their bulky size and frequency limitations. For on-chip applications, the insertion loss of a transformer is an important metric. Since the insertion loss of a transformer depends highly on the source and load impedance, it's convenient to find the minimum insertion loss under ideal conditions, a bi-conjugate source and load match. Under such conditions, we know that the gain of the transformer peaks at

$$G_{max} = \frac{y_{21}}{y_{12}}(K_s - \sqrt{K_s^2 - 1}) = K_s - \sqrt{K_s^2 - 1} \tag{10.30}$$

since $y_{21} = y_{12}$ for a passive device. The gain is thus related to the stability factor K_s, which is by definition $K_s \geq 1$ for a passive device. When $K_s = 1$, the insertion loss is 0 dB. The K_s factor is easily computed from the Z parameters of the transformer

$$Z = \begin{pmatrix} R_p + j\omega L_p & j\omega M \\ j\omega M & R_s + j\omega L_s \end{pmatrix} \tag{10.31}$$

where

$$K_s = \frac{2\Re(z_{22})\Re(z_{11}) - \Re(z_{21}z_{12})}{|z_{21}z_{12}|} \tag{10.32}$$

One particularly simple case of interest is a 1 : 1 transformer, where each winding has a resistance R_x

$$K_s = \frac{2R_x^2 + \omega^2 M^2}{\omega^2 M^2} = \frac{2R_x^2 + \omega^2 k^2 L^2}{\omega^2 k^2 L^2} \tag{10.33}$$

The maximum gain can be parameterized by the unitless $Q = \omega L/R_x$ factor and the transformer coupling factor $K = M/L_x$

$$G_{max}(Q, k) = 1 + \frac{2}{Q^2 k^2} - 2\sqrt{\frac{1}{Q^4 k^4} + \frac{1}{Q^2 k^2}} \tag{10.34}$$

A plot of G_{max} is shown in Fig. 10.15, which shows that reasonably low insertion loss can be obtained with relatively modest Q factor and coupling factor k. For instance, a $Q = 10 - 15$ is easily achieved in the $1 - 5$ GHz frequency range. On-chip planar transformers have $k \approx 0.75$, which implies an insertion loss of about 0.8, or about 1 dB insertion loss. While the above result is derived for a simple 1 : 1 transformer, the result can be generalized if we simply notice that N windings driven in shunt on the primary and driven in series on the secondary will transform the impedance by a factor of N, as shown in Fig. 10.16. Furthermore, the insertion loss is independent of the number of sections. This is a powerful result because for LC-matching networks we found a strong dependence on the matching ratio (Eq. (7.113)). This was also true for transmission line quarter wave-matching sections (Eq. (9.116)).

Another important figure of merit for the transformer is the maximum frequency that one can safely call a transformer by its name, before it resonates and degenerates into another structure. Consider a pair of coupled inductors forming a transformer with equal capacitive primary and secondary loads, as shown in Fig. 10.17. Let's calculate the Z matrix of this structure directly by solving the three mesh currents $i_1 - i_3$ for a source current i_s at port 1

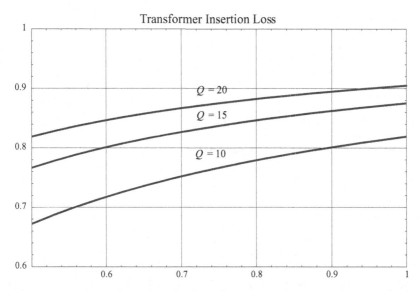

Figure 10.15 The minimum insertion loss of a 1 : 1 transformer as a function of magnetic coupling factor K and winding quality factor Q.

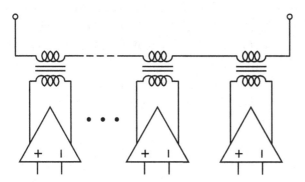

Figure 10.16 Transformer power combiner employing core 1 : 1 sections. For N stages the impedance is transformed down by N for each driver.

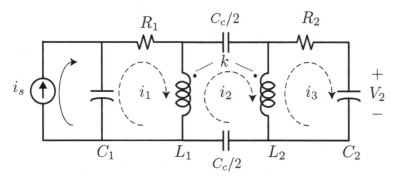

Figure 10.17 Lumped equivalent circuit model for a transformer including the winding capacitance and the coupling capacitance C_c.

and an open-circuited port 2. The mesh equations are given by

$$0 = (i_1 - i_s)\frac{1}{sC_1} + i_1 R_1 + (i_1 - i_2)s L_1 + (i_2 - i_3)s M \tag{10.35}$$

$$0 = (i_2 - i_1)s L_1 - (i_2 - i_3)s M + i_2 \frac{1}{sC_c} + (i_2 - i_3)s L_2 + (i_1 - i_2)s M \tag{10.36}$$

$$0 = (i_3 - i_2)s L_2 - (i_1 - i_2)s M + i_3 R_2 + i_3 \frac{1}{sC_2} \tag{10.37}$$

To find Z_{11}, we calculate the voltage at port 1

$$Z_{11} = \left.\frac{V_1}{I_1}\right|_{I_2=0} = \frac{i_s - i_1}{sC_1 i_s} \tag{10.38}$$

Likewise, we compute Z_{21}

$$Z_{21} = \left.\frac{V_2}{I_1}\right|_{I_2=0} = \frac{i_3}{sC_2 i_s} \tag{10.39}$$

Since the structure is symmetric, Z_{2k} can be obtained by simply interchanging subscripts. The resulting solutions, while straightforward to derive, are too long to derive by hand. Instead, a symbolic analysis program is used.

To build up to the complexity of the full solution, let's start with a simple case with $C_c = 0$, or no capacitive coupling. In particular, the fully symmetric case, with $L_1 = L_2 = L$, $C_1 = C_2 = C$, and $R_1 = R_2 = R$, is easy to solve (only two equations) and particularly insightful. In this case we have

$$Z_{11} = \frac{C(L - M)(L + M)s^3 + 2CLRs^2 + CR^2s + Ls + R}{(Cs(R + (L - M)s) + 1)(Cs(R + (L + M)s) + 1)} \tag{10.40}$$

and

$$Z_{21} = \frac{sM}{(Cs(R + (L - M)s) + 1)(Cs(R + (L + M)s) + 1)} \tag{10.41}$$

A plot of the magnitude of Z_{11} is shown in Fig. 10.18a. Notice that there two resonant frequencies are evident. A plot of the phase of Z_{21}, which is $90°$ for an ideal transformer, is shown in Fig. 10.18b, showing a large deviation from ideal behavior near the first resonant frequency.

The poles of the system are given by roots of the denominator of the common term $(Cs(R + (L - M)s) + 1)(Cs(R + (L + M)s) + 1)$. For $R = 0$, we see immediately that these frequencies are given by

$$\omega^+ = \frac{1}{C(L + M)} = \frac{1}{LC(1 + k)} \tag{10.42}$$

and

$$\omega^- = \frac{1}{C(L - M)} = \frac{1}{LC(1 - k)} \tag{10.43}$$

Since this structure has two degrees of freedom, it has two natural modes of oscillation. These fundamental modes are defined by resonance and anti-resonance. These modes are

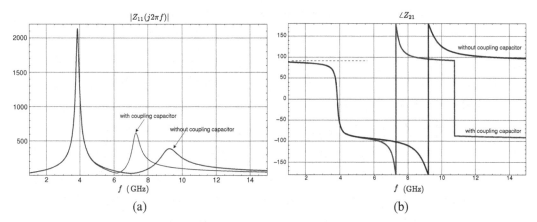

Figure 10.18 (a) The magnitude of Z_{11} for a lumped transformer with winding and coupling capacitance. (b) The phase of Z_{12} for a lumped transformer. The curve with the higher second resonance frequency corresponds to the case with $C_c = 0$, or no coupling capacitance. As expected, the coupling capacitor does not disturb the even mode.

also called the *even* and *odd* mode, due to the symmetry of the excitation. Yet a third name for this is the *differential* or *balanced* mode versus *common mode*.

The lower-frequency resonance occurs in the even mode as the primary and secondary currents flow in phase so as the maximize the magnetic energy in the system. The net inductance, therefore, is $L_1 + L_2 + 2M$. The capacitance in even mode is the winding capacitance to ground, $C_e = C$. It is interesting to note that in even mode we can place a coupling capacitor C_c across the coupled inductors without disturbing the mode voltages and currents, since the net voltage across this capacitor is zero. Thus the lower-frequency resonance mode of the transformer is given by the *self*-winding capacitance and parasitic capacitance to ground, rather than the coupling capacitance. We can thus predict the even mode resonant frequency ω_e as

$$\omega^+ = \sqrt{\frac{1}{C_0(L_1 + L_2 + 2M)}} \tag{10.44}$$

In anti-resonance, or odd mode, the magnetic energy is minimized by circulating current into the windings out of phase. The capacitance of this mode, though, includes the interwinding capacitance $C_o = C + 2C_c$. The factor of two accounts for the Miller effect. The frequency of this mode is therefore

$$\omega^- = \sqrt{\frac{1}{(C + 2C_c)(L_1 + L_2 - 2M)}} \tag{10.45}$$

If $C_c \ll C_0$, we see that the even mode resonance has a lower cutoff frequency. For most integrated transformers, this is indeed the case and so the culprit is the interwinding capacitance.

One of the limitations of the above analysis is the coupling capacitor is treated as a lumped element. In reality, the capacitive coupling is distributed and intermingled with the

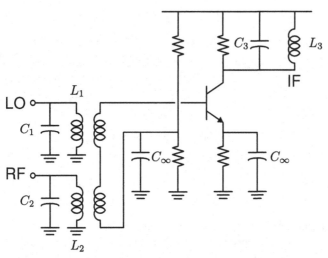

Figure 10.19 A bipolar single transistor mixer employs a transformer to sum the RF and LO signals at the base of the transistor.

magnetic coupling. A full analysis of the problem, therefore, requires distributed analysis, a topic we will cover in Chapter 11.

10.10   Circuits with transformers

Mixers

Integrated transformers are increasingly employed in integrated circuits to solve a variety of problems. The natural differential operation of the transformer is a big advantage in the design of mixers, where balanced operation results in fewer spurious tones and higher power gain. Additionally, the inherent power combining of a transformer can be used to combine the LO and RF signals in a single transistor mixer, as shown in Fig. 10.19. While a simple capacitive transformer is more compact, the transformer can be used as a hybrid to provide LO and RF isolation.

In a Gilbert cell single-balanced mixer, shown in Fig. 10.20a, the transformer can be used to boost the effective transconductance of the input stage M1 by the turns ratio of the device. Furthermore, since the switching pair operates with a DC grounded source, there is plenty of headroom on the drain, which can be employed to boost the gain of the stage by cascoding the stage or by simply increasing the load resistance. In the double-balanced mixer shown in Fig. 10.20b, a balun converts the single-ended RF signal into a balanced signal, eliminating the transconductance stage altogether [36]. Since the transformer is a linear device, the distortion of the mixer is low, especially the $IIP2$ due to the balanced drive. At the output the differential signal is also converted back to a single-ended signal. Single-ended drive is often required to drive an off-chip component, such as a filter.

The classic diode ring mixer shown in Fig. 10.21, is still the preferred choice for extremely linear applications. Furthermore, since diodes work into the THz regime, this mixer can operate at extremely high frequencies. As a passive mixer, the mixer has an insertion loss

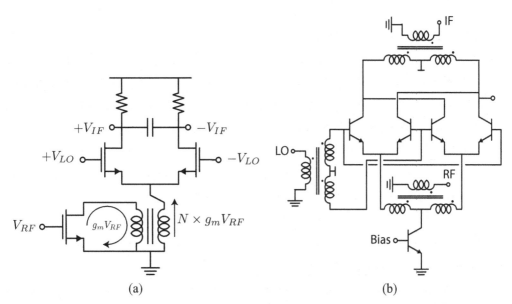

Figure 10.20 (a) A transformer is employed to boost the transconductance of a mixer. At the same time it allows the circuit to operate on a lower supply voltage. (b) A fully balanced version of the Gilbert mixer employs transformers to convert single-ended RF, LO, and IF to a differential circuit. Center taps are used to deliver bias voltages.

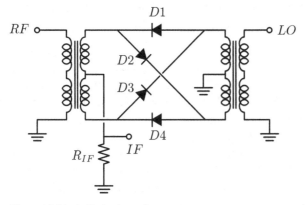

Figure 10.21 A diode ring mixer.

of about 6 dB. The mixer, though, has good isolation, where the $LO \longleftrightarrow RF$ isolation comes from the transformers. The balanced operation also results in RF and LO rejection. The downside is that the LO signal drive must be large since it must turn on/off diodes. Thus the power consumption in the LO port can be quite high.

The ring mixer is a fully balanced mixer but we have the option of driving the LO and RF single ended. As we shall see, the LO signal alternating turns on diodes D1/D2 or D3/D4, thereby connecting the RF voltage to the IF with alternative polarity. During a positive LO voltage, the secondary of the LO transformer applies a positive voltage across D1 in series with the load R, thereby forward biasing the diode. Likewise, the secondary terminal of the

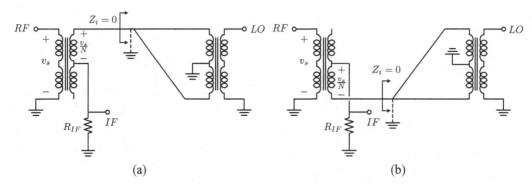

Figure 10.22 Equivalent circuit for the diode ring mixer during the (a) positive and (b) negative LO.

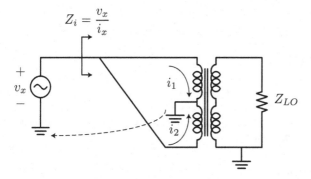

Figure 10.23 Calculation of the input impedance seen by the RF port of the diode ring mixer.

LO transformer applies a negative voltage to the cathode of D2, thereby forward biasing it in series with the load. Diodes D3/D4 are reverse biased and therefore open circuits (ideally).

The equivalent circuit for the positive LO cycle is shown in Fig. 10.22a, where the diode on-resistance is assumed to be zero for simplicity. The RF signal is applied to the load through the center tap of the transformer. We shall show that the impedance looking into the transformer is low (ideally zero), and thus the entire RF signal at the secondary terminal is applied to the load $v_o = -v_s/N$.

Note the input impedance looking into the LO transformer is ideally short circuit, as shown in Fig. 10.23. To see this, apply a test voltage v_x to this node. If we ignore the effect of the LO signal for now, we see that two equal currents $i_1 = i_2 = i_x/2$ flow into the secondary of the LO transformer. The return current comes from the center tap ground. The induced voltage at the primary is given by

$$v_1 = j\omega(M_{12} - M_{13})i_x/2 \equiv 0 \tag{10.46}$$

This is because $M_{13} = M_{12}$ and the currents on the secondary are out of phase. The voltages induced on the secondary side are likewise zero (assuming perfect coupling)

$$v_2 = j\omega L_2 i_x/2 + j\omega M_{23}(-i_x/2) \equiv 0 \tag{10.47}$$
$$v_3 = j\omega L_3(-i_x/2) + j\omega M_{32}i_x/2 \equiv 0 \tag{10.48}$$

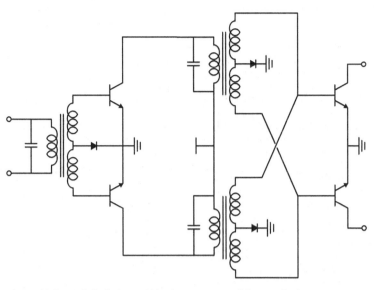

Figure 10.24 A fully balanced bipolar power amplifier employing resonant transformers [65].

By symmetry, during the negative LO cycle, the mixer simplifies to the equivalent circuit shown in Fig. 10.22b. During this cycle, the RF signal is applied to the load through the center tap and the bottom of the transformer, thus producing an output signal in phase with the RF

$$v_o = +v_s/N \qquad (10.49)$$

The operation is thus similar to the double-balanced Gilbert cell mixer in voltage mode. The RF voltage is multiplied by ± 1 with a rate of the LO signal. The lack of DC means that the RF is rejected at the IF port. Likewise, the LO/RF signals are isolated by the switches. The only feed-through occurs due to reverse isolation of the diodes. The diode ring mixer is very linear and quite attractive in applications where linearity reigns supreme over power consumption. At microwave frequencies, the transformers can be replaced by couplers (see Chapter 15). Since diodes can operate up to extremely high frequencies (THz), the entire circuit can work up to THz region.

Interstage power matching

In the balanced power amplifier shown in Fig. 10.24, transformers are used for the inter-stage matching [65] and power combining. The output stage consists of large bipolar devices capable of driving more than 5 W of power into a 50Ω (transformed) load impedance. Large devices have high-input capacitance and require a large base current drive. Two parallel transformer windings at the output provide the current drive from two windings at the input. This allows a high-voltage swing to be maintained at the driver for maximal efficiency. By employing a capacitor in shunt with the transformer, the current gain of the transformer is increased by the Q factor of the network. To see this, consider the equivalent circuit shown

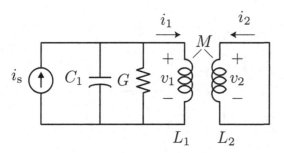

Figure 10.25 A resonant transformer boosts the current gain by the Q factor of the structure.

in Fig. 10.25, where for simplicity the secondary has been shorted. The current gain in the transformer is simply given by solving

$$v_2 = 0 = j\omega M i_1 + j\omega L_2 i_2 \tag{10.50}$$

$$\frac{i_2}{i_1} = -\frac{M}{L_2} = -k\sqrt{\frac{L_1}{L_2}} \tag{10.51}$$

The current i_1 is the source current minus the current in the capacitor

$$i_1 = i_s - Gv_1 - j\omega Cv_1 = i_s - (G + j\omega C)(j\omega M i_2 + j\omega L_1 i_1) \tag{10.52}$$

The conductor G represents the total loss of the transformer reflected to the primary. In terms of the coupling factor k

$$i_1 = i_s - (G + j\omega C)j\omega L_1(1 - k^2) \tag{10.53}$$

The current i_s is therefore given by

$$i_s = i_1(1 - \omega^2 L_1 C_1(1 - k^2) + j\omega L_1 G) \tag{10.54}$$

At the resonant frequency $\omega_0 = 1/\sqrt{L_1 C_1(1 - k^2)}$, the first two terms cancel and we have

$$i_1 = i_s \frac{R}{j\omega L_1} = -jQi_s \tag{10.55}$$

If the Q factor of the transformer is large, the effective current boost is given by the turns ratio *times* the Q factor, greatly reducing the drive current.

Power combining

As already noted, 1:1 transformers are also good elements for power combining on-chip. As shown in Fig. 10.26a, the pseudo-differential pairs drive the primary coils in a balanced fashion on a low supply voltage. The output loop is magnetically coupled to differential pair loops and has N times the flux, and hence N times the voltage. This voltage step-up is a very useful property as the breakdown voltage of high-frequency transistors is low, approaching sub-1 V in a modern CMOS process. A potential layout of such a structure is shown in Fig. 10.27a. Two thick overlapping metal layers are used to build the transformer to minimize the losses and to maximize the coupling factor k.

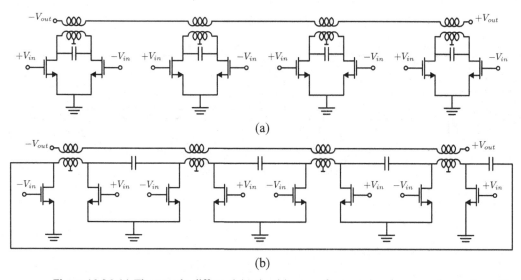

(a)

(b)

Figure 10.26 (a) The pseudo-differential pairs drive transformer primaries over a low voltage. The secondaries are connected in series to drive a high voltage on to the load. (b) In the DAT technique [1], the adjacent cells are driven differentially to form virtual grounds, allowing one to AC short the transformer connections and thus eliminate the lead inductance.

Notice that the inner turns of Fig. 10.27a do not contribute magnetic flux to the output since the currents are flowing in opposite directions, whereas they contribute to extra loss due to the resistivity of the metal layers. If possible, then, these extra segments should be eliminated. This is precisely the approach shown in Fig. 10.26b, and in the layout of Fig. 10.27b, where the input devices are connected in a circular fashion in order to form a closed loop [1]. By virtue of the virtual ground nodes created from the differential drive, the extra lead inductance in driving each sub-loop is eliminated. This results in enhanced power, combining efficiency and a more symmetric layout.

Magnetic feedback

In Fig. 10.28, the transformer is embedded in a differential LNA [36]. Here the transformer serves two functions. The primary windings are used to degenerate the input device and aid matching. The secondary winding inductance serves as a tuned load for the LNA. Since a cascode device is not used, the linearity of the LNA is very good due to the high voltage swing at the output. Without a cascode, though, the stability is an issue due to the feedback capacitance C_{gd}. Here the magnetic coupling is used to neutralize this feedback, improving the stability of the amplifier. Since the amplifier is inverting, the transformer windings are used in inverting configuration in order to produce a signal in phase with the input.

Another application of magnetic feedback is in the voltage controlled oscillator (VCO) shown in Fig. 10.29 [38] [44]. In a typical cross-coupled VCO, shown Fig. 10.29a, capacitors commonly form the feedback. Since the BJT devices need base current, a separate DC biasing scheme is necessary to bias the devices at the right voltage and to provide DC current. To avoid noise from the biasing circuitry, large resistors are often employed. But

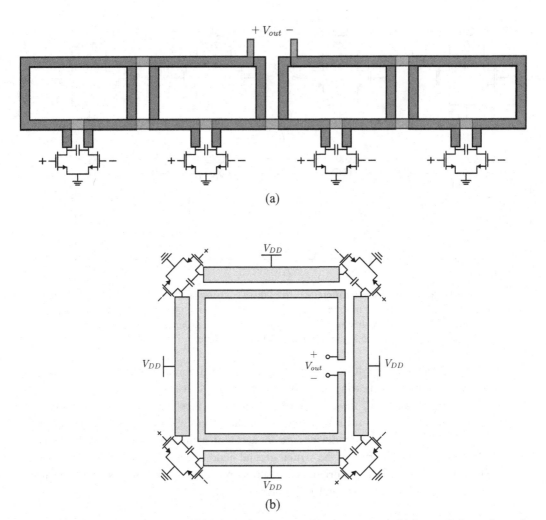

Figure 10.27 (a) A potential layout for the transformer power-combining technique shown in Fig. 10.26a. (b) The ring layout of the DAT structure of Fig. 10.26b.

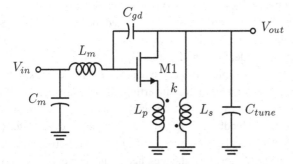

Figure 10.28 AC schematic of LNA neutralized by the transformer magnetic feedback [36].

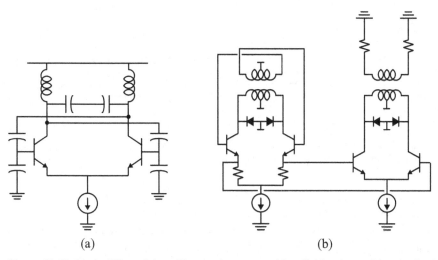

Figure 10.29 (a) A differential oscillator using a capacitive divider for positive feedback. (b) A differential oscillator using a transformer for positive feedback [44].

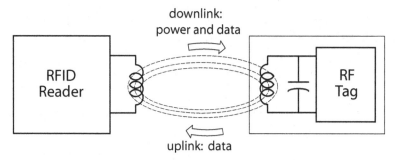

Figure 10.30 A typical RFID system employing magnetic coupling.

since the base current of a BJT varies a lot over process and temperature, to provide a steady voltage at the base is challenging [37]. The transformer feedback circuit of Fig. 10.29b solves this problem since a DC voltage can be easily delivered to the center tap of the primary and secondary. Since the resistance of the windings is small, the variation in base current has a negligible impact on the DC bias voltage. A second differential pair driven by the emitters of the VCO core provides a buffered output. The buffer uses a tuned balun to drive the off-chip load.

RFID

In the RFID system shown in Fig. 10.30, the RF tag is usually a flexible small label that can be attached to an object. The tag "antenna" is usually a printed square spiral inductor which forms the secondary winding of a transformer. The primary coil is a larger coil on a "reader" device. In the near field of the coil, the magnetic field of the "reader" generates sufficient flux to induce a sufficiently large AC voltage at the secondary winding. This voltage is

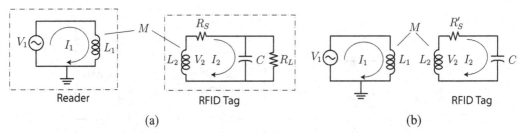

Figure 10.31 (a) Equivalent circuit for RFID tag and reader communication and power link. (b) Simplified circuit where the tag load resistance has been absorbed into the winding resistance.

rectified and used to power the RFID tag, and so the voltage needs to be large enough to turn on an on-chip diode.

To maximize the range, we usually employ a resonant circuit at the secondary. To see the benefit of a resonant secondary, consider the voltage induced on the secondary winding as shown in Fig. 10.31a

$$V_2 = j\omega M I_1 + j\omega L_2 I_2 = I_2(R_S + R_L || j\omega C)$$

It's convenient to transform the load from a shunt RC circuit to a series RC circuit, as shown in Fig. 10.31b. The load resistance is now easily absorbed into the winding resistance R'_S

$$V_2 = j\omega M I_1 + j\omega L_2 \frac{V_2}{R'_S + \frac{1}{j\omega C}}$$

The transfer function from the primary current to the secondary voltage is therefore given by

$$\frac{V_2}{I_1} = \frac{j\omega M\left(R'_S + \frac{1}{j\omega C}\right)}{R_S + \frac{1}{j\omega C} - j\omega L_2}$$

If the load R_L is relatively large, then the load voltage without resonance ($C = 0$) is given by $j\omega M I_1$. In the resonant circuit, though, the maximum load voltage is at a frequency $\omega_0 = \frac{1}{L_2 C}$

$$\frac{V_2}{I_1} = j\omega_0 M\left(1 + \frac{1}{j\omega C R'_S}\right) = j\omega_0 M(1 + Q')$$

If a large Q' factor can be maintained, there is a substantial improvement in the range of the system. One disadvantage of the resonant system is the potential loss in range due to frequency misalignment. Due to process variations, the resonant frequency of the tags will vary. The reader device can "search" for the tag by sweeping its frequency across a wide range and observing the drive current, shown for a particular case in Fig. 10.32a. The frequency of minimum drive current corresponds to the tag resonant frequency.

Once a power link has been established, the tag can communicate with the reader in various ways. A common approach places a switch across the terminals of the tag to modulate the load resistance of the tag. Since the reader current is a function of the tag resistance R_L,

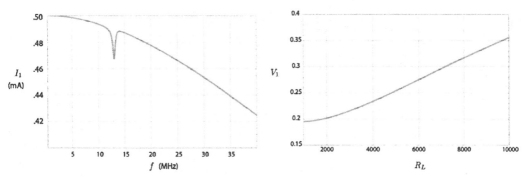

Figure 10.32 (a) The current drive of the primary reader as a function of frequency. Note the dip at resonance. (b) The reader current as a function of the load resistance.

the variation in reader current can be amplified and detected. A plot of the primary current as a function of R_L is shown in Fig. 10.32b.

Example 21

RFID using magnetic coupling

A popular band for RFID is 13.56 MHz. At a distance of 30 cm, the coupling factor k between the primary and secondary is $k = 0.01$. Assume the RFID tag requires 1 mA of current and 1 V DC to operate. The average equivalent resistance of the tag is then 1 kΩ. Let's assume this factor dominates the quality factor. For a tag with $L = 0.5\,\mu$H, the Q factor at resonance is $Q = 1000/(13.56 \times 0.5 \times 2\pi) = 23.5$. If we assume the reader device has ten times more inductance, then

$$M = k\sqrt{L_1 L_2} = 0.01 \times \sqrt{0.5 \times 5}\,\mu\text{H} = 0.016\,\mu\text{H}$$

For a reader current of $I_1 = 100$ mA, the induced voltage at the secondary is given by

$$V_2 \approx I_1 j\omega M(1 + Q') = 100\,\text{mA} \times 13.56 \times 2\pi \times 0.016(1 + 23.5)\,\text{V} = 3.3\,\text{V}$$

which is sufficient for the tag to operate. Note that without resonance the induced voltage is only 136 mV, below the diode turn-on voltage.

10.11   References

Some of the material of this chapter comes from *Modern Communicaiton Circuits* [58]. The application of transformers for power combining and impedance matching is drawn from seminal papers by the Cal Tech team [1] and from Simburger [65], as well as our own research efforts at Berkeley [35]. For an in-depth discussion of transformers in an integrated circuit environment, I highly recommend the thesis and research papers by John Long [36].

11 Distributed circuits

This chapter brings together many of the concepts from this book in order to analyze distributed circuits. In particular, we will analyze common integrated passive elements such as capacitors and resistors as distributed structures. We will also see how even a tiny transistor can act like a distributed circuit at high frequency. Next we consider transmission line transformers, structures that incorporate transformers in a distributed fashion for enhanced bandwidth. Finally, we'll outline the design of distributed structures incorporating active elements, such as distributed amplifiers.

11.1 Distributed RC circuits

The analysis technique for lossy transmission lines can be applied to many practical problems. In Fig. 11.1, an IC resistor using a diffusion layer or thin film material is seen to couple to the substrate in a distributed fashion. A lumped model employing finite Cs and Rs can only work up to a certain frequency point.

Likewise, an IC capacitor Q factor is determined largely by the series resistance. Since the current through the plates is non-uniform in the direction along the plates, the resistance is not simply the total resistance of the plates due to the distributed effects.

Distributed resistor

The distributed resistor can be analyzed as a transmission line formed between the diffusion layer and the substrate. The capacitance per unit length is determined by the reverse-biased pn junction. At any frequency of interest, we assume that the resistance per unit length R' dominates over the inductance per unit length L'. This is a reasonable assumption because the return current flows in the substrate at very close proximity to the current in the resistor. Likewise, both the diffusion region and the substrate have high sheet resistance, and thus for all practical purposes $Z'_s = j\omega L' + R' \approx R'$.

The line propagation constant, $\gamma = \sqrt{R'j\omega C'}$, has a phase of $45°$. The Y matrix for the resistor is easily calculated

$$Y = Y_0 \begin{pmatrix} \coth \gamma \ell & \operatorname{sech} \gamma \ell \\ \operatorname{sech} \gamma \ell & \coth \gamma \ell \end{pmatrix} \tag{11.1}$$

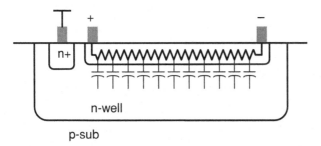

Figure 11.1 An IC diffusion resistor has distributed capacitance to the substrate arising from the reverse-biased pn-junction.

As should be expected, $Y_{11} = Y_{22}$ due to symmetry, and $Y_{12} = Y_{21}$ due to the passivity of the structure. Since $R' = \frac{R_\square}{W}$ and $C' = \frac{W\epsilon_{Si}}{t_{dep}}$, we have

$$Z_0 = \frac{1}{Y_0} = \sqrt{\frac{R'}{j\omega C'}} = \frac{1}{W}\sqrt{\frac{R_\square}{j\omega C_x}} \qquad (11.2)$$

where $C_x = \frac{\epsilon_{Si}}{t_{dep}}$ and

$$\gamma = \sqrt{j\omega R'C'} = \sqrt{j\omega R_\square C_x} \qquad (11.3)$$

The propagation constant is independent of the width of the resistor. The characteristic impedance, though, drops with W.

The optimal size of a resistor can be analyzed at a given frequency. Then for a grounded one-terminal resistor, the input impedance is given by

$$Z_{in} = Z_0 \tanh \gamma \ell = \frac{1}{W}\sqrt{\frac{R_\square}{j\omega C_x}} \tanh \left(\sqrt{j R_\square \omega C_x}\,\ell\right) \qquad (11.4)$$

For a given desired resistance $\frac{\ell}{W} R_\square = R_0$, we can substitute for ℓ

$$Z_{in} = Z_0 \tanh \gamma \ell = \frac{1}{W}\sqrt{\frac{R_\square}{j\omega C_x}} \tanh \left(\sqrt{j R_\square \omega C_x}\,\frac{R_0}{R_\square} W\right) \qquad (11.5)$$

For example, let $R_\square = 100\,\Omega/\square$ and

$$C_x = \epsilon_{SiO_2}/t_0 = 3.45 \times 10^{-5}\,\mathrm{F/m^2}$$

The plot of $|Z_{in}|$ for a nominally $10\,\mathrm{k\Omega}$ resistor versus frequency is shown in Fig. 11.2. The $W = 1\,\mu m$ resistor has a relatively flat frequency response up to $1\,\mathrm{GHz}$, whereas the $W = 5\,\mu m$ resistor rolls off quickly and is about half of its nominal size at $1\,\mathrm{GHz}$. The variation of the impedance magnitude versus W is shown Fig. 11.2. Larger W resistors have better precision and matching, but clearly the extra capacitance hurts at high frequency.

We can also plot the voltage and current along a resistor structure as shown in Fig. 11.3. The resistor is grounded at one end ($x = 0$) and driven by a unit source voltage at the other terminal ($x = -\ell$). The voltage variation at low frequency is linear but at high frequency it decays exponentially. More interestingly, though, at low frequency the current is uniformly

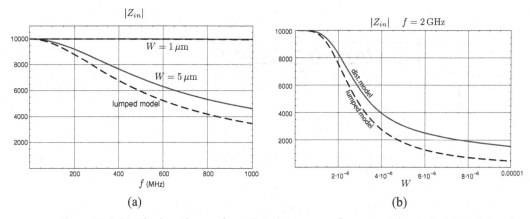

Figure 11.2 (a) The impedance of an IC resistor versus frequency. The dashed line shows the impedance of a lumped RC model of the line. (b) The impedance of the IC resistor at $f = 2\,\text{GHz}$ versus the width (length adjusted to achieve same resistance).

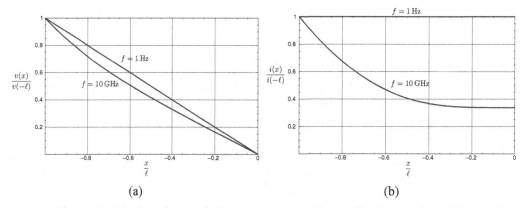

Figure 11.3 The (a) voltage and (b) current variation along a distributed resistor with $R = 3\,\text{k}\Omega$, $W = 10\,\mu$, $C_x = 3.9\frac{\epsilon_0}{1\,\mu}$, $R_\square = 300\,\Omega$ at low frequency and at $f = 10\,\text{GHz}$.

.

distributed through the resistor but at high frequency the current deviates significantly, with much more current concentrated in the first half of the structure.

Metal-insulator-metal (MIM) capacitor

A common IC structure is a MIM capacitor built by sandwiching two conducting plates in close proximity, separated by a thin sheet of insulator, as shown in Fig. 11.4a. The plates are constructed with metal layers (aluminum or copper) or polysilicon, or a combination of metal/poly. In either case, we can treat this circuit as a distributed RC structure to determine the effective loss and quality factor of the structure.

For most practical circuits, we would like to determine the effect of the plate resistance on the structure under a low loss assumption. In other words, the resistance of the metal

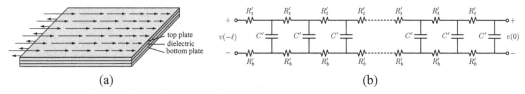

Figure 11.4 (a) A metal-insulator-metal (MIM) capacitor with uniform current flow into the top plane and out of the bottom plate. (b) A distributed RC model for the MIM capacitor.

plates is assumed to be very small. For simplicity we adopt a simple one-dimensional model shown in Fig. 11.4b.

Single contact structure

It is critical to specify the fashion in which we drive the capacitor. We assume that a contact is made along the edge of the capacitor. For a single-sided contact, we contact one edge of the capacitor and leave the other side open. We shall also simplify the analysis by ignoring the substrate parasitics of the MIM capacitor, or the bottom-plate capacitance.

We immediately recognize this as a simple distributed RC structure as the resistance of the bottom plate can be lumped into the top plate to form an effective resistance per unit length $R' = R'_T + R'_B$. The input impedance of the open-ended structure is readily calculated

$$Z_{in} = \frac{\gamma}{j\omega C'} \coth(\gamma \ell) \tag{11.6}$$

Using a Taylor series expansion, we see that the circuit can be represented by a lumped equivalent circuit, consisting of a resistor in series with a capacitor

$$Z_{in} \approx \frac{R_T + R_B}{3} + \frac{1}{j\omega C} \tag{11.7}$$

The above form has an intuitive appeal since we should have expected the effective resistance to be lower for a distributed circuit since not all of the current flows through the plates uniformly. In fact, we could completely anticipate the above results with the following assumption. For a lossless capacitor, the current in the top plate drops linearly due to the uniform displacement current

$$\nabla \cdot J = -j\omega\rho = J_d \tag{11.8}$$

or more simply

$$\frac{dJ}{dx} = J_d = \text{constant} \tag{11.9}$$

implying that $J(x) = J_0(1 - x/\ell)$. The ohmic loss in each metal plate is given by

$$P_L = \int_V \frac{|J|^2}{2\sigma} dV = \frac{1}{2\sigma} \int_0^W \int_{-\delta}^0 \int_0^\ell |J_0|^2 \left(1 - \frac{x}{\ell}\right)^2 dx\,dy\,dz \tag{11.10}$$

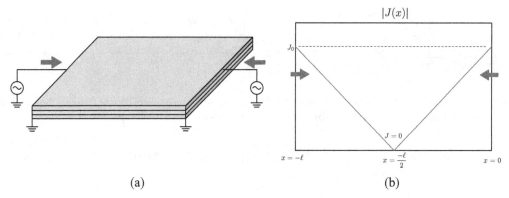

(a)   (b)

Figure 11.5 (a) A double contacted MIM capacitor is driven with an equal signal on both sides in order to reduce the loss. (b) The current distribution in a double contact MIM cap.

where δ is the effective plate thickness (or skin depth) at the frequency of interest. Note that $J_0 = I/(W\delta)$ so that

$$P_L = \frac{1}{2} \frac{W\delta\ell^3}{3\sigma} |J_0|^2 = \frac{1}{2} \times \frac{1}{3} \times \frac{\ell}{W} \frac{1}{\sigma\delta} |I|^2 = \frac{1}{2} \times \frac{1}{3} R_{plate} |I|^2 \qquad (11.11)$$

From circuit theory we have

$$P_L = \frac{1}{2} R_{eq} |I|^2 \qquad (11.12)$$

Showing that the effective resistance is a factor $1/3$ of the plate resistance.

Double contact structure

An important variation of the above result is to contact the structure from both sides, as shown in Fig. 11.5a. This is a common layout trick that will reduce the effective resistance of the structure. We can solve this problem in an analogous fashion but now we apply the voltage boundary conditions

$$v(0) = V_s = V^+ + V^- \qquad (11.13)$$
$$v(-\ell) = V_s = V^+ e^{\gamma\ell} + V^- e^{-\gamma\ell} \qquad (11.14)$$

The current leaving the voltage source is the sum of the currents entering the structure on both sides

$$i_T = i(-\ell) - i(0) \qquad (11.15)$$

where

$$i(0) = (V^+ - V^-)Y_0 \qquad (11.16)$$
$$i(-\ell) = (V^+ e^{\gamma\ell} - V^- e^{-\gamma\ell})Y_0 \qquad (11.17)$$

These equations are readily solved for V^+ and V^- and it is easy to show that

$$i_T = 2Y_0 V_s \tanh\left(\frac{\gamma\ell}{2}\right) \qquad (11.18)$$

Expanding the hyperbolic term $\tanh(x/2) \approx x/2$, we find that the equivalent capacitance is of course the total capacitance

$$\frac{\gamma\ell}{2} 2Y_0 = Y_0\gamma\ell = j\omega C \tag{11.19}$$

To find the equivalent impedance including loss, we expand the input impedance

$$Z_{in} = \frac{Z_0}{2} \frac{1}{\tanh(\gamma\ell/2)} \approx \frac{Z_0}{\gamma\ell} + \frac{Z_0\gamma\ell}{12} \tag{11.20}$$

The second term is the equivalent resistance

$$\gamma\ell Z_0 = \sqrt{j\omega R'C'}\sqrt{\frac{R'}{j\omega C'}}\ell = R'\ell = R \tag{11.21}$$

So that the equivalent circuit has a resistance four times lower than the single contact case

$$Z_{in} \approx \frac{R}{12} + \frac{1}{j\omega C} \tag{11.22}$$

This result can also be anticipated from the low-loss linear current distribution model developed in the previous section. Since for each current source, the current flows to the middle of the capacitor plate as shown in Fig. 11.5b, the resistance can be estimated to be

$$R_{eq} = \frac{1}{2} \times \frac{1}{3} \times \frac{L/2}{W} R_\square = \frac{R}{12} \tag{11.23}$$

where the first factor of $1/2$ accounts for the fact that the resistance of each half plate is in parallel, the factor of $1/3$ accounts for the distributed nature of the current on each half plate, and the final term is simply the resistance of the half plate.

11.2 Transmission line transformers

Transformers pose several problems at high-frequencies. As we have seen, the distributed winding capacitance limits the high-frequency application. The windings cannot be physically isolated since the coupling factor reduces. At low frequency a magnetic core can be used to couple transformers, but at high frequency the magnetic losses preclude magnetic cores. Furthermore, for most integrated circuit processes, no magnetic materials are available.

Thus for coreless integrated inductors, the overall coupling factor of about $k \sim 0.8 - 0.9$ can be achieved. The effect of the non-unity coupling factor and high-winding capacitance spells disaster for high-frequency transformers.

On the other hand, if we view a transformer as a transmission line, as shown in Fig. 11.6, then we see that the winding capacitance is in fact a critical part of the structure. Thus transformers double as transmission lines. They are very convenient building blocks, especially at "lower" frequencies when ordinary transmission lines are too bulky to be employed.

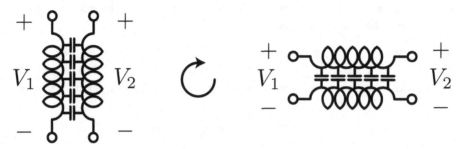

Figure 11.6 The parasitic winding capacitance in a transformer can be exploited if we employ the structure as a transmission line.

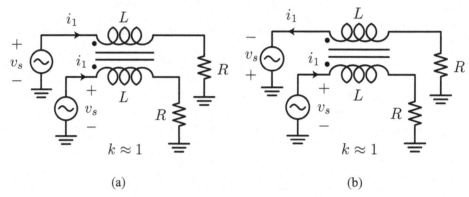

(a)   (b)

Figure 11.7 (a) A transformer excited with a common-mode signal source presents a large series inductance and thus acts like a common-mode "choke." (b) A transformer excited with a differential signal presents a small series inductance if $k \approx 1$.

Low frequencies: a common-mode choke

We begin by noting the operation of a simple $1:1$ transmission line transformer at low frequency. The input impedance for a *common-mode* signal, shown in Fig. 11.7a, is simply given by

$$Z_{in} = \frac{v_s}{i_1} = j\omega L + j\omega M + R = j\omega(L + M) + R = j\omega L(1 + k) + R \approx j\omega L(1 + k)$$

(11.24)

assuming $\omega L \gg R$. Thus the common-mode input impedance is very high, especially at low frequency when a core can boost the inductance to a large value. For a differential mode input, though, shown in Fig. 11.7b, the input impedance is much lower

$$Z_{in} = \frac{v_s}{i_1} = j\omega L - j\omega M + R = j\omega(L - M) + R = j\omega L(1 - k) + R \approx R \quad (11.25)$$

assuming $k \approx 1$. We see that such a structure by its very nature prevents common-mode AC currents from flowing. Furthermore, since a common-mode signal energizes the

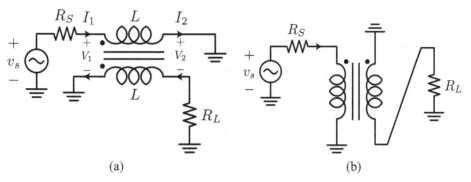

(a)   (b)

Figure 11.8 (a) A properly matched transmission line can function as a broadband inverter. (b) At low frequency, the magnetic coupling alone is enough to produce inversion.

core,[1] whereas a differential signal does not, the loss will be much higher for a common-mode signal at high frequencies.

Broadband inverter

We now excite the transformer as a transmission line with a differential (balanced) signal, as shown in Fig. 11.8a. Note that at low frequency the circuit simply inverts the input signal and applies it to the load R_L, as shown in Fig. 11.8b. At high frequency, though, we employ the transmission line equations. From the $ABCD$ matrix

$$V_1 = \cosh \gamma \ell V_2 + Z_0 \sinh \gamma \ell I_2 \tag{11.26}$$

$$V_2 = I_2 R_L \tag{11.27}$$

Assume that the differential characteristic impedance $Z_0 = \sqrt{\frac{L-M}{C}} = R_L$, so that no reflections occur at the load. Then we have

$$V_1 = (\cosh \gamma \ell + \sinh \gamma \ell) V_2 = e^{\gamma \ell} V_2 \tag{11.28}$$

We see that the output signal is given by

$$v_L = -V_2 = -e^{-\gamma \ell} V_1 \tag{11.29}$$

For the lossless case, $\gamma \ell = jk\ell$, and thus the circuit behaves like

$$v_L = -e^{-jk\ell} V_1 \tag{11.30}$$

Thus the output signal is equal to the input over a broad frequency range

$$|v_L| = |V_1| \tag{11.31}$$

The phase of the transfer function is $180°$ as long as $k\ell \ll 1$. The input impedance $Z_{in} = Z_0 = R_L$ is likewise matched over a broad frequency range.

---

[1] By this we mean that the magnetic field is stronger for a common-mode signal, since the current in the primary and secondary reinforce the magnetic flux. For a differential signal, the magnetic flux cancels to produce a smaller field.

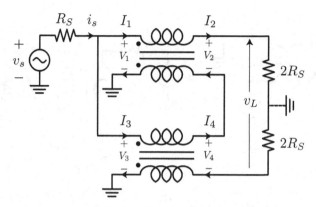

Figure 11.9 A transmission line balun consists of the input shunt connection of an inverter and a delay line, with the outputs connected in series to boost the voltage.

Transmission line transformer balun

Consider now the circuit shown in Fig. 11.9, where the input drives two transformers, the top configured as a non-inverting delay line and the bottom acting like an inverting delay line, as derived in the last section.

Using the notation form the figure, by observation we have

$$I_2 = I_4 \tag{11.32}$$

$$V_L = V_2 + V_4 \tag{11.33}$$

$$i_s = I_1 + I_3 \tag{11.34}$$

$$I_2 = I_4 = \frac{v_L}{4R_S} \tag{11.35}$$

Now apply the transmission line equations (assuming a lossless line)

$$v_L = V_2 + V_4 = \cos k\ell\, V_s - j Z_0 \sin k\ell\, I_1 + \cos k\ell\, V_s - j Z_0 \sin k\ell\, I_3 \tag{11.36}$$

$$= 2\cos k\ell\, V_s - j Z_0 \sin k\ell\, \underbrace{(I_1 + I_3)}_{i_s} \tag{11.37}$$

$$i_s = I_1 + I_3 = j Y_0 \sin k\ell\, \underbrace{(V_2 + V_4)}_{v_L} + 2\cos k\ell\, \underbrace{I_2}_{=I_4} \tag{11.38}$$

Combining the above equations we have

$$i_s = \left(j Y_0 \sin k\ell + \frac{2\cos k\ell}{4R_S} \right) v_L \tag{11.39}$$

$$v_L + j Z_0 \sin k\ell\, i_s = 2\cos k\ell\, v_s \tag{11.40}$$

$$v_L + j Z_0 \sin k\ell \left(j Y_0 \sin k\ell + \frac{\cos k\ell}{2R_S} \right) v_L = 2\cos k\ell\, v_s \tag{11.41}$$

Finally, we can solve for the output voltage

$$v_L = v_s \frac{2\cos k\ell}{1 - \sin^2 k\ell + j \sin k\ell \cos k\ell \frac{Z_0}{2R_S}} \tag{11.42}$$

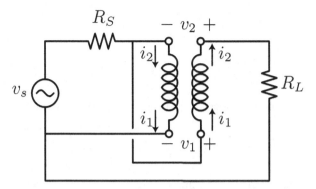

Figure 11.10 A 4:1 Guannela transformer.

The voltage gain is given by

$$G_v = \frac{v_L}{v_s} = \frac{2}{\cos k\ell + j \sin k\ell \frac{Z_0}{2R_S}} = \frac{2}{e^{jk\ell}} = 2e^{-jk\ell} \tag{11.43}$$

where the last equality holds if we select $Z_0 = 2R_S$. We see that the output voltage is twice the input voltage plus a delay. This relation is a broadband relationship for a low-loss circuit

$$|G_v| = 2 \tag{11.44}$$

Using Eq. (11.36), we can also derive the input impedance

$$2v_s e^{-jk\ell} + jZ_0 \sin k\ell i_s = 2 \cos k\ell v_s \tag{11.45}$$

$$v_s(2 \cos k\ell - 2e^{-jk\ell}) = j \underbrace{Z_0}_{2R_s} \sin k\ell i_s \tag{11.46}$$

Simplifying we have

$$\frac{v_s}{i_s} = R_s \tag{11.47}$$

which shows that the circuit behaves like a 4 : 1 impedance matching circuit.

4:1 Unbalanced Guannella transformer

Consider the Guannella transformer shown in Fig. 11.10. Intuitively, when the transmission line is electrically short, we can see that the source current is twice as large as the load current. We thus expect that an impedance match occurs for a load four times as large as the source. We can verify this at high frequency by applying KVL around the source and load loop

$$v_s = i_s R_S - v_2 + i_2 R_L = (i_1 + i_2)R_S - v_2 + i_2 R_L \tag{11.48}$$

We can also take a KVL loop around the source

$$v_s = i_s R_S + v_1 = (i_1 + i_2)R_S + v_1 \tag{11.49}$$

Assuming the transformer is acting like a differential transmission line, the current and voltage are related by

$$v_1 = v_2 \cos \beta \ell + j i_2 Z_0 \sin \beta \ell \qquad (11.50)$$

$$i_1 = i_2 \cos \beta \ell + j v_2 Y_0 \sin \beta \ell \qquad (11.51)$$

The above four equations contain four unknowns v_1, v_2, i_1, and i_2. Solving for the load current i_2 we have the output power

$$P_o = \frac{1}{2} |i_2|^2 R_L \qquad (11.52)$$

To find the optimal T-line characteristic impedance Z_0 we differentiate the output power to find that

$$\frac{\partial P_o}{\partial Z_0} = 0 \qquad (11.53)$$

when

$$Z_0 = \sqrt{R_S R_L} \qquad (11.54)$$

The above result holds independent of the T-line length ℓ. The optimal load impedance is given by solving

$$\frac{\partial P_o}{\partial R_L} = 0 \qquad (11.55)$$

which yields

$$R_L = \frac{2 R_S (1 + \cos \beta \ell)}{\cos \beta \ell} \approx 4 R_S \qquad (11.56)$$

The last equality holds if the line length $\beta \ell \ll 1$ is sufficiently small. The ratio of the load power to the available power is given by

$$P_a = \frac{v_s^2}{8 R_S} \qquad (11.57)$$

$$\frac{P_o}{P_a} = \frac{1 + \cos \beta \ell}{\frac{5}{4} \left(1 + \frac{6}{5} \cos \beta \ell + \cos^2 \beta \ell\right)} \qquad (11.58)$$

The load power drops to zero if $\beta \ell = \pi$ or if the transmission line approaches a half-wavelength, $\ell = \lambda/2$. Since

$$v_L = v_1 + v_2 \qquad (11.59)$$

we can see that the load voltage is zero when the phase shift from input to output is 180°. So this transformer is only effective when $\ell \ll \lambda/2$. Typically we keep $\ell < \lambda/10$.

At low frequencies the equivalent circuit of the transformer is shown in Fig. 11.11a. Note that the transmission line behavior is replaced by magnetic flux coupling of the auto-transformer. An equivalent circuit for the transformer, shown in Fig.11.11b, can be used to find the low-frequency cutoff. In general we see that the low-frequency signal is shunted to ground by the inductance $k^2 L$.

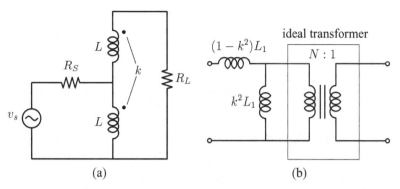

Figure 11.11 (a) Low-frequency model of Guannella transformer. The energy transfer is primarily due to magnetic coupling. (b) Equivalent model for center tapped transformer.

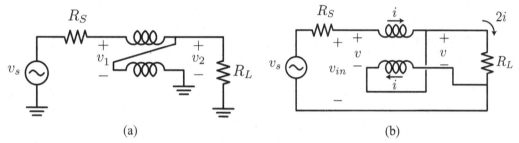

Figure 11.12 (a) A step-up Guannella transformer. (b) Redrawn for analysis.

Since the transformer only works when the electrical T-line length is small, we can simplify the analysis by assuming that

$$i_1 = i_2 = i \tag{11.60}$$

$$v_1 = v_2 = v \tag{11.61}$$

The following relations follow readily from the above conditions

$$v_L = v_1 + v_2 = 2v \tag{11.62}$$

$$i_L = i \tag{11.63}$$

$$R_L = \frac{v_L}{i_L} = \frac{2v}{i} \tag{11.64}$$

$$Z_{in} = \frac{v}{2i} \tag{11.65}$$

$$\frac{v}{i} = 2Z_{in} \tag{11.66}$$

$$R_L = 4Z_{in} \tag{11.67}$$

$$Z_{in} = \frac{R_L}{4} \tag{11.68}$$

The circuit shown in Fig. 11.12a performs the dual operation. The above circuit doubles the current to the load and thus the input impedance is boosted to $Z_{in} = 4R_L$. Let's redraw the circuit, as shown in Fig. 11.12b, and analyze it using the ideal T-line

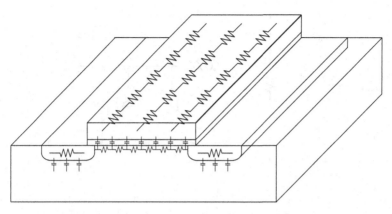

Figure 11.13 Cross section of FET device highlighting the distributed RC lines along the gate width and channel length of the device.

equations (11.60)–(11.61)

$$i_s = i \tag{11.69}$$
$$v_{in} = v + 2i\,R_L \tag{11.70}$$
$$v_L = v = 2i\,R_L \tag{11.71}$$
$$v_{in} = 2i\,R_L \times 2 = 4i\,R_L \tag{11.72}$$
$$Z_{in} = \frac{v_{in}}{i} = 4R_L \tag{11.73}$$

11.3   FETs at high frequency

The MOS structure is a distributed RC network by its very nature. But since high-frequency transistors are made with extremely short channel lengths L, we must work the transistor at very high frequencies to reveal its distributed behavior. A schematic representation of the MOSFET is shown in Fig. 11.13. Note the presence of distributed resistance along the *width* of the gate due to the resistance of the gate material (usually polysilicon), and the distributed resistance along the *length* or the channel of the device. These resistances are intermingled with the device C_{ox}. There is of course distributed inductance at play as well, but the device is so small that the resistance dominates over the inductance and thus it can be safely ignored for scale devices. The drain and source also couple to the substrate through distributed resistances in the device, namely the substrate impedances.

In this section we focus on the distributed gate resistance of the device. We divide the gate resistance of an FET into two terms, an extrinsic resistance, due to the resistance of the contacts, metal layers, and polysilicon gate structure, and the intrinsic or channel portion.

Extrinsic gate resistance

The extrinsic resistance is usually dominated by the polysilicon resistance, since the metal layers are much more conductive. The polysilicon gate structure is equivalent to an RC gate

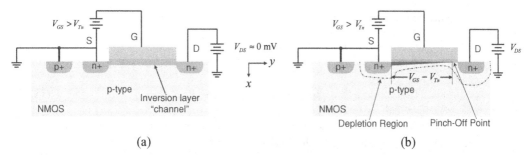

Figure 11.14 (a) A MOSFET in the linear region behaves like a distributed RC circuit. (b) A MOSFET in saturation region has a non-uniform channel resistance due to the non-uniform channel inversion layer.

transmission line where the capacitance per unit length is given the oxide capacitance and the resistance is simply the distributed resistance of the gate material.

For now let's ignore the corresponding intrinsic bottom plate component of the resistance, which is in fact the channel resistance. Since the FET is really a two-dimensional structure, it's important to note that the bottom plate current does not flow in the same direction but perpendicular to the gate current as it flows into the source and drain of the transistor.

From the previous section we know that the extrinsic gate resistance can be modeled by an equivalent resistor with a value of either $1/3$ or $1/12$ of the gate resistance, where the $1/3$ case corresponds to a single-sided gate contact and $1/12$ corresponds to a double-sided gate contact.

$$R_{g,ext} = \frac{1}{3|12} \frac{W}{L} R_{\square} \tag{11.74}$$

In practice long width transistors are avoided precisely to minimize the gate resistance. The layout of a large transistor is therefore made of parallel fingers as shown in Fig. 8.34. For N fingers the resistance is reduced by N^2

$$R_{g,ext} = \frac{1}{N} \frac{1}{3|12} \frac{W/N}{L} R_{\square} = \frac{1}{N^2} \frac{1}{3|12} \frac{W}{L} R_{\square} \tag{11.75}$$

Intrinsic gate NQS resistance

To find the intrinsic portion of the gate resistance, it's important to notice that the current flow is orthogonal to the gate width current. Now the loss is due to the channel resistance between the source and drain. Since in general this resistance is non-uniform (due to a non-zero V_{ds}), the analysis is more complicated than a uniform transmission line.

For simplicity consider a transistor in a linear region with $V_{ds} \approx 0$. Now this structure, shown in Fig. 11.14a, is similar to the capacitor structure we analyzed before so the input capacitance is given by approximately C_{gs} in series with a resistance R_i, the distributed resistance of the channel. In the triode region the channel resistance is given by[2]

$$g_{ds} = \frac{\partial I_{ds}}{\partial V_{ds}} = \mu C_{ox} \frac{W}{L} \frac{\partial}{\partial V_{ds}} \left((V_{gs} - V_T)V_{ds} - \frac{V_{ds}^2}{2} \right) \tag{11.76}$$

$$g_{ds} \approx \mu C_{ox} \frac{W}{L} (V_{gs} - V_T) \tag{11.77}$$

But notice that this RC line is grounded on two sides, the source and the drain. Thus the equivalent resistance is $1/12$ of the channel resistance.

For a transistor biased with $V_{ds} \neq 0$, the analysis is more complicated. The channel is non-uniform, since the effect of the drain voltage is to counter the gate voltage and thus the amount of inversion is lower on the drain side, as shown in Fig. 11.14b. A practical approach is to simply assume that the intrinsic gate resistance is some fraction of the channel resistance. In pinch-off the channel resistance is given by the voltage drop V_{ds} across the channel, excluding the voltage drop along the pinch-off region

$$R_{ch} = \frac{V_{ds} - V_{pinch-off}}{I_{ds}} = \frac{V_{gs} - V_T}{I_{ds}} \tag{11.78}$$

A detailed analysis shows that in the pinch-off region this effective resistance is given by

$$R_i = \frac{1}{5g_{ds}} \tag{11.79}$$

where g_{ds} is the small-signal channel conductance. Note that in the pinch-off region, for a square-law device, $g_m = g_{ds}$. This means that the Q factor of the input intrinsic device is given by

$$Q_{gs} = \frac{1}{\omega R_i C_{gs}} = \frac{5g_m}{\omega C_{gs}} = \frac{5\omega_T}{\omega} \tag{11.80}$$

The quality factor at frequencies far removed from the transistor ω_T is indeed very large. Only as we approach the device ω_T does this loss term become an important.

Gate induced noise

Looking into the gate of a transistor we "see" a fraction of the channel. Since the channel is resistive, it has associated thermal noise. Thus we expect that some of this noise will leak into the gate. Naively we may assume that the noise at the gate of the transistor should be $4kT\Re(Z_{gs}) = 4kT\frac{1}{5g_m}$. But this is naive because the transistor is not a simple passive two-port. A more detailed analysis must include the correlation between the drain and gate noise since they arise from a common source. But fortuitously, if we simply assume that the gate resistor is noisy and independent from the channel thermal noise, we actually can predict the transistor noise performance quite well. This noise model is attributed to Pospieszalski [40].

---

[2]   For simplicity we assume a square law device.

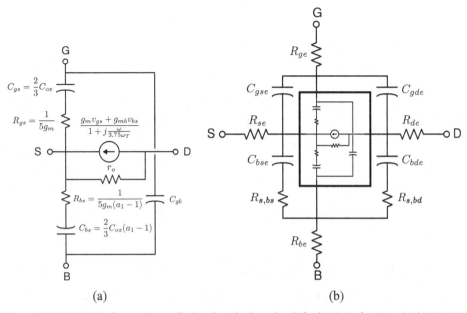

Figure 11.15 (a) High-frequency small-signal equivalent circuit for intrinsic four-terminal MOSFET device. (b) Equivalent small-signal circuit for extrinsic four-terminal MOSFET device.

FET equivalent circuit

A commonly used high-frequency equivalent circuit for the FET is shown in Fig. 11.15a [61]. The distributed nature of the transistor is captured by lumped resistances placed judiciously into the model. The most important deviation from a standard low-frequency model is the loss of the input network due to the intrinsic channel resistance R_{gs}. Note that the device transconductance g_m is complex so as to capture the delay in the drain current. Alternatively, the g_m can be controlled by the voltage across C_{gs}, which introduces a delay as well. Even though this approach is not strictly accurate, since the time constant is different, it does do a fairly decent job of capturing the non-quasistatic effect.

A more sophisticated model, which includes the distributed width dependence of the transistor, has been developed by [16]. The above model is an approximation to a complete y-parameter based on the non-quasistatic model presented in [61]. The complete model of Fig. 11.15b includes the external parasitics, including the drain and source substrate resistances and various overlap capacitors.

11.4 Distributed amplifier

The distributed amplifier, also called a *Traveling Wave Amplifier*, is a key broadband building block. The concept of a distributed amplifier is actually quite simple and illustrated in Fig. 11.16.

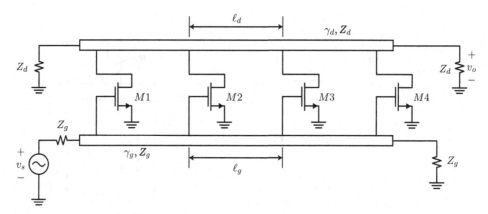

Figure 11.16 A distributed amplifier consists of N parallel transistors connected using gate and drain transmission lines.

Ideal lossless distributed amplifier

Here we assume for simplicity that the transistors have no parasitics and N transistors are connected in parallel along transmission lines. The gates of each transistor are placed periodically along a "gate" transmission line, whereas the drains are connected to a "drain" transmission line. Each line is terminated into Z_0, the characteristic impedance of the line.

Now at low frequency, we see that the output current into the load $Z_L = Z_0$ is simply given by

$$i_o = \frac{1}{2}N \times g_m v_{gs} = \frac{1}{4}N \times g_m v_s \tag{11.81}$$

Note that the first factor of $1/2$ occurs since the output currents split into each side of the transmission line equally. Likewise, the second factor of $1/2$ is simply due to voltage division at the source of the gate line.

What is so special about this structure? The key observation is that the ideal distributed amplifier has infinite bandwidth. To see this, recall that an ideal lossless terminated TEM line operates with a simple frequency response from DC out to infinite frequency. Any voltage excited at the input of the amplifier travels as a wave towards the gate termination resistance. Thus each transistor is excited turn by turn at a fixed interval. The same argument applies to the drain transmission line as the output current of each transistor is injected on to the line and travels towards the load with a fixed delay. Now by its very nature, the distributed amplifier is delay balanced. In other words, the input transistor closest to the source is furthest from the load. The path from source to each transistor to load has the same delay. So the currents sum in phase and this is true at any frequency.

Let us make this simple argument more rigorous with a simple derivation. The ith transistor is excited with an input voltage

$$v_{gs,i} = \frac{v_s}{2}e^{-j(i-1)\beta_g \ell_g} \tag{11.82}$$

where β_g is the propagation constant on the gate line. The load current is also a summation of N currents each coming from the input transistors

$$I_d = \frac{1}{2} \sum_{i=1}^{N} i_{d,i} e^{-(N-i)j\beta_d \ell_d} \tag{11.83}$$

where $i_{d,i} = -g_m v_{gs,i}$ resulting in

$$I_d = -\frac{g_m}{4} v_s \sum_{i=1}^{N} e^{-(i-1)j\beta_g \ell_g} e^{-(N-i)j\beta_d \ell_d} \tag{11.84}$$

$$= -\frac{g_m}{4} v_s e^{-Nj\beta_d \ell_d} e^{j\beta_g \ell_g} \sum_{i=1}^{N} e^{-ij(\beta_g \ell_g - \beta_d \ell_d)} \tag{11.85}$$

The above equation applies for any arbitrary line, but obviously we would like to synchronize the delay on the gate and drain line

$$\beta_g \ell_g = \beta_d \ell_d = \theta \tag{11.86}$$

which results in the following obvious conclusion

$$I_d = -\frac{g_m}{4} v_s e^{-(N-1)j\theta} \cdot N \tag{11.87}$$

Which confirms the frequency independence of the current gain. The power gain is given by

$$G = \frac{P_{out}}{P_{in}} = \frac{\frac{1}{2}|I_d|^2 Z_d}{\frac{1}{8}|v_s|^2 / Z_g} = \frac{g_m^2 N^2 Z_d Z_g}{4} \tag{11.88}$$

Notice the penalties for "cheating" nature! Our amplifier gain grows only like N^2 with the number of stages, unlike a cascade where the gain grows like g_m^N. There is also a built-in delay in the amplifier.

How does this work with real transistors? Remember that up to now we have ignored the frequency variation of the transistor. Well, that's the beauty of the distributed amplifier. The transistor input capacitance and output capacitance can be *absorbed* into the gate and drain line! In other words, we can form artificial transmission lines where the cutoff frequency of the line is a function of how aggressively we distribute the transistors along the line. This means that in theory we have a broadband amplifier up to the cutoff frequency of the artificial line, a big improvement over the bandwidth of simply putting N transistors in parallel. But of course there is a catch. Unfortunately our transistors also have input/output resistance, and this introduces increased losses on to the input and output transmission lines.

Lossy distributed amplifier

Let's now consider the losses in our transmission lines due to the transistor parasitics. Using a hybrid-π model for each transistor results in the gate and drain lines shown in Fig. 11.17. Our previous analysis holds if we simply consider the *complex* propagation constant on the gate and drain line. If we ignore the lumped nature of the transistors and simply assume that a

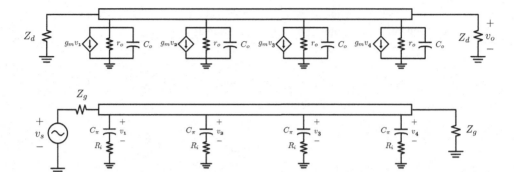

Figure 11.17 A distributed-lumped equivalent circuit for the distributed amplifier using a simple hybrid-π model. Note that the Miller capacitance C_μ has been neglected.

distributed admittance is added to the gate and drain line, we can use our previous formulas for transmission line propagation constant and characteristic impedance. This argument holds if the spacing between the transistors is much less than the wavelength, $\ell \ll \lambda$.

For the gate line, assume a low-loss line so that the the impedance per unit length is given by

$$Z = j\omega L'_g \tag{11.89}$$

where as the admittance per unit length is lossy

$$Y = j\omega C'_g + \frac{j\omega C_{gs}/\ell_g}{1 + j\omega R_i C_{gs}} \tag{11.90}$$

Under a low-loss approximation the characteristic impedance is given by

$$Z_g = \sqrt{\frac{Z}{Y}} \approx \sqrt{\frac{L'_g}{C'_g + C_{gs}/\ell_g}} \tag{11.91}$$

and the propagation constant is given by

$$\gamma_g = \alpha_g + j\beta_g = \sqrt{-\omega^2 L'_g \left(C'_g + \frac{C_{gs}}{\ell_g}(1 - j\omega R_i C_{gs}) \right)} \tag{11.92}$$

For the drain line, a similar calculation for the drain leads to

$$Z_d = \sqrt{\frac{L_d}{C_d + C_{ds}/\ell_d}} \tag{11.93}$$

$$\gamma_d = \sqrt{j\omega L_d \left(\frac{1}{R_{ds}\ell_d} + j\omega(C'_d + C_{ds}/\ell_d) \right)} \tag{11.94}$$

Let's now redo our gain calculation taking the losses into account. If we assume that $\omega \ll \frac{1}{R_i C_{gs}}$, then

$$v_{gs,i} = \frac{v_s}{2} e^{-(i-1)\gamma_g \ell_g} \left(\frac{1}{1 + j\omega R_i C_{gs}} \right) \approx \frac{v_s}{2} e^{-(i-1)\gamma_g \ell_g} \tag{11.95}$$

Allowing us to reuse our calculation from above

$$I_d = -\frac{g_m v_s}{4} e^{-N\gamma_d \ell_d} e^{\gamma_g \ell_g} \sum_{i=1}^{N} e^{-i(\gamma_g \ell_g - \gamma_d \ell_d)} \tag{11.96}$$

Now phase matching alone is not sufficient since we cannot in general cancel the argument of the exponential. Taking the summation results in

$$I_d = -\frac{g_m v_s}{4} \frac{e^{-N\gamma_g \ell_g} - e^{-N\gamma_d \ell_d}}{e^{-\gamma_g \ell_g} - e^{-\gamma_d \ell_d}} \tag{11.97}$$

$$G = \frac{g_m^2 Z_d Z_g}{4} \left| \frac{e^{-N\gamma_g \ell_g} - e^{-N\gamma_d \ell_d}}{e^{-\gamma_g \ell_g} - e^{-\gamma_d \ell_d}} \right|^2 \tag{11.98}$$

Under a low-loss condition, the denominator is further simplified

$$G = \frac{g_m^2 Z_d Z_g}{4} \left| \frac{e^{-N\gamma_g \ell_g} - e^{-N\gamma_d \ell_d}}{\gamma_g \ell_g - \gamma_d \ell_d} \right|^2 \tag{11.99}$$

The key observation is that the gain is no longer a monotonic function of the number of stages N. In other words, as we add more sections in parallel eventually we reach a point of diminishing returns. Why does this happen? Simply because an exponential function will always beat a polynomial. In a short sprint, a polynomial may win, but give it enough distance (or time), and the exponential will always conquer. Since the loss on the transmission line is an exponential function of the length, and adding more stages increases the length, the gain of an extra transistor cannot overcome the loss on the line.

To find the optimum number of stages, we can take the derivative of G with respect to N to find N_{opt}

$$N_{opt} = \frac{\ln\left(\frac{\alpha_g \ell_g}{\alpha_d \ell_d}\right)}{\alpha_g \ell_g - \alpha_d \ell_d} \tag{11.100}$$

In a real design, the optimal number of stages may turn out to be quite small, say 2–3 stages, and thus the achievable gain is limited. Adding insult to injury the gain is not flat due to the frequency dependence of α_g and α_d. Furthermore, under a moderate to high-loss case, the characteristic impedances Z_g and Z_d are also frequency dependent, causing further problems.

Artificial distributed amplifier

To understand the frequency dependence of the gain we must do a more careful analysis of the lumped/distributed line. In fact, in the analysis that follows, we will replace the transmission line with a lumped inductor, as shown in Fig. 11.18. This is commonly employed to decrease the area of the distributed amplifier. This forms an *artificial* transmission line that behaves very much like a real transmission line until we get close to the cutoff frequency.

The analysis of an artificial transmission line can be undertaken by the Image Parameter Method [47]. This method is equivalent to the analysis of Bloch Waves [30] in a periodic structure. To begin let's consider the problem of calculating the input impedance of an

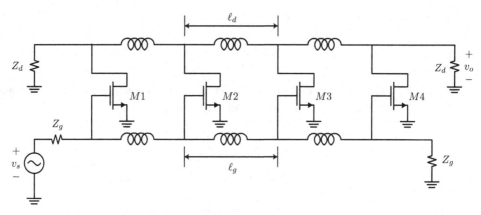

Figure 11.18 An artificial distributed amplifier consists of N parallel transistors connected using lumped gate and drain artificial transmission lines.

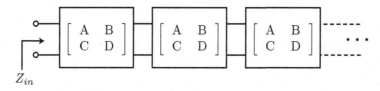

Z_{in}

Figure 11.19 An infinite cascade of two-ports form an artificial transmission line.

infinite cascade of two-port networks, shown in Fig. 11.19. As you may recall, we did this calculation when we started our analysis of transmission lines. In terms of the two-port $ABCD$ parameters, the input impedance of a two-port is given by

$$Z_i = \frac{DZ_L + B}{CZ_L + A} \tag{11.101}$$

For an infinite cascade, the input impedance at any point can be computed by observing that $Z_L = Z_i$

$$Z_i = \frac{DZ_i + B}{CZ_i + A} \tag{11.102}$$

Solving the above equation leads to the following equation

$$CZ_i^2 + (A - D)Z_i - B = 0 \tag{11.103}$$

Suppose that the two-port in question is symmetric, so that $A = D$. Then we simply have

$$Z_i = \sqrt{\frac{B}{C}} \tag{11.104}$$

Now let us consider wave propagation along this infinite two-port cascade. The voltage transfer function is easily computed

$$\frac{V_2}{V_1} = \sqrt{AD} - \sqrt{BC} = e^{-\gamma} \tag{11.105}$$

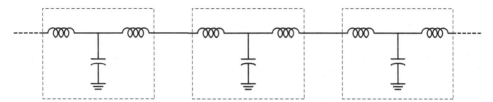

Figure 11.20 An artificial LC transmission line. The basic unit cell is a T network.

For wave propagation, the transfer function is given by the exponential propagation factor. If we can show that γ has an imaginary component, then wave propagation can occur. We can simplify the above equation as follows. Note that for any reciprocal two-port $AD - BC = 1$ so that

$$e^{\gamma} = \frac{1}{\sqrt{AD} - \sqrt{BC}} = \frac{AD - BC}{\sqrt{AD} - \sqrt{BC}} = \sqrt{AD} + \sqrt{BC} \qquad (11.106)$$

Using $2\cosh\gamma = e^{\gamma} + e^{-\gamma}$ gives us

$$\cosh\gamma = \sqrt{AD} = A \qquad (11.107)$$

In other words, the propagation characteristics of a two-port cascade is determined by the magnitude and phase of A.

Consider the artificial LC transmission line shown in Fig. 11.20. Let's compute the characteristic impedance and propagation constant. For the single T section, it's easy to show that

$$B = j\omega L_g \left(1 - \frac{\omega^2 L_g C_g}{4}\right) \qquad (11.108)$$

$$C = j\omega C_g - \frac{j\omega L_g \omega^2 C_g^2}{4} \qquad (11.109)$$

so that

$$Z_i = \sqrt{\frac{L}{C}} \sqrt{1 - \left(\frac{\omega}{\omega_c}\right)^2} \qquad (11.110)$$

where we have defined the cutoff frequency $\omega_c = \frac{2}{\sqrt{LC}}$.

Note that for frequencies far removed below ω_c, the line characteristic impedance is flat but as we approach ω_c, as shown in Fig. 11.21a, there is significant deviation from $\sqrt{L/C}$. Now consider propagation constant γ. You can show that

$$e^{\gamma} = 1 - \frac{2\omega^2}{\omega_c^2} + \frac{2\omega}{\omega_c}\sqrt{\frac{\omega^2}{\omega_c^2} - 1} \qquad (11.111)$$

We now see that for $\omega < \omega_c$, the propagation constant γ is imaginary, which shows that wave propagation occurs at these frequencies. On the other hand, above ω_c, γ is a real quantity and the transmission line cannot propagate waves. This is shown in Fig. 11.21b, where $\gamma = \alpha + j\beta$ is broken into real and imaginary parts. Therefore in practice we are limited to the bandwidth given by the cutoff frequency of the line. There are techniques to

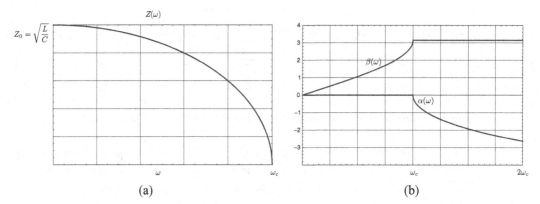

Figure 11.21 (a) The characteristic impedance of an artificial transmission line begins at Z_0 but drops to zero at ω_c. (b) The real and imaginary part of the propagation constant $\gamma = \alpha + j\beta$ versus frequency for an artificial transmission line. Note that $\alpha = 0$ for $\omega < \omega_c$.

improve the gain flatness of the distributed structure by terminating the gate and drain line in the correct frequency-dependent image impedance. This requires matching networks to properly synthesize the image impedance from a fixed load [47] [43].

11.5 References

The material on transmission line transformers originates in several courses I have taught at UC Berkeley. In preparing material on transmission line transformers, I drew from *Modern Communication Circuits* [58] and Hagen's book [26]. The transistor high-frequency equivalent circuits material are from Tsividis [61] and the BSIM4 manual [68]. The material on distributed amplifiers is an extension of material a previous book [43] and Pozar [47].

12 High-speed switching circuits

A lot of this book has been concerned with time-harmonic excitation and behavior (frequency domain) of passive and distributed elements. This is convenient – as the differential equations simplify to algebraic ones – and it is also a very common situation. Most RF circuits, for example, are narrowband, and therefore for all practical purposes the input signal is a sinusoid at the carrier frequency. But there are a host of situations where the signals of interest are wideband. This includes digital switching circuits, high-speed data circuits, and ultra-wideband circuits and systems. Furthermore, digital circuits are switching increasingly faster, now into the multi-GHz regime. There are also many high-speed links, such as chip-to-chip communication on a PCB. These high-speed interface circuits drive pulses or spectrally rich waveforms on long board traces. We will treat these different cases in a uniform manner by generically calling these circuit applications "high-speed" switching circuits.

12.1 Transmission lines and high-speed switching circuits

The focus of the chapter is switching waveforms on transmission lines. You may wonder why we would use transmission line analysis for switching circuits, especially small on-chip or on-board circuits. Let's take a practical example to get a feel for the problem. Let's say a digital chip has a dimension of less than 1 cm. For a time-harmonic circuit, we know that we may treat most structures as lumped as long as their dimension does not exceed $1/10\lambda$. For a SiO_2 transmission line structure, we equate the chip dimension to one tenth the wavelength $\lambda = 10 \times 1\,\text{cm} = 10\,\text{cm}$, in order to solve for the maximum operating frequency

$$f = c/\lambda = \frac{3}{2} \times 10^8/10\,\text{cm} = 1.5\,\text{GHz} \tag{12.1}$$

For signals with center frequency below about 1 GHz, therefore, we may use lumped analysis.[1]

For a digital waveform, or a switching waveform, we cannot simply use the switching or clock frequency F_{clk} to determine the boundary between lumped and distributed analysis. Recall that a deterministic square wave has a discrete spectrum that decays like $1/N$ with

---

[1] We should be careful to not excite the circuit in *slow wave* modes that have significantly reduced velocity. This occurs, for instance, when waves flow between a conductor with a return current flowing in the Si substrate.

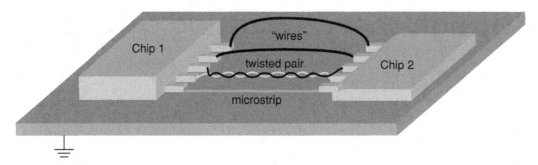

Figure 12.1 Two chips can be connected with wires using the backplane ground reference, or more explicitly with a twisted pair connection or a microstrip transmission line.

the Nth harmonic frequency. For a randomly switching waveform, we must evaluate the Fourier transform of the autocorrelation function [48], which results in the well known sinc characteristic. This spectrum is rich in harmonics and therefore a good fraction of the energy will appear at frequencies above the switching rate. A more realistic waveform would have a non-zero rise and fall time. For such a waveform, it can be shown that most of the energy is contained in the spectrum from DC to the knee frequency, $F_{knee} = 0.5/T_r$, where T_r is the waveform risetime [29].

From this we conclude that a digital waveform with knee frequency significantly below 1.5 GHz can also be treated as a lumped circuit. This corresponds to a risetime of

$$T_r = 0.5/F_{knee} = 0.5/1.5\,\text{ns} = \frac{1}{3}\,\text{ns} \tag{12.2}$$

If we take the risetime to be 10% of the clock period, then the minimum clock period is $T = 10 \times T_r \approx 3\,\text{ns}$, or $F_{clk} = 300\,\text{MHz}$. Digital circuits are already exceeding this frequency by an order of magnitude. But since most on-chip circuits are a small fraction of the chip dimension, with the exception of the clock distribution network or a signal that must propagate over a long distance, most on-chip structures can be treated in a lumped manner. But now we see that off-chip structures, the package, board, connectors, and especially cables, are large enough to require full distributed analysis.

Advantages of transmission lines

There are many situations as a circuit designer where we must decide between using transmission lines and simple point to point connections [29]. In Fig. 12.1 we show a typical example where two integrated circuits are communicating on a PCB. Let's say that the two chips are close enough where simple wires can be used to connect the ICs rather than transmission lines. Is there any reason to use transmission lines at all in this case?

If we use a "wire" to connect the two chips, then we would keep the length as short as possible to minimize the inductance. If the input impedance of the integrated circuit is capacitive, a common situation, the wire can be modeled by an LC circuit, where the capacitance is due to the wire capacitance summed with the output and input capacitance of the circuit. From chapter 7, we know that the step response of this circuit will exhibit ringing if the $Q > 1/2$. Ringing is undesirable because it forces us to slow the clock down to allow the waveforms to settle to the correct values. Alternatively, we can introduce loss into the circuit to lower the Q to eliminate the ringing.

But there are other problems with this circuit as shown. The high inductance wire tends to generate large magnetic fields that will generate a lot of EMI (electromagnetic interference). This means that the circuit may fail to qualify for a production environment. More importantly, the circuit has a large tendency to interfere due to mutual inductance. Any nearby wire experiences cross-talk due to the mutual inductance. Charging and discharging currents dI/dt couple the flux leading to potential problems. The crude solution, as always, is to slow down the clock. This of course also means increasing the risetime.

Alternative interconnections are shown in Fig. 12.1. We reduce the wire inductance by placing it closer to the ground plane, thus naturally creating a transmission line structure. The reduction in inductance occurs since ground return current is closer to the signal. If the ground plane is far away, we can use a ground wire in a twisted pair fashion, forming a pseudo-transmission line. While two parallel wires would form a real transmission line, the twisted pair has better isolation. If a ground plane is available, we can use a PCB trace to form a microstrip transmission line. This naturally cuts down the area of the loop and minimizes the magnetic coupling and EMI. But what about the ringing?

In the next sections, we shall show that the circuit with the transmission line will also ring unless we properly *terminate* the transmission lines. These transmission line terminations require us to drive resistive loads, which increases the power dissipation of the circuits. But in return our circuits are robust and can be driven to very high clock frequencies.

12.2   Transients on transmission lines

Time domain voltage/current waveforms

In Chapter 9 we found that the currents and voltages on the transmission line satisfy the one-dimensional wave equation. This is a partial differential equation. The solution depends on the boundary and the initial conditions. For the lossless case, we found

$$\frac{\partial^2 f}{\partial z^2} = L'C'\frac{\partial^2 v}{\partial t^2} \tag{12.3}$$

It's easy to show that the function $f(z,t) = f(z \pm vt) = f(u)$ satisfies the wave equation.

$$\frac{\partial f}{\partial z} = \frac{\partial f}{\partial u}\frac{\partial u}{\partial z} = \frac{\partial f}{\partial u} \tag{12.4}$$

$$\frac{\partial^2 f}{\partial^2 z} = \frac{\partial^2 f}{\partial u^2} \tag{12.5}$$

$$\frac{\partial f}{\partial t} = \frac{\partial f}{\partial u}\frac{\partial u}{\partial t} = \pm v\frac{\partial f}{\partial u} \tag{12.6}$$

$$\frac{\partial^2 f}{\partial t^2} = \pm v\frac{\partial}{\partial u}\left(\frac{\partial f}{\partial t}\right) = v^2\frac{\partial^2 f}{\partial u^2} \tag{12.7}$$

$$\frac{\partial^2 f}{\partial z^2} = \frac{1}{v^2}\frac{\partial^2 f}{\partial t^2} \tag{12.8}$$

We see that this is a very general solution. In theory, any waveform can be excited onto a lossless transmission line and it will travel along the line without dispersion.

The most general voltage solution is therefore two waves, $v(z, t) = f^+(z - vt) + f^-(z + vt)$, where $v = \sqrt{\frac{1}{L'C'}}$. The speed of motion can be deduced if we observe the speed of a point on the waveform

$$z \pm vt = \text{constant} \tag{12.9}$$

To follow this point as time elapses, we must move the z coordinate in step. This point moves with velocity

$$\frac{dz}{dt} \pm v = 0 \tag{12.10}$$

This is the speed at which we move, $\frac{dz}{dt} = \pm v$, and v is the velocity of wave propagation. The current also satisfies the wave equation $i(z, t) = g^+(z - vt) + g^-(z + vt)$. Recall that on a transmission line, current and voltage are related by

$$\frac{\partial i}{\partial z} = -C' \frac{\partial v}{\partial t} \tag{12.11}$$

For the general function this gives

$$\frac{\partial g^+}{\partial u} + \frac{\partial g^-}{\partial u} = -C' \left(-v \frac{\partial f^+}{\partial u} + v \frac{\partial f^-}{\partial u} \right) \tag{12.12}$$

Since the forward waves are independent of the reverse waves

$$\frac{\partial g^+}{\partial u} = C'v \frac{\partial f^+}{\partial u} \tag{12.13}$$

$$\frac{\partial g^-}{\partial u} = -C'v \frac{\partial f^-}{\partial u} \tag{12.14}$$

Within a constant we have

$$g^+ = \frac{f^+}{Z_0} \tag{12.15}$$

$$g^- = -\frac{f^-}{Z_0} \tag{12.16}$$

where $Z_0 = \sqrt{\frac{L'}{C'}}$ is the "Characteristic Impedance" of the line.

12.3   Step function excitation of an infinite line

Consider exciting a step function on to a transmission line. Note that a finite width pulse can be constructed as a sum of two step functions, $p_\tau(t) = u(t) - u(t - \tau)$. The line is assumed uncharged: $Q(z, 0) = 0$, $\psi(z, 0) = 0$ or equivalently $v(z, 0) = 0$ and $i(z, 0) = 0$. By physical intuition, we would only expect a forward traveling wave since the line is infinite

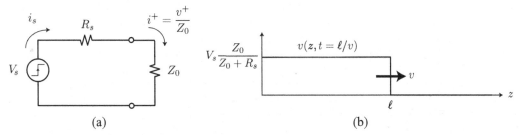

Figure 12.2 (a) Equivalent circuit for a voltage source driving an infinite transmission line. (b) The waveform on the transmission line at time $t = \ell/v$ due to a step function a the source.

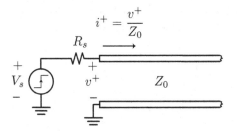

Figure 12.3 The power delivered to the infinite transmission line by a step voltage source.

in extent. The general form of current and voltage on the line is given by

$$v(z, t) = v^+(z - vt) \tag{12.17}$$

$$i(z, t) = i^+(z - vt) = \frac{v^+(z - vt)}{Z_0} \tag{12.18}$$

The T-line looks like a resistor of Z_0 ohms! We may therefore model the line with the equivalent circuit shown in Fig. 12.2a.

Since $i_s = i^+$, the excited voltage wave has an amplitude of

$$v^+ = \frac{Z_0}{Z_0 + R_s} V_s \tag{12.19}$$

It's surprising that the voltage magnitude on the line is not equal to the source voltage. As shown in Fig. 12.2b, the voltage on the line is a delayed version of the source voltage.

Energy on a transmission line

Let's calculate the energy to "charge" a transmission line. The setup is shown in Fig. 12.3. The power flow into the line is given by

$$P_{line}^+ = i^+(0, t)v^+(0, t) = \frac{(v^+(0, t))^2}{Z_0} \tag{12.20}$$

Or in terms of the source voltage

$$P_{line}^+ = \left(\frac{Z_0}{Z_0 + R_s}\right)^2 \frac{V_s^2}{Z_0} = \frac{Z_0}{(Z_0 + R_s)^2} V_s^2 \tag{12.21}$$

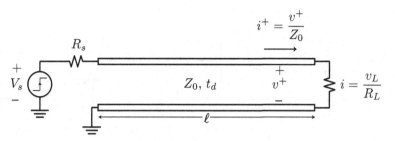

Figure 12.4 A finite section of transmission line terminated with a load resistance R_L.

But where is the power going? The line is lossless! The energy stored by a capacitor and inductor is $\frac{1}{2}CV^2 / \frac{1}{2}LI^2$. At time t_d, a length of $\ell = vt_d$ has been "charged"

$$\frac{1}{2}CV^2 = \frac{1}{2}\ell C' \left(\frac{Z_0}{Z_0 + R_s} \right)^2 V_s^2 \tag{12.22}$$

$$\frac{1}{2}LI^2 = \frac{1}{2}\ell L' \left(\frac{V_s}{Z_0 + R_s} \right)^2 \tag{12.23}$$

The total energy is thus

$$\frac{1}{2}LI^2 + \frac{1}{2}CV^2 = \frac{1}{2}\frac{\ell V_s^2}{(Z_0 + R_s)^2} \left(L' + C' Z_0^2 \right) \tag{12.24}$$

Recall that $Z_0 = \sqrt{L'/C'}$. The total energy stored on the line at time $t_d = \ell/v$

$$E_{line}(\ell/v) = \ell L' \frac{V_s^2}{(Z_0 + R_s)^2} \tag{12.25}$$

And the energy delivered onto the line in time t_d:

$$P_{line} \times \frac{\ell}{v} = \frac{\frac{1}{v}Z_0 V_s^2}{(Z_0 + R_s)^2} = \ell \sqrt{\frac{L'}{C'}} \sqrt{L'C'} \frac{V_s^2}{(Z_0 + R_s)^2} \tag{12.26}$$

As expected from conservation of energy, the results match.

12.4   Terminated transmission line

Consider a finite transmission line with a termination resistance. This setup, shown in Fig. 12.4, is perhaps the most common practical situation. At the load we know that Ohm's Law applies, $I_L = V_L/R_L$. So at time $t = \ell/v$, our step reaches the load. Since the current on the T-line is $i^+ = v^+/Z_0 = V_s/(Z_0 + R_s)$ and the current at the load is V_L/R_L, a discontinuity is produced at the load.

Thus a reflected wave must be created at discontinuity to satisfy the boundary condition

$$V_L(t) = v^+(\ell, t) + v^-(\ell, t) \tag{12.27}$$

$$I_L(t) = \frac{1}{Z_0}v^+(\ell, t) - \frac{1}{Z_0}v^-(\ell, t) = V_L(t)/R_L \tag{12.28}$$

Solving for the forward and reflected waves

$$2v^+(\ell, t) = V_L(t)(1 + Z_0/R_L) \tag{12.29}$$

$$2v^-(\ell, t) = V_L(t)(1 - Z_0/R_L) \tag{12.30}$$

And therefore the reflection from the load is given by

$$\Gamma_L = \frac{V^-(\ell, t)}{V^+(\ell, t)} = \frac{R_L - Z_0}{R_L + Z_0} \tag{12.31}$$

The reflection coefficient (same as the time harmonic case) plays a very important concept for transmission lines, and for a real termination resistor, $-1 \le \Gamma_L \le 1$. Some noteworthy cases are

- $\Gamma_L = -1$ for $R_L = 0$ (short)
- $\Gamma_L = +1$ for $R_L = \infty$ (open)
- $\Gamma_L = 0$ for $R_L = Z_0$ (match)

Most often, we strive to impedance match to provide the proper termination to avoid reflections. Otherwise if $\Gamma_L \ne 0$, a new reflected wave travels toward the source and unless $R_s = Z_0$, another reflection also occurs at source! To see this consider the wave arriving at the source. Recall that since the wave equation is a linear partial differential equation, a superposition of any number of solutions is also a solution. At the source end the boundary condition is as follows

$$V_s - I_s R_s = v_1^+ + v_1^- + v_2^+ \tag{12.32}$$

The new term v_2^+ is used to satisfy the boundary condition. The current continuity requires $I_s = i_1^+ + i_1^- + i_2^+$

$$V_s = (v_1^+ - v_1^- + v_2^+)\frac{R_s}{Z_0} + v_1^+ + v_1^- + v_2^+ \tag{12.33}$$

Solve for v_2^+ in terms of known terms

$$V_s = \left(1 + \frac{R_s}{Z_0}\right)(v_1^+ + v_2^+) + \left(1 - \frac{R_s}{Z_0}\right)v_1^- \tag{12.34}$$

But $v_1^+ = \frac{Z_0}{R_s + Z_0}V_s$

$$V_s = \frac{R_s + Z_0}{Z_0}\frac{Z_0}{R_s + Z_0}V_s + \left(1 - \frac{R_s}{Z_0}\right)v_1^- + \left(1 + \frac{R_s}{Z_0}\right)v_2^+ \tag{12.35}$$

So the source terms cancel out and

$$v_2^+ = \frac{R_s - Z_0}{Z_0 + R_s}v_1^- = \Gamma_s v_1^- \tag{12.36}$$

The reflected wave bounces off the source impedance with a reflection coefficient given by the same equation as before

$$\Gamma(R) = \frac{R - Z_0}{R + Z_0} \tag{12.37}$$

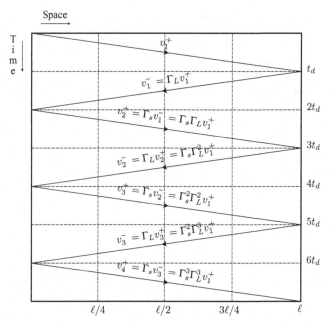

Figure 12.5 The bounce diagram.

The source appears as a short for the incoming wave. Invoke superposition! The term v_1^+ took care of the source boundary condition so our new v_2^+ only needed to compensate for the v_1^- wave. The reflected wave is only a function of v_1^-.

The bounce diagram

We can track the multiple reflections with a "bounce diagram," shown in Fig. 12.5. If we freeze time, using the bounce diagram we can figure out how many reflections have occurred. For instance, at time $2.5t_d = 2.5\ell/v$ three waves have been excited (v_1^+, v_1^-, v_2^+), but v_2^+ has only traveled a distance of $\ell/2$. To the left of $\ell/2$, the voltage is a summation of three components

$$v = v_1^+ + v_1^- + v_2^+ = v_1^+(1 + \Gamma_L + \Gamma_L\Gamma_s) \tag{12.38}$$

To the right of $\ell/2$, the voltage has only two components

$$v = v_1^+ + v_1^- = v_1^+(1 + \Gamma_L) \tag{12.39}$$

We can also pick an arbitrary point on the line and plot the evolution of voltage as a function of time. For instance, at the load, assuming $R_L > Z_0$ and $R_S > Z_0$, so that $\Gamma_{s,L} > 0$, the voltage at the load will will increase with each new arrival of a reflection, as shown in Fig. 12.6.

Physical intuition: shorted Line

To gain physical insight into the transient behavior of a transmission line, consider Fig. 12.7, an LC ladder model. The initial step charges the "first" capacitor through the "first" inductor

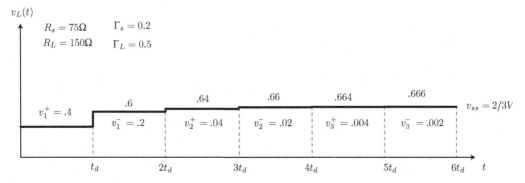

Figure 12.6 The voltage waveform on a distributed transmission line at a given instant of time.

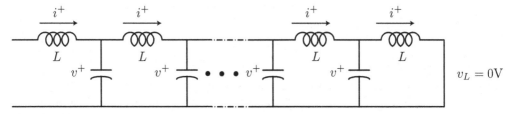

Figure 12.7 An LC ladder model of a shorted transmission line.

since the line is uncharged. There is a delay since on the rising edge of the step, the inductor is an open. Each successive capacitor is charged by "its" inductor in a uniform fashion ... this is the forward wave v_1^+.

The voltage on the line goes up from left to right due to the delay in charging each inductor through the inductors. The last inductor, though, does not have a capacitor to charge. Thus the last inductor is "discharged," with extra charge flowing through discharging the "last" capacitor. As this capacitor discharges, so does its neighboring capacitor to the left. Again there is a delay in discharging the caps due to the inductors. This discharging represents the backward wave v_1^-.

Steady-state waveform

To find steady-state voltage on the line, we sum over all reflected waves

$$v_{ss} = v_1^+ + v_1^- + v_2^+ + v_2^- + v_3^+ + v_3^- + v_4^+ + v_4^- + \cdots \qquad (12.40)$$

Or in terms of the first wave on the line

$$v_{ss} = v_1^+(1 + \Gamma_L + \Gamma_L\Gamma_s + \Gamma_L^2\Gamma_s + \Gamma_L^2\Gamma_s^2 + \Gamma_L^3\Gamma_s^2 + \Gamma_L^3\Gamma_s^3 + \cdots \qquad (12.41)$$

Notice geometric sums of terms like $\Gamma_L^k\Gamma_s^k$ and $\Gamma_L^{k+1}\Gamma_s^k$. Let $x = \Gamma_L\Gamma_s$

$$v_{ss} = v_1^+(1 + x + x^2 + \cdots + \Gamma_L(1 + x + x^2 + \cdots)) \qquad (12.42)$$

The sums converge since $|x| < 1$

$$v_{ss} = v_1^+ \left(\frac{1}{1 - \Gamma_L\Gamma_s} + \frac{\Gamma_L}{1 - \Gamma_L\Gamma_s} \right) \qquad (12.43)$$

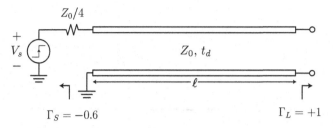

Figure 12.8 A step function voltage source drives an open line.

or more compactly

$$v_{ss} = v_1^+ \left(\frac{1 + \Gamma_L}{1 - \Gamma_L \Gamma_s} \right) \qquad (12.44)$$

Substituting for Γ_L and Γ_s gives

$$v_{ss} = V_s \frac{R_L}{R_L + R_s} \qquad (12.45)$$

In steady state, the equivalent circuit shows that the transmission line has disappeared. This happens because if we wait long enough, the effects of propagation delay do not matter. Conversely, if the propagation speed were infinite, then the T-line would not matter. But the presence of the T-line will be felt if we disconnect the source or load! That's because the T-line *stores* reactive energy in the capacitance and inductance. Note that every real circuit behaves this way! Lumped circuit theory is an abstraction that only applies to infinitesimally small circuits.

Ringing for open/short Loads

Consider a source driving an open line shown in Fig. 12.8. Suppose the source impedance is $Z_0/4$, so $\Gamma_s = -0.6$, and the load is open so $\Gamma_L = 1$. As before a positive going wave is launched $v_1^+ = 4V_s/5$. Upon reaching the load, a reflected wave of equal amplitude is generated and the load voltage overshoots $v_L = v_1^+ + v_1^- = 1.6\text{V}$. Note that the current reflection is negative of the voltage

$$\Gamma_i = \frac{i^-}{i^+} = -\frac{v^-}{v^+} = -\Gamma_v \qquad (12.46)$$

This means that the sum of the currents at load is zero (open).

At the source a new reflection is created $v_2^+ = \Gamma_L \Gamma_s v_1^+$, and note $\Gamma_s < 0$, so $v_2^+ = -.6 \times 0.8 = -0.48$. At a time $3t_p$, the line charged initially to $v_1^+ + v_1^-$ drops in value

$$v_L(3t_p) = v_1^+ + v_1^- + v_2^+ + v_2^- = 1.6 - 2 \times .48 = .64 \qquad (12.47)$$

So the voltage on the line undershoots < 1. And on the next cycle $5t_p$ the load voltage again overshoots. We observe ringing with frequency $2t_p$, as shown in Fig. 12.9. When we drive an unterminated line, especially and open-circuited or a short-circuited line, we observe ringing. In a digital circuit with a long propagation delay t_p, if the ringing is severe enough, it may require us to slow down the clock to allow the voltage to properly

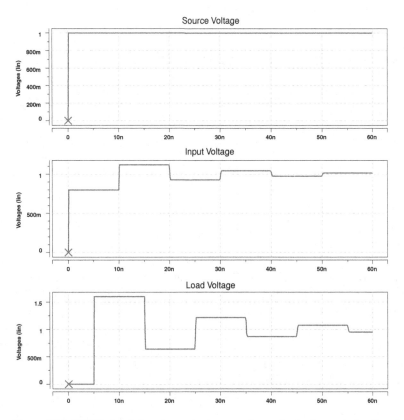

Figure 12.9 SPICE simulation shows ringing waveforms for the transmission line shown in Fig. 12.8. The time delay of the transmission line is $t_p = 5\,\text{ns}$.

settle. This is one of the main advantages of a terminated transmission line in such a system.

Transmission line resonator

In Chapter 9 we found that transmission lines resonate when the electrical length becomes a multiple of the quarter wavelength. At odd multiples the line behaves like a parallel RLC circuit, whereas at even multiples it behaves like a series RLC circuit. In the time domain, we shall see that a transmission line can be used to generate periodic waveforms with fast transitions and short periods.

The setup is shown in Fig. 12.10. We assume that the transmission is initially charged to a voltage V_0 and then at time $t = 0^+$ we short the transmission line to ground. Since the transmission line is lossless, the energy stored in the line, $CV^2/2$, cannot disappear. Instead as the line is shorted, it will be converted into magnetic energy $LI^2/2$, and then back again to electrical energy. So we see a resonance occurring on the line but instead of sinusoidal waves, the line will have a square wave voltage and current waveform. From a frequency domain perspective, we see that the initial closure of the switch excites multiple

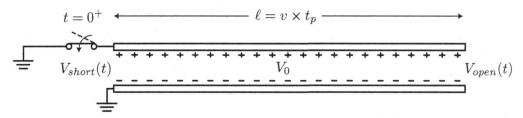

Figure 12.10 A transmission line initially charged to a voltage V_0 is discharged by closing the switch at time $t = 0^+$.

harmonics on the transmission line, which sum to form a square wave. Since the initial step is rich in harmonics, it will in theory impart energy on to an infinite number of harmonics frequencies.

Let's solve this problem in the following manner. At time $t = 0$, the voltage on the line is given by V_0. When the switch is closed, we generate a wave $V_1^+ = -V_0$ at the position of the switch, which discharges the voltage there. This voltage wave travels towards the open end of the transmission line at the speed of light in the medium, $v = \sqrt{1/L'C'}$, and as it travels it discharges the voltage on the line to zero. In a time $t = t_p = \ell/v$, this wave will reach the open end and the voltage on the entire line is reduced to zero. Now all the energy of the line is contained in the current, or the magnetic energy. Due to the open boundary condition, though, the voltage V^+ is reflected and now a new wave $V_1^- = V_1^+ = -V_0$ is generated. This wave now travels leftward towards the shorted end of the transmission line. As it travels, it reduces the voltage on the line to $-V_0$. Now we can see that this voltage, upon reaching the shorted end, will generate a new voltage, $V_2^+ = -V_2^- = V_0$, which travels forward towards the open end. At the shorted end of the voltage, each subsequent wave generated always cancels the incoming voltage to produce a net zero voltage.

In this way we see that the voltage on the transmission line oscillates between V_0 and $-V_0$ as a function of time at the load. At the short end, it drops to zero as the switch is closed. At the open end, though, the voltage will oscillate. The period of this oscillation is given by the time delay of the transmission line, or $T = 2t_p$. A plot of the voltage at the open and short end is shown in Fig. 12.11. Interestingly, at the midpoint of the transmission line, the voltage waveform is more complicated, as partial reflections drive the line to zero volts, then negative, and then positive again. Since the line is lossless, this oscillation will ensue until the line is somehow disturbed again. Very fast oscillations can be produced in this way. Unfortunately every real transmission line has loss and so the voltage waveform will decay towards zero. If we can somehow introduce energy back onto the line, we can in fact produce a distributed oscillator.

Cascade of transmission lines

Consider the junction between two transmission lines Z_{01} and Z_{02}, shown in Fig. 12.12a. At the interface $z = 0$, the boundary conditions are that the voltage/current have to be

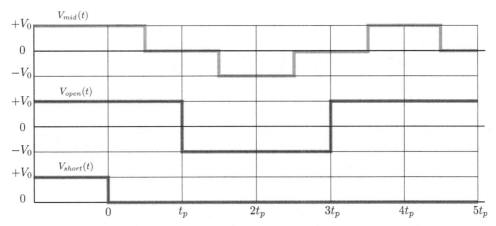

Figure 12.11 The voltage waveforms on the initially charged transmission line at points $z = 0$ (shorted end), $z = \ell$ (open end), and at the midpoint.

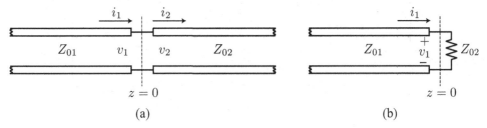

Figure 12.12 (a) The cascade of two transmission lines. (b) The equivalent load presented by the second transmission line.

continuous

$$v_1^+ + v_1^- = v_2^+ \tag{12.48}$$

$$(v_1^+ - v_1^-)/Z_{01} = v_2^+/Z_{02} \tag{12.49}$$

Solve these equations in terms of v_1^+. The reflection coefficient has the same form (easy to remember)

$$\Gamma = \frac{v_1^-}{v_1^+} = \frac{Z_{02} - Z_{01}}{Z_{01} + Z_{02}} \tag{12.50}$$

As shown in Fig. 12.12b, the second line looks like a load impedance of value Z_{02}. The wave launched on the new transmission line at the interface is given by

$$v_2^+ = v_1^+ + v_1^- = v_1^+(1 + \Gamma) = \tau v_1^+ \tag{12.51}$$

This "transmitted" wave has a coefficient

$$\tau = 1 + \Gamma = \frac{2Z_{02}}{Z_{01} + Z_{02}} \tag{12.52}$$

Notice that the sum of τ^2 and Γ^2 is unity

$$\tau^2 + \Gamma^2 = 1 \tag{12.53}$$

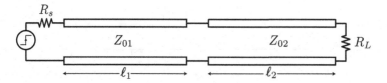

Figure 12.13 The load is connected to the source by a cascade of two sections of transmission lines.

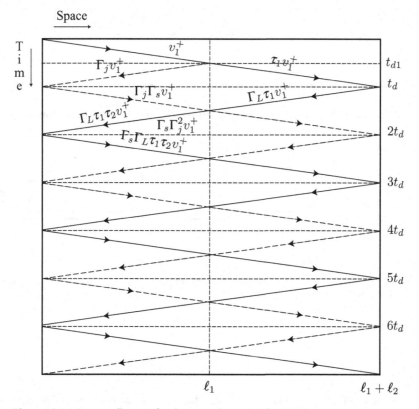

Figure 12.14 Bounce diagram for the setup shown in Fig. 12.13.

This follows from conservation of energy. We can construct a bounce diagram for the scenario shown in Fig. 12.13. Thus reflections occur at multiple interfaces, shown in Fig. 12.14.

Junction of parallel T-lines

Consider the junction of three transmission lines, shown in Fig. 12.15. Using the same approach as before, we invoke voltage/current continuity at the interface

$$v_1^+ + v_1^- = v_2^+ = v_3^+ \tag{12.54}$$

$$\frac{v_1^+ - v_1^-}{Z_{01}} = \frac{v_2^+}{Z_{02}} + \frac{v_3^+}{Z_{03}} \tag{12.55}$$

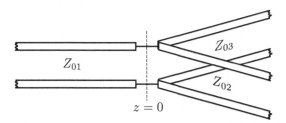

Figure 12.15 A transmission line drives two parallel lines.

But $v_2^+ = v_3^+$, so the interface just looks like the case of the junction of two transmission lines, Z_{01} and a new line, with characteristic impedance $Z_{01} || Z_{02}$.

Example 22

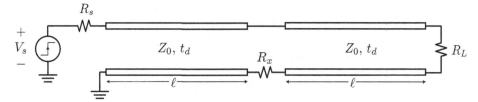

Figure 12.16 A layout error introduces a ground resistance between two transmission lines.

Consider the following circuit consisting of two transmission lines of characteristic impedance Z_0. Due to a layout error, the ground connection is not good and presents a resistance of $R_x = .5Z_0$. The circuit is excited by a pulse at the generator with amplitude 1V and source impedance $R_s = 2Z_0$ (zero rise-time). The load is matched $R_L = Z_0$.

We analyze this problem by drawing the bounce diagram, shown in Fig. 12.17, for the circuit. We only include the action on the first transmission line. From the bounce diagram, we can sketch the voltage waveform at the load and at an arbitrary point, such as $z/\ell = .25$ as a function of time on the first transmission line. These graphs are shown in Fig. 12.18.

12.5 Reactive terminations

Consider a reactive termination. This situation is common in practice because a shorted or open load has non-zero reactance. Let us analyze the problem of an inductive load first. When a pulse first "sees" the inductance at the load, it looks like an open, so $\Gamma_0 = +1$. As time progresses, the inductor looks more and more like a short! So $\Gamma_\infty = -1$.

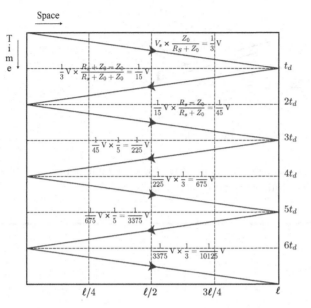

Figure 12.17 Bounce diagram for discontinuity shown in Fig. 12.16.

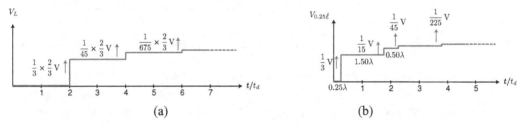

(a)   (b)

Figure 12.18 (a) The voltage variation at the load. (b) The voltage variation at a distance of 0.25λ on the first transmission line. Note drawings are not drawn to scale (vertical).

So intuitively we might expect the reflection coefficient to look like Fig. 12.19. The graph starts at $+1$ and ends at -1. In between we'll see that it goes through exponential decay (first-order ODE).

Do equations confirm our intuition?

$$v_L = L\frac{di}{dt} = L\frac{d}{dt}\left(\frac{v^+}{Z_0} - \frac{v^-}{Z_0}\right) \tag{12.56}$$

And the voltage at the load is given by $v^+ + v^-$

$$v^- + \frac{L}{Z_0}\frac{dv^-}{dt} = \frac{L}{Z_0}\frac{dv^+}{dt} - v^+ \tag{12.57}$$

The right-hand side is known, it's the incoming waveform. For the step response, the derivative term on the RHS is zero at the load

$$v^+ = \frac{Z_0}{Z_0 + R_s}V_s \tag{12.58}$$

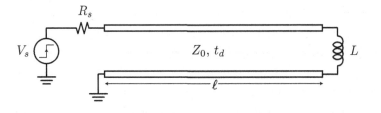

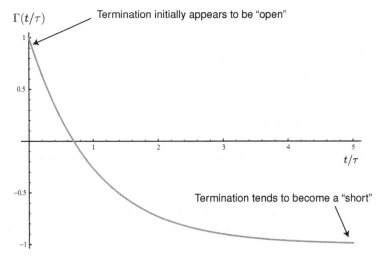

Figure 12.19 The reflection coefficient versus time for an inductive load.

So we have a simpler case $\frac{dv^+}{dt} = 0$. We must solve the following equation

$$v^- + \frac{L}{Z_0}\frac{dv^-}{dt} = -v^+ \qquad (12.59)$$

For simplicity, assume at $t = 0$ the wave v^+ arrives at load. In the Laplace domain

$$V^-(s) + \frac{sL}{Z_0}V^-(s) - \frac{L}{Z_0}v^-(0) = -v^+/s \qquad (12.60)$$

Solve for reflection $V^-(s)$

$$V^-(s) = \frac{v^-(0)L/Z_0}{1 + sL/Z_0} - \frac{v^+}{s(1 + sL/Z_0)} \qquad (12.61)$$

Break this into basic terms using partial fraction expansion

$$\frac{-1}{s(1 + sL/Z_0)} = \frac{-1}{s} + \frac{L/Z_0}{1 + sL/Z_0} \qquad (12.62)$$

Invert the equations to get back to time domain $t > 0$

$$v^-(t) = (v^-(0) + v^+)e^{-t/\tau} - v^+ \qquad (12.63)$$

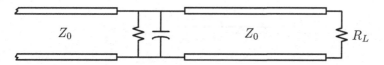

Figure 12.20 A shunt RC discontinuity.

Note that $v^-(0) = v^+$, since initially the inductor is an open. So the reflection coefficient is

$$\Gamma(t) = 2e^{-t/\tau} - 1 \tag{12.64}$$

The reflection coefficient decays with time constant L/Z_0.

In fact, we can simplify the above procedure by using Laplace analysis directly. We can find the reflection coefficient $\Gamma(s)$ directly in the Laplace domain without any equations. For a periodic excitation, such as a square wave, we can use Fourier analysis directly. This is illustrated by the next example.

Example 23

The transmission line shown in Fig. 12.20 has a discontinuity connected in shunt as shown. The discontinuity consists of a capacitor and a parallel resistor. The load is matched to the line. Derive the equations governing the reflection v^- at the discontinuity.

The load seen by the first transmission line is given by $R||C||Z_0$, or

$$Z_L = \left(\frac{1}{R} + \frac{1}{Z_0} + sC \right)^{-1}$$

$$= \frac{RZ_0}{R + Z_0 + sRZ_0C}$$

The complex reflection coefficient seen by the first transmission line is given by

$$\Gamma(s) = \frac{Z_L - Z_0}{Z_L + Z_0} = \frac{-(Z_0^2 + RZ_0^2sC)}{2RZ_0 + Z_0^2 + RZ_0^2sC}$$

We can find $V^- = \rho V^+$ by using Fourier or Laplace transform analysis.

12.6   Transmission line dispersion

In reality all transmission lines deviate from ideal behavior in two important ways. First, all transmission lines have loss, and thus a wave on a transmission line decays in magnitude as it travels down the line. Second, most real transmission lines have non-linear phase delay versus frequency, leading to waveform dispersion. In Chapter 9 we found that we could quantify the loss in terms of the real part $\alpha(\omega)$ of the complex propagation constant $\gamma = \alpha + j\beta$ and the phase constant is determined by the imaginary component $\beta(\omega)$.

Signal decay can be compensated for by placing amplifiers, or repeaters, along the transmission line. But the waveforms do not shrink by simple scaling as magnitude distortion occurs since α is not a constant function of frequency. The different frequency components of the waveform experience different attenuation, with the attenuation typically an increasing function of frequency. The high-frequency components of a signal, such as the sharp edges, tend to be softened, as they travel along a transmission line.

Frequency-dependent losses occur for multiple reasons. In most transmission line structures, for instance, the conductive losses increase like $\sqrt{f}$ due to the skin effect. Other loss mechanisms, such as dielectric loss, also increase with frequency, due to dielectric resonance.

In theory we can compensate for the frequency-dependent losses by introducing the correct compensation network. For instance, an amplifier can boost high frequencies (with a zero in the transfer function) to clean up the waveform. But in a typical transmission line we have to also compensate for the waveform distortion due to the non-linear phase constant $\beta(\omega)$. Recall that a non-distorting transfer function should ideally scale and delay a waveform. An ideal transmission line has this ideal characteristic as the phase constant is a linear function of frequency. For a lossless transmission line we have

$$\gamma = \sqrt{j\omega L j\omega C} = j\omega\sqrt{LC} \tag{12.65}$$

a linear relation between the phase delay versus frequency. The group delay for any waveform is therefore a constant. Since switching waveforms are broadband in nature, we need to ensure that over the bandwidth of the waveform, the phase delay has a linear response. Intuitively it is clear that a non-uniform group delay implies that different parts of the waveform arrive at different times and thus corrupt the waveform. For a lossy transmission line where the losses are dominated by the conductive losses of the material, we have

$$\gamma = \sqrt{(j\omega L + R)j\omega C} = j\omega\sqrt{LC}\sqrt{1 - j\frac{R}{\omega L}} \tag{12.66}$$

For low frequencies, or if $\omega L \ll R$, the line is an RC distributed line that we studied in Chapter 11. For this case the attenuation and group delay are frequency dependent, leading to distortion. For high frequencies, though, $\omega \gg R/L$, the line behaves like an ideal low-loss transmission line. Unfortunately for a real line the conductive losses increase with frequency and lead to dispersion. Furthermore, inclusion of dielectric losses leads to further distortion.

To demonstrate the effects of dispersion, consider the following wavepacket traveling on a transmission line

$$p(x,t) = \sum_{k=1}^{100} 0.07\left(e^{-(0.1k-3)^2} + e^{-(0.1k+3)^2}\right) \cos\left(0.1kx - \frac{0.1kt}{\sqrt{1 + 0.02(0.1k)^2}}\right) \tag{12.67}$$

The waveform dispersion occurs since the phase delay decreases non-linearly for each successive harmonic k of the signal, as shown in Fig. 12.21. A plot of the waveform shape

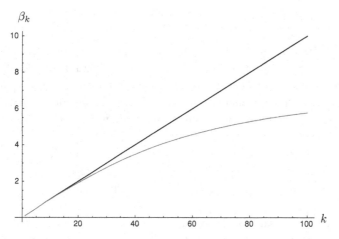

Figure 12.21 The dispersion curve shows that increasing harmonics k of the wavepacket have smaller propagation constants β_k.

traveling down the line is shown in Fig. 12.22. In Fig. 12.22a we have removed the dispersion to show an ideal case of a wavepacket traveling to the right. In Fig. 12.22b we see severe distortion occurring. Since higher frequencies travel slower, the wavepacket broadens and distorts along the line.

We found from Eq. 9.77 that if $R/L = G/C$, then the lossy transmission line is also dispersionless. This is why old telephone networks included artificial transmission lines with lumped inductor sections, used to raise the inductance per unit length to satisfy this equation. In a modern communication system, we solve this problem using the power of digital electronics. In essence, if we could discover the phase and magnitude response of the channel, the transmission line in this case, then we could construct a digital filter to compensate for the phase non-linearity of the channel. This is known as equalization. Alternatively, we can break our signals into multiple narrowband carriers, and send these carriers through the channel. Since each carrier is narrowband, phase distortion does not occur, since over a narrow bandwidth the group delay is approximately constant. Inter-carrier interference can be minimized by using orthogonal carrier tones. We can also distribute energy into the tones in such a way as to maximize the overall performance of the system. Since different tones will experience different levels of attenuation, it is intuitively clear that we should inject more power into weak tones to maximize the overall signal to noise ratio. This is in fact the manner in which modern communication systems compensate for the channel properties. Digital subscriber line (DSL) uses an approach similar to this to pack significantly more data on to a telephone wire than otherwise possible.[2] A similar scheme is employed in OFDM broadband wireless systems such as 802.11a/g wireless LAN.

---

[2]   Recall the days of a 300 baud modem? Well at least you recall the 56K modem. A DSL link can pump 1.5 Mbps through the same line.

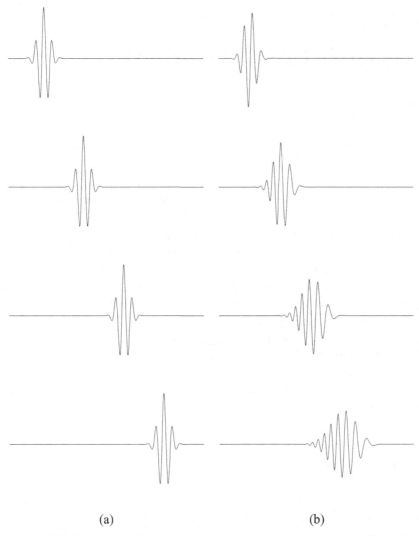

(a)   (b)

Figure 12.22 (a) A wavepacket travels at constant speed along a transmission line with no dispersion. (b) A wavepacket experiences severe dispersion as it travels through a transmission line.

12.7 References

This chapter has been adapted from my lecture notes prepared for an electromagnetics course. I have used several references, including *Fields and Waves in Communication Electronics* [52], Cheng [6], and Inan and Inan [62]. Other important references include Johnson and Graham [29].

13 Magnetic and electrical coupling and isolation

Coupling of electric, magnetic, and electromagnetic fields from one part of a circuit to another is usually undesirable and can lead to many problems, such as noise injection, interference, instability through parasitic feedback, and transient ringing. Ultimately this leads to reduced sensitivity in a system such as a radio receiver or a fast digital circuit. In this chapter we explore various coupling mechanisms and techniques for isolation. Entire books have been dedicated to this subject, such as the excellent book by Ott [45]. Much of the material for this chapter has been inspired from this source and the interested reader is highly encouraged to read Ott for further details.

In practice coupling problems are very difficult to solve, since the source of the coupling is often difficult to identify. The techniques of this chapter should give the necessary tools to properly diagnose and solve such coupling problems. As we shall see, there is no straightforward way to do this, but applying fundamental principles of electromagnetics, with the right approximations, can lead to tractable solutions.

13.1 Electrical coupling

Electrical coupling occurs due to stray capacitance in a circuit. A typical scenario is shown in Fig. 13.1. In this example the source of interference is from the first circuit, and the *receptor* circuit picks up the interference or noise due to the coupling capacitor C_c. A simplified equivalent circuit calculation shows that the magnitude of the noise pickup is given by the voltage divider

$$v_n = v_1 \frac{Z_2}{Z_2 + \frac{1}{j\omega C_c}} \tag{13.1}$$

where Z_2 is the effective impedance to ground. The effect of current (magnetic field) in the first circuit is neglected. If we assume that the magnitude of coupling is relatively weak, then the currents and voltages induced in the receptor circuit will not substantially influence the voltage in the first. A common situation occurs when $R_2 \ll 1/(j\omega C_2)$, and so the magnitude of voltage induced in the second circuit is given by

$$v_n = j\omega R_2 C_c v_1 = R_2 \times i_n \tag{13.2}$$

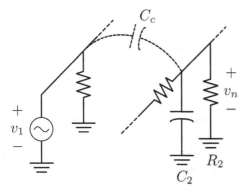

Figure 13.1 Electrical coupling occurs between circuits due to stray capacitive coupling.

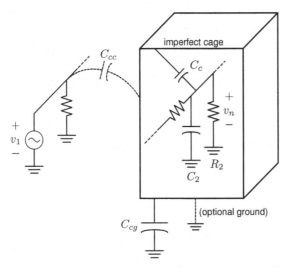

Figure 13.2 The electrical isolation between circuits can be improved considerably by enclosing the circuit inside of a metallic enclosure (cage). Unless the cage completely encloses the circuit, the cage must be grounded to reduce the coupling.

where the noise current is given by $i_n = j\omega C_c v_1$. Obviously, the way to reduce this noise pickup is to reduce C_c. This can be accomplished in several ways. The most direct approach is to increase the distance between the source of the interference and the receptor circuit, resulting in a lower value of C_c. This is not always possible, and in practice the capacitance C_c tends to decrease only logarithmically after the spacing between the conductors is increased substantially. Operating at lower impedance also tends to minimize the electrical noise coupling, although it leads to an increase in magnetic coupling (see next section). A much more effective approach is to shield the circuits from one another.

Electrical shielding

If the receptor circuit is surrounded by a metallic *cage*, commonly known as a Faraday Cage, then the coupling can be eliminated. This is shown in Fig. 13.2a. The metallic cage must completely surround the circuit in order to shield it. In the same way, the source of the noise can be shielded to minimize interference and radiation, and this is in fact the preferable course of action. Highly sensitive circuits are often caged in a metal enclosure.

It is vital that the cage be completely closed, with no holes. For such a case, the inside of the cage is surrounded by a closed equipotential surface. Since the electrostatic potential satisfies Laplace's equation in this case, we can argue simply that the potential inside must be constant. This is so because a constant potential satisfies the boundary conditions and it also satisfies Laplace's equation. Since the solution is unique, it is then the only possible solution. Since the potential inside is a constant, no differential noise voltage can be induced and no external interference can disturb the circuit inside the structure. The moment the structure is punctuated with a hole, though, this argument falls apart and external fields can leak inside. Thus, if signals must enter and exit the cage, it is important to keep the openings small. Often the cage will be grounded. While this is not strictly necessary, it has many advantages, especially when the cage is non-ideal.

For a cage with an opening, the shielding is non-ideal. Assume that all the field lines from the source terminate on the cage. Also assume that the cage is not grounded. Then the noise voltage pickup on the cage can be calculated with the aid of Fig. 13.2

$$v_{nc} = \frac{C_{cc}}{C_{cc} + C_{cg}} v_1 \tag{13.3}$$

where C_{cg} is the capacitance of the shield to ground. The noise pickup by the circuit is therefore given by

$$v_n = \frac{R}{R + \frac{1}{j\omega C_c}} v_{nc} \approx v_{nc} \tag{13.4}$$

Note that as the physical size of the cage is increased, the noise pickup decreases. In the limit that $C_{cg} \to \infty$, the noise pickup disappears. In essence, the cage itself is acting like a "ground," since it has a very large supply of charge that can easily respond to an external disturbance. If the cage is not physically large but grounded, then the coupling is precisely zero, $v_n = 0$, since charge flows on and off of the shield to maintain zero voltage on the shield, and since the potential of the shield does not change, the voltage pickup on the circuit is also zero. In practice, the shield can take on many forms, such as a coaxial cable (Fig. 13.3) or on-chip approximation thereof (Fig. 13.4).

Ground plane

If the circuit cannot be physically covered with a metallic cage, then the electrical coupling is reduced if the circuits are placed near a ground plane, as shown in Fig. 13.5. Ground planes are often employed on PCBs and integrated circuits. Since the majority of the electric fields from the noise source terminate on the ground plane, the coupling capacitance C_c is reduced substantially. It is important for the ground plane to be truly grounded. An ungrounded plane can actually enhance the coupling, as shown in Fig. 13.6. Here the effective coupling capacitance is given by the series combination of C_{1g} and C_{2g} (assuming a small resistance on the ground plane), which can be larger than C_c, the capacitance in the absence of the

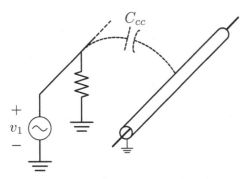

Figure 13.3 A coaxial structure with a grounded outside conductor acts as an electrical shield.

Figure 13.4 In an IC environment, a sensitive trace can be enclosed in a rectangular cage structure.

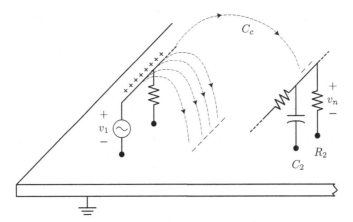

Figure 13.5 A ground plane can help isolate structures by providing a termination to electrical field lines.

ground plane. Since the ground plane is not grounded, it responds to the external field by charge separation, maintaining the equipotential in the bulk of the plane. This charge separation, though, effectively transfers the noise disturbance to the receptor circuit.

13.2 Magnetic coupling

Magnetic coupling occurs due to mutual inductance between circuit elements. In Fig. 13.7, current from the source wire generates an emf in the receptor wire due to this magnetic

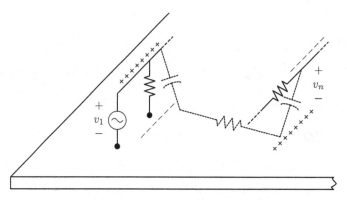

Figure 13.6 An ungrounded floating "ground plane" can result in an unintended increase in the electrical coupling from one circuit to another.

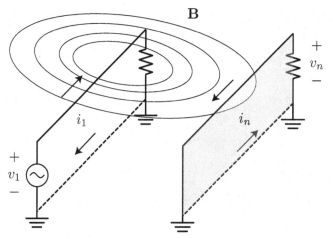

Figure 13.7 Magnetic coupling occurs between wires or traces in a circuit due to magnetic flux leakage.

coupling. Again, if we assume that the magnetic coupling is weak, then the noise disturbance current and fields is only altered in a minor way due to the coupling and we can assume that

$$v_n = j\omega M i_1 \tag{13.5}$$

We have learned many techniques to calculate the mutual inductance M, and we know that to reduce M we must minimize the flux pickup. If we assume that the flux crossing the receptor circuit is uniform, then the only way to reduce the magnetic pickup is by reducing the effective area of the second circuit.

Magnetic isolation

In the problem of two wires, note that placing a conductive *non-magnetic* cylindrical shield around the receptor wire, shown in Fig. 13.8a, is absolutely ineffective at shielding the magnetic field. This is because no current flows in the shield and thus the flux pickup by

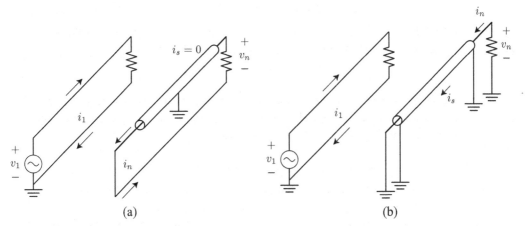

Figure 13.8 (a) A cable employing a non-magnetic conductive coaxial ground layer cannot intrinsically shield against magnetic fields. (b) By allowing current to flow in the shield, induced currents flow and shield the inner conductor.

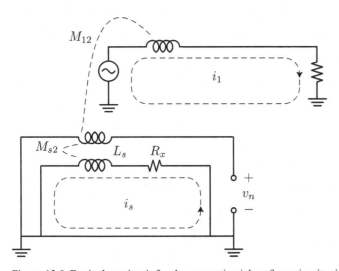

Figure 13.9 Equivalent circuit for the magnetic pickup for a circuit with a shield conductor.

the wire is the same as before. This is true even if we ground the shield at a one-point, as shown. On the other hand, if we ground the shield at two points, preferably the ends of the shield as shown in Fig. 13.8b, then the induced current flowing in the shield generates a magnetic field, which tends to cancel the flux of the source, providing isolation. Note that it is the induced current that effectively shields the wire, not the shield itself. To calculate the shielding effectiveness, refer to Fig. 13.9, where M_s is the mutual inductance between the shield and the source, M_{12} is the mutual inductance between the source and the wire, and finally M_{s2} is the mutual inductance between the shield and the receptor wire. The voltage induced on the receptor wire is given by

$$v_n = j\omega M_{12} i_1 + j\omega M_{s2} i_s \tag{13.6}$$

If the shield current flows uniformly around the receptor wire, then the mutual inductance $M_{12} = M_{s2}$. This is because the magnetic field inside the cylindrical tube is precisely zero (see Chapter 4) and so the flux inside the wire is also zero. Then we have

$$v_n = j\omega M_{s2}(i_1 - |i_s|) \tag{13.7}$$

where we have explicitly noted that the phase of the shield current is opposite to the source current by Lenz's Law. Then the shielding effectiveness depends very much on the induced current i_s. For an ideal shield with zero resistance

$$v_s = 0 = j\omega M_{12}i_1 + j\omega L_s i_s \tag{13.8}$$

So the induced shield current is simply $i_s = -i_1\frac{M_{12}}{L_s}$. But since the flux inside the shield conductor is precisely zero, then $L_s = M_{12}$ and the current is precisely equal to $-i_1$. Under this circumstance the response of the shield is perfect and no voltage is induced on the inner wire. Unfortunately, there is some shield resistance, which upsets this delicate balance. If we insert the shield resistance into the above equation, we have

$$v_s = 0 = R_x i_s + j\omega L_s i_s + j\omega L_s i_1 \tag{13.9}$$

or

$$i_s = -\frac{j\omega L_s}{R_x + j\omega L_s}i_1 \tag{13.10}$$

and

$$v_n = j\omega L_s(i_1 + i_s) = j\omega L_s \frac{R_x}{R_x + j\omega L_s}i_1 \tag{13.11}$$

The efficacy of the shield is determined by the shield cutoff frequency, $\omega_c = R_x/L_s$. Asymptotically, the noise pickup approaches a constant given by

$$v_{2,\infty} = R_x i_1 \tag{13.12}$$

This means that at high frequency we must strive to reduce the shield resistance as much as possible. At low frequencies, or frequencies below the cutoff, the noise pickup increases with frequency such as

$$v_{2,0} = j\omega L_s i_1 \tag{13.13}$$

which is the same coupling as the case without the shield. In such a case, it may be advantageous to only ground the shield at a single point to prevent shield currents altogether. The reason for this will be demonstrated in the next section.

Shield and ground together

As we have seen many times in this book, the closer the return current, the less magnetic pickup and radiation, as illustrated in Fig. 13.10a–c. Since return current flows through the coaxial ground shield (or through the ground of a microstrip), as opposed to the ground plane, the area of the circuit is much smaller, therefore the magnetic pickup/radiation is

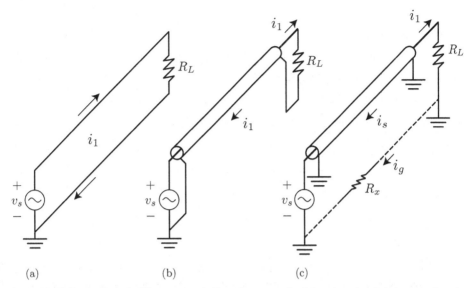

Figure 13.10 For uniform incident magnetic flux, the magnetic pickup in a circuit is a strong function of circuit area. In (a) the area is quite large, making the circuit very sensitive to external disturbance. In (b) the return current flows close to the wire in the ground shield, making the circuit very insensitive to external disturbance. (c) The return current can flow either through the ground plane or through the shield. Case (b) is preferable for magnetic shielding.

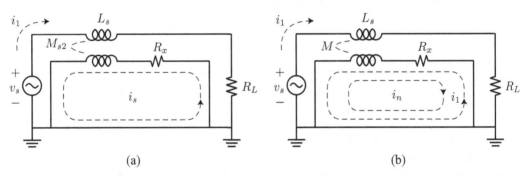

Figure 13.11 (a) Equivalent circuit for calculation of return current for shielded conductor, where the shield has resistance R_x. (b) If the shield is grounded at both the source and load, an extra loop between the shield and ground can pick up a noise i_n due to incident magnetic flux.

reduced substantially. In Fig. 13.10b, the shield is grounded at the source and so all the return current must flow through the shield. This is the desired situation, since in Fig. 13.10c the current can flow partly through the ground plane. At high frequency, virtually all the current will flow in the shield as opposed to the ground, since it forms the path of least impedance. We can find the fraction of current flowing in the shield by writing KVL around the two loops shown in Fig. 13.11a

$$v_s = j\omega L_s i_1 + i_1 R_L - j\omega M i_s \tag{13.14}$$

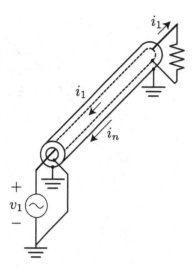

Figure 13.12 A triaxial conductor structure is used to prevent the flow of ground loop currents inside the signal conductor. Coaxial structures act as triaxial shields at very high frequencies due to skin effect.

and

$$0 = j\omega L_s i_s + i_s R_x - j\omega M i_1 \tag{13.15}$$

or

$$i_s = \frac{j\omega M}{j\omega L + R_x} i_1 \tag{13.16}$$

where R_x is the resistance in the ground shield. For the case of uniform current in a circular shield, we have that $M = L_s$ so that the shield current is essentially the same as the center conductor well above the cutoff frequency.

But a disadvantage of this circuit is that even though the return current in the shield reduces the loop area of the primary loop, a secondary loop is formed between the shield and the ground plane. Thus a noise current, shown in Fig. 13.11b, will flow in this ground loop and cause noise coupling into the circuit. Thus it is in general preferable to avoid this scenario by only grounding the outer conductor at one point.

It is interesting to note that this sort of noise coupling does not occur at very high frequencies due to the skin effect. This is so because the noise voltage induced in the ground loop flows on the outer surface of the shield, whereas the return current flows on the inner surface of the shield, thus providing natural isolation in the circuit. At low frequencies, the same effect can be obtained if we use a triaxial cable, as shown in Fig. 13.12. The outer cable shield provides a capacitive and inductive shield against noise voltage and currents.

Twisting the signal and return current wire together is also a very effective way to reduce the area of the loop, as shown in Fig. 13.13a. It has a further benefit that the voltage is induced differently on successive sections of the circuit. In fact, we can see that the induced flux angle varies 360° over successive sub-loops and thus there is flux cancellation between loops. For a uniform incident flux, therefore, we expect very good isolation. The same

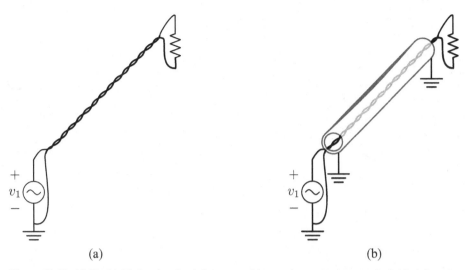

(a)   (b)

Figure 13.13 (a) Unshielded twisted pair has a good immunity against magnetic field pickup due to the small circuit area and the twisting of the loops. (b) A shielded twisted pair is a very effective way to reduce electrical and magnetic coupling. The grounded shield is also useful for eliminating ground loop current pickup in the signal conductors.

precaution against grounding both ends of the twisted pair applies, since a ground loop can form between the ground plane and the return wire of the twisted pair, creating a new coupling mechanism for the noise. In practice, the main limitation of the twisted pair wiring scheme is the frequency cutoff and non-constant Z_0, making it harder to terminate the twisted pair compared with a true transmission line. The shielded twisted pair, shown in Fig. 13.13b, provides very good isolation.

13.3 Ground noise coupling

Supply and ground bounce

All real circuits have finite inductance to supply and ground. Thus, if a circuit draws a varying current from the supply, due to dI/dt, it will experience what is commonly referred to as ground "bounce" and supply "bounce," as shown in Fig. 13.14a. In this particular example, we show two gates driving off-chip loads. Since the loads for most digital circuits are capacitive, any variation in output voltage results in a variation in the drive current. For a given load transition waveform, $V_L(t)$, we can estimate the current drawn from the supply or ground by $I = -CdV_L/dt$. In this case we demonstrate a high-to-low transition, which draws current from the load into the ground terminal. Since the power and ground terminals of the chip have inductance, due to bond-wires, package leads, and board traces, any variation in the supply current will cause a voltage shift in the supply, given by $V_{gnd} = LdI/dt$. This results in the "bounce" waveform, obtaining a second derivative characteristic shape shown in Fig. 13.14b. In the worse case, if the bounce is sufficiently large (set by the noise margins of the digital gate), we may experience a false transition on another output. As the on-chip

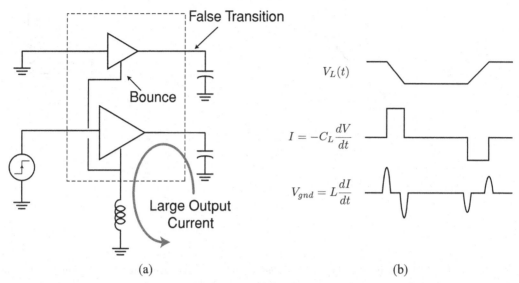

False Transition

Bounce

Large Output Current

$V_L(t)$

$I = -C_L \dfrac{dV}{dt}$

$V_{gnd} = L \dfrac{dI}{dt}$

(a)   (b)

Figure 13.14 (a) An example setup demonstrating ground bounce. Likewise, the low-to-high transition creates supply bounce. (b) The current and voltage waveforms during a ground or supply bounce event.

ground is a shared connection, the ground bounce can alter the V_{gs} and threshold V_T of other transistors, resulting in possible false transitions.

For a real digital circuit, there are thousands (or possibly millions) of gates transitioning in any given cycle. This will result in an apparent noise-like shape for the ground and supply waveforms, which create a pseudo-random current. Since the voltage and ground is shared by a lot of other circuits, this shift injects supply and ground noise into other circuits. Besides possible false transitions, the noise will disturb sensitive analog and RF circuitry, sharing the same substrate. Since the amount of bounce is proportional to L and dI/dt, the faster the current variations occur, the more bouncing that occurs.

Ground bounce is alleviated by placing a large capacitor on-chip, often referred to as a bypass capacitor. This capacitor acts like a local battery, supplying current to on-chip circuitry and bypassing the supply. Intuitively, the bypass capacitor acts as a good battery if its impedance is low, or if $1/\omega C$ is sufficiently low. If we can make this much lower than the supply impedance, proportional to ωL, then we would expect current variations to come from this capacitor. A DC current flowing through the inductors will then compensate for the average charge lost on the bypass capacitor. In Fig. 13.15, we see that connecting a large bypass capacitance has a beneficial impact on the supply bounce.

In particular, consider the case outlined in Fig. 13.14, where a high-to-low transition at the load discharges through an on-chip inverter. For the idealized current and voltage waveforms shown, ground bounce occurs during the current transition, since $V_L = L\,dI/dt$. In a like manner, supply bounce occurs during a low-to-high transition. Let's estimate the amount of bounce to see if it poses a problem. In a typical scenario there might be around 2 nH of cumulative ground or supply inductance. Imagine switching a 50 Ω load between $0 - 1$ V at a rate of 1 GHz, with rising/falling edges of 50 ps. Then $dI/dt = 1\,\text{V}/50\Omega/10\,\text{ps} = 0.8 \times 10^9\,A/s$, and $L\,dI/dt = 800\,\text{mV}$, a very substantial bounce! Clearly this is intolerable.

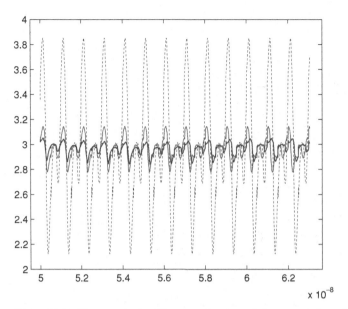

Figure 13.15 The magnitude of the ground and supply bounce can be reduced by increasing the size of the bypass capacitance. The bypass capacitor is changed from 10 pF, to 100 pF, and finally 400 pF to minimize the supply bounce. The package has 10 nH of inductance and the chip draws about 150 mA of current and operates at 1 GHz clock frequency.

Notice that the problem is exacerbated by faster switching, a common scenario in today's high-speed electronics. This example may seem contrived, but consider a digital circuit with one million transistors switching. At any given time only a small fraction may be switching, but that is still a substantial current drawn from the supply. On-chip most loads are capacitive, so the current draw is also proportional to the switching rate $I = C dV/dt$. A typical gate might drive $C = 10$ fF of capacitance, so the peak current is $I_C = 10$ fF $\cdot$ 1 V/50 ps $= 200\,\mu$A. If 10% of the million gates switch at any clock edge, we have a huge 20 A of current drawn from the supply!

The problem with ground bounce is not exclusive to digital circuits. In Fig. 13.16 we see that an amplifier (say a power amplifier or buffer) drives an off-chip load. Due to the large current drawn into the load, large variations occur. For instance, to drive 100 mW into an off-chip 50 Ω load at 5 GHz requires a drive current 63 mA. Only 100 pH of ground inductance will cause a parasitic swing of 200 mV! For internal circuits biased using the on-chip references, this does not pose a severe problem. But for an off-chip signal at the input referenced to the external ground, this signal will compete and de-sensitize the circuit. For instance, the low-noise amplifier (LNA) will have problems co-existing with this power amplifier if they share the same ground. The solution, of course, is to separate the ground connections on-chip and to use decoupling (next section) to reduce the pickup.

Bypass and decoupling

As outlined in the previous section, supply bypassing is critical in order to reduce ground bounce and noise in high-speed circuits. The bypass capacitor must be large enough to

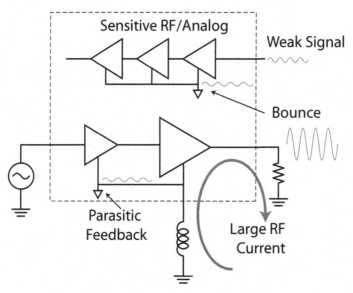

Figure 13.16 Ground and supply bounce in an analog/RF circuit.

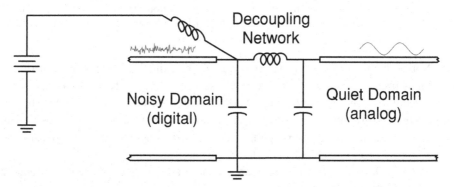

Figure 13.17 A decoupling network is a filter to reject high-frequency noise signals from entering a sensitive circuit from the supply lead.

supply the necessary switching current, else it will introduce another source of ground bounce. Furthermore, the effective series resistance (ESR) and inductance (ESL) of the capacitor needs to be small enough in order to minimize this secondary ground bounce. This is why several capacitors in parallel are preferred over a single large capacitor. This is also why a capacitor should be placed on-chip. But on-chip bypass is not enough and we typically must also place bypass capacitors on the PC board as close as possible to the chip. The proximity is important because the further we place the capacitor, the more inductance will be introduced, creating undesirable bounce.

While bypass can minimize the noise, it cannot eliminate it completely. A sensitive analog or RF circuit might be sensitive to μV level signals, and such noise could be intolerable. *Decoupling* circuits, shown in Fig. 13.17, can minimize the noise injection into a sensitive analog circuit. Here the LC filter acts like a low-pass filter, only allowing low-frequency

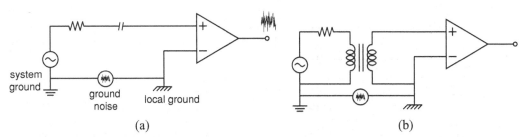

(a)   (b)

Figure 13.18 (a) Ground noise in a circuit can be modeled by a voltage source between two ground points in a circuit. (b) A transformer breaks the ground loop and eliminates the ground noise from entering a sensitive circuit.

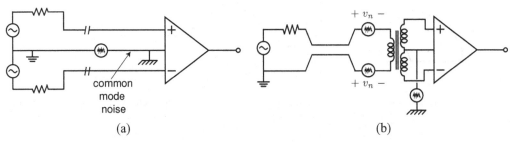

(a)   (b)

Figure 13.19 (a) Balanced circuit operation is naturally immune to ground noise. (b) A balun can be employed at the input of a single-ended circuit to reject ground noise.

signals to couple to the analog circuit. Note that separate supply pins are used for the analog circuitry for this purpose.

The design of power supply networks may seem trivial at first, but we already see that it can potentially involve many LC circuits. The danger of LC circuits is that they can resonate, creating new problems near the resonance frequency.

Ground noise rejection techniques

In circuits with ground noise, the "ground" is not constant and varies from point to point in the circuit. This results in noise injection from the ground into sensitive circuits, as shown in Fig. 13.18a. An easy way to remedy the situation is to employ a transformer, as shown in Fig. 13.18b. The ground now is applied across the primary and secondary of the transformer, and not at the input of the sensitive circuit. Another technique uses a common-mode choke, as described in Section 11.2. Since the ground voltage induces a common-mode current, this is rejected by the transformer. For ultra-low noise applications, optical couplers provide a useful way of isolation, especially when the DC voltage levels are markedly different. Finally, balanced or differential circuit operation, shown in Fig. 13.19a, is a very easy and natural way to combat common-mode noise in circuits, and most sensitive integrated circuits employ differential operation for precisely this reason. The common-mode rejection ratio ($CMRR$) is an important metric that specifies how much of the noise can be rejected. Since noise can also enter the circuit through the power supply, the power supply rejection ratio

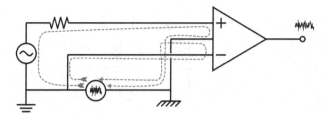

Figure 13.20 Any imbalance in a circuit can lead to a noisy differential input voltage. Since the path through the source includes the source resistance, it has a different impedance than the return current path. A fully balanced source has a symmetric impedance to ground from either terminal.

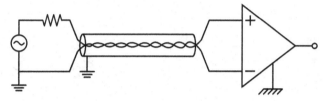

Figure 13.21 Shielding can be used in conjunction with differential circuitry to improve the isolation. The point where the shield is grounded should be carefully considered.

$(PSSR)$ is also important. If the circuit components are single ended, then a balun can be used to reject the common-mode signal, as shown in Fig. 13.19b.

Care must be taken to ensure fully balanced operation for perfect common-mode cancellation to occur. For instance, in the situation shown in Fig. 13.20, the ground voltage induces a differential voltage since the path through the source has an additional resistance. The solution is to use a fully balanced source, so that the ground noise generates only the common-mode signal. Alternatively, a balun can be used to convert a single ended signal to a differential signal. Here two signals are used to route a single-ended signal to the input of the differential circuit. Since the two wires in close proximity (such as a twisted pair or transmission line), the noise pickup is common-mode. Likewise, the noise pickup from the common ground of the balun is injected into the amplifier as common-mode noise.

Differential or balanced operation can be used in conjunction with shielding for very effective isolation. An example is shown in Fig. 13.21, where a balanced amplifier is shielded. Here we assume that the amplifier is floating and the best point to tie a single ground is at the source [45]. For a floating source, the opposite is true.

13.4   Substrate coupling

Substrate injection mechanisms

In standard Si integrated circuit environments, all devices reside on a common substrate. Direct injection into the substrate occurs through the substrate contact and guard rings, since they form ohmic contacts. Indirect injection occurs through devices. As illustrated generally in Fig. 13.22, and more specifically in Fig. 13.23, while devices are isolated from the

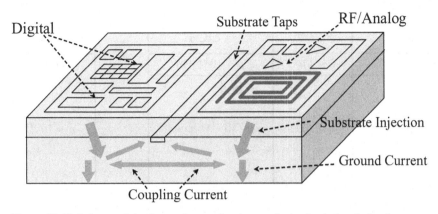

Figure 13.22 Substrate injection and reception in a typical mixed signal circuit can couple noisy digital circuits to sensitive analog and RF circuitry.

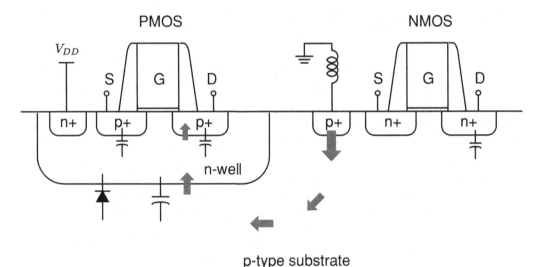

Figure 13.23 Substrate injection and reception in a typical p-type CMOS substrate occurs through substrate taps and device parasitic capacitance.

substrate at DC, AC currents can readily flow into the substrate through parasitic capacitors. In passive integrated capacitors, the back-plate couples to the substrate. A shield can reduce this coupling at the expense of enhancement of the back-plate parasitics. Spiral inductors and transformers likewise inject signals into the substrate through the oxide capacitance. This can be shielded to a certain extent by employing pattered shields (in order to prevent eddy currents). Magnetically induced substrate currents, though, are nearly impossible to shield unless magnetic materials are employed. Since passive devices are typically physically large, they are good sources and receptors for substrate noise.

Active devices inject current into the substrate through their junction source and drain capacitance, and through the bulk connection. Noise in the substrate is likewise coupled through the device through the back-gate effect, which modulates the threshold and thus

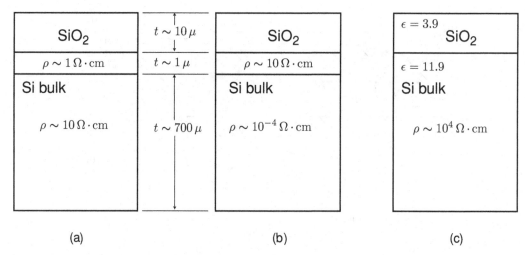

Figure 13.24 Typical CMOS substrate profiles. (a) A lightly doped substrate. (b) A heavily doped substrate. (c) An undoped substrate.

current of the device. Impact ionization current also injects charges into the substrate [61]. Junction leakage currents and gate tunneling current provide other conducting paths to the substrate, but are typically orders of magnitude smaller than the capacitive currents.

Since the semiconducting substrate is a moderately good conductor, substrate currents flow from one part of the circuit to another. The injection mechanisms act as a noise reception mechanism, generating substrate noise. In mixed-signal circuits, the fast and pseudo-random digital switching signals generate significant dI/dt, which modulate the ground and supply nodes of the circuit due to ground and supply inductance. This modulation is coupled through the substrate through substrate taps and well connections. Sensitive analog transistors therefore will pick up this "noise" leading to lower sensitivity, enhanced phase noise, and other undesirable phenomena. Since nMOS transistors are in the bulk substrate, or the common p-substrate, they pick up the noise directly from the substrate. pMOS devices, on the other hand, pick up noise from the well junction capacitance. For this reason, a triple-well technology is an attractive solution for reducing the noise coupling (see next section).

Substrate isolation

The best way to reduce substrate noise is through careful system design and planning. For instance, current mode logic generates substantially less noise than CMOS, since the differential circuit draws less current from the common supply, generating less ground bounce. Isolation of digital and analog grounds is another vital step in preventing crosstalk.

Layout techniques can ameliorate substrate noise problems by physical separation of the interference source from the receptor and by shielding. Shielding is often in the form of a guard ring or "substrate wall." The efficacy of these techniques is a strong function of the substrate technology. Three typical substrate profiles are shown in Fig. 13.24. In

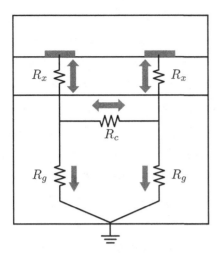

Figure 13.25 The equivalent circuit for substrate coupling. In a lightly doped substrate, R_c is a significant portion of the coupling resistance and increases with separation. In a heavily doped substrate, R_c is insignificant and separation is not effective in reducing the resistance.

Fig. 13.24a, a moderately resistive substrate of $\rho \sim 10\Omega/\mathrm{cm}$ is employed. A surface layer of higher conductivity, about $\sim 1\Omega/\mathrm{cm}$, is often placed to form the body for transistors. In Fig. 13.24b, a highly doped substrate is used, often in digital circuits. This helps in preventing latch-up but is very bad for analog and RF circuits. The substrate resistivity is about $\sim 0.1\mathrm{m}\Omega/\mathrm{cm}$, or about 1000 to 10 000 times more conductive than the first case. A few layers of higher resistivity appear at the surface to house the transistor wells. Finally, in Fig. 13.24c we have a extremely resistive substrate, with $\sim 10\mathrm{k}\Omega/\mathrm{cm}$. This is usually undoped intrinsic Si, or perhaps an insulator such as sapphire. The substrate is only used for mechanical support, and electrical isolation from the transistors is formed through a thick buried oxide layer. This substrate profile is common in SOI technology.

In a lightly doped substrate, the isolation improves with physical separation. The equivalent circuit of Fig. 13.25 shows that increasing the distance increases the resistance between the source and receptor. For small separations the depth of current flow is an important factor, but for far-away coupling, resistance increases nearly linearly. Moreover, at high frequencies the current flows along the skin of the substrate. Unfortunately, in a heavily doped substrate, the isolation does not improve with distance. This is because the heavily doped substrate acts like a ground plane, and R_C in Fig. 13.25 shorts all points on the chip to each other. The resistance is therefore dominated by the vertical path and so the coupling tends to flatten out with separation.

Guard rings

A guard ring, shown in Fig. 13.26, is typically a square ring of substrate contacts or well trenches surrounding a transistor, or a set of transistors. The guard ring can surround the noisy circuit and act as a sink for interference generated by circuits (such as the digital logic portion of a chip), and it can likewise surround a sensitive circuit, in order to sink the noisy

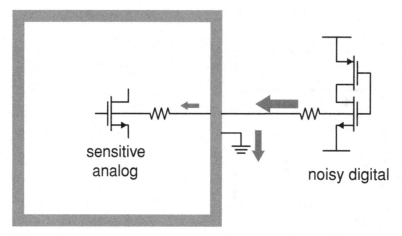

Figure 13.26 A guard ring is used to improve the substrate isolation.

currents and shunt them to ground. As such it acts like a capacitive shield. Unfortunately the ring lies at the surface of the substrate, limiting the efficiency of the shield. In some process technologies we may employ deep n-well or p-well layers to isolate circuitry. The deeper the contact, the more current it can intercept, thus improving the efficiency. The efficiency of the guard ring also depends strongly on how well it is grounded. For a guard ring to be effective, the inductance to ground must be smaller than the receptor circuit impedance to ground. This means that the guard ring should be placed as close as possible and surrounding the receptor to minimize the impedance to ground.

FET substrate network

In addition to acting as a noise injection and reception mechanism, the Si conductive substrate also acts as an additional loss mechanism. In some instances, such as high-frequency on-chip inductors, it is in fact one of the dominant sources of loss. We have seen this in Chapter 6, when constructing equivalent circuits for inductors and capacitors.

In CMOS FETs, the substrate impedance has a significant impact on the output impedance of the device, since the drain injects current directly into the transistor body. The actual substrate loss depends on the layout, especially the number of transistor fingers and the placement of the body contact. This is shown in Fig. 13.27, where two components of the substrate current can be identified. The "vertical" current flows to the substrate contacts on either side of the transistor fingers. This resistance component tends to increase as we increase the transistor finger width W. The "horizontal" component, on the other hand, tends to lower as we increase W. The overall resistance is difficult to calculate analytically, due to the three-dimensional nature of the layout. Practice shows that surrounding the transistor with as many contacts as possible is the best way to minimize the substrate resistance. For a high-frequency transistor employing many short fingers, W is small and the vertical component of the resistance dominates for all but the outer edge fingers. Thus the vertical contacts should be moved as close as possible to the transistor.

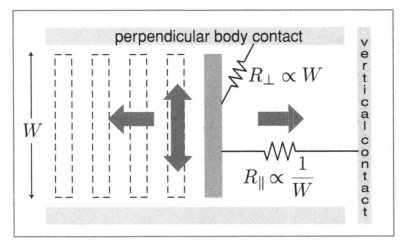

Figure 13.27 The layout of a FET has a strong influence on the body substrate resistance. Currents can flow in parallel to the fingers to a "vertical" ground or perpendicular to the fingers for an edge ground.

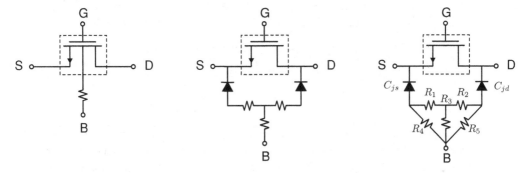

Figure 13.28 A substrate resistance network with a single body resistance is quite effective for most applications. For more accuracy, three or five resistors should be used.

A simple drain resistor to body is often sufficient to model the substrate loss in a transistor. An improvement is the three element network, shown in Fig. 13.28b. It has been shown that the substrate network with five elements, shown in Fig. 13.28c, is necessary to obtain an extremely good match to 2D and 3D simulations of the substrate parasitics of a transistor [15]. These substrate networks have been implemented into the popular BSIM4 transistor model [68].

13.5 Package coupling

In many cases the parasitic coupling through the package is the main source of coupling in an IC. The coupling on-board may also present problems. This is because relatively long

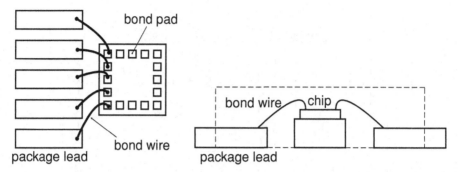

Figure 13.29 (a) Topview and (b) sideview of the bondwires in an integrated circuit. For rapid calculations, these bondwires can be approximated as piecewise linear structures.

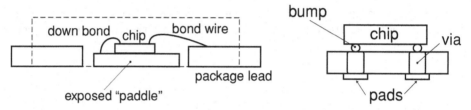

Figure 13.30 (b) The inductance of bondwires can be reduced by using short "downbonds" to the package ground. An exposed paddle allows currents to flow through the package ground on to the board ground. (a) Flip-chip technology is very effective in reducing the bondwire inductance.

bond wires make connections to the package leads where substantial inductive and capacitive coupling occurs. As we have seen, the package lead inductance is also responsible for ground bounce, which causes substrate noise injection. Full wave EM simulation is necessary to model the complex three-dimensional package and bondwire environment, but in practice simple hand calculations and careful planning can greatly reduce package coupling. The formulas from Grover (see Chapter 6) are very handy in these calculations. Recall that the inductance of a path is only defined when the entire loop is identified. The presence of a ground plane should also be considered in cases that the die (chip) is mounted onto a ground. This is sometimes called "an exposed paddle," since the back side of the package has a hole to allow it to be soldered on to a ground plane. In this case "down-bonds" (see Fig. 13.30a) can be used to form a low-inductance path to ground. The actual topology of the bond wire can be approximated by a piecewise curve.

Low inductance packages include flip-chip mounted packages, shown in Fig. 13.30b, eliminating the bond wires entirely. The inductance of a typical "path" can be cut down from $\sim 1\,\text{nH}$ to about $\sim 100\,\text{pH}$, a factor of ten reduction or more. The inductance is dominated by the stub bumps from the chip down to the package substrate. One disadvantage of flip-chip technology is that we cannot probe the top surface of the chip, which is often convenient when debugging. Furthermore, most of the heat removal must occur through the conductive bumps as opposed to the Si substrate.

13.6   References

As noted in the introduction, the first few sections were inspired by reading Ott's classic work [45]. Ott dedicates an entire book to the topic of this chapter. I also relied on some useful discussions in Johnson and Graham [29], papers posted on the *Designer's Guide* [32]. A more detailed discussion of substrate coupling and simulation can be found in Gharpurey [19].

14 Electromagnetic propagation and radiation

In this chapter we explore the propagation of the simplest wave solutions of Maxwell's equations. We begin with the one-dimensional case and generalize to a planar wave in vector coordinates. We shall find that many properties of plane waves are identical to the flow of voltages and currents along a transmission line. We derive equations for wave reflection and refraction into various media. Finally, we introduce the Poynting vector to facilitate calculations involving the power flow of an electromagnetic field.

14.1 Maxwell's equations in source-free regions

One-dimensional waves

In a source-free region $\rho = 0$ and $\mathbf{J} = 0$, which simplifies Maxwell's equations to

$$\nabla \cdot \mathbf{D} = 0 \tag{14.1}$$

$$\nabla \cdot \mathbf{B} = 0 \tag{14.2}$$

$$\nabla \times \mathbf{E} = -\frac{\partial \mathbf{B}}{\partial t} = -\mu \frac{\partial \mathbf{H}}{\partial t} \tag{14.3}$$

$$\nabla \times \mathbf{H} = \frac{\partial \mathbf{D}}{\partial t} = \epsilon \frac{\partial \mathbf{E}}{\partial t} \tag{14.4}$$

Assume that $\mathbf{E}$ and $\mathbf{H}$ are uniform in the $x - y$ plane so that $\frac{\partial}{\partial x} = 0$ and $\frac{\partial}{\partial y} = 0$. For this case the $\nabla \times \mathbf{E}$ simplifies to

$$\nabla \times \mathbf{E} = \begin{pmatrix} \hat{\mathbf{x}} & \hat{\mathbf{y}} & \hat{\mathbf{z}} \\ 0 & 0 & \frac{\partial}{\partial z} \\ E_x & E_y & E_z \end{pmatrix} \tag{14.5}$$

$$(\nabla \times \mathbf{E})_x = -\frac{\partial E_y}{\partial z} = -\mu \frac{\partial H_x}{\partial t} \tag{14.6}$$

$$(\nabla \times \mathbf{E})_y = -\frac{\partial E_x}{\partial z} = -\mu \frac{\partial H_y}{\partial t} \tag{14.7}$$

$$(\nabla \times \mathbf{E})_z = 0 = -\mu \frac{\partial H_z}{\partial t} \tag{14.8}$$

Similarly, writing out the curl of $\mathbf{H}$ in rectangular coordinates

$$\nabla \times \mathbf{H} = \epsilon \frac{\partial \mathbf{E}}{\partial t} \tag{14.9}$$

$$-\frac{\partial H_y}{\partial z} = \epsilon \frac{\partial E_x}{\partial t} \tag{14.10}$$

$$\frac{\partial H_x}{\partial z} = \epsilon \frac{\partial E_y}{\partial t} \tag{14.11}$$

$$0 = \epsilon \frac{\partial E_z}{\partial t} \tag{14.12}$$

Time variation in the $\hat{\mathbf{z}}$ direction is zero. Thus the fields are entirely transverse to the direction of propagation. We call such fields *TEM* "waves."

Polarized TEM fields

For simplicity, assume $E_y = 0$. We say the field is polarized in the $\hat{\mathbf{x}}$-direction. This implies that $H_x = 0$ and $H_y \neq 0$

$$\frac{\partial E_x}{\partial z} = -\mu \frac{\partial H_y}{\partial t} \tag{14.13}$$

$$-\frac{\partial H_y}{\partial z} = \epsilon \frac{\partial E_x}{\partial t} \tag{14.14}$$

$$\frac{\partial^2 E_x}{\partial z^2} = -\mu \frac{\partial^2 H_y}{\partial z \partial t} \tag{14.15}$$

$$\frac{\partial^2 H_y}{\partial z \partial t} = -\epsilon \frac{\partial^2 E_x}{\partial t^2} \tag{14.16}$$

We finally have it, a one-dimensional wave equation

$$\frac{\partial^2 E_x}{\partial z^2} = \mu \epsilon \frac{\partial^2 E_x}{\partial t^2} \tag{14.17}$$

Notice the similarity between this equation and the wave equation we derived for voltages and currents along a transmission line (see for instance Eq. (9.12)). As before, the wave velocity is $v = \frac{1}{\sqrt{\mu \epsilon}}$ and the general solution to this equation is

$$E_x(z, t) = f_1 \left(t - \frac{z}{v} \right) + f_2 \left(t + \frac{z}{v} \right) \tag{14.18}$$

Let's review why this is the general solution

$$\frac{\partial E_x}{\partial t} = f_1' + f_2' \tag{14.19}$$

$$\frac{\partial^2 E_x}{\partial t^2} = f_1'' + f_2'' \tag{14.20}$$

$$\frac{\partial E_x}{\partial z} = -\frac{1}{v} f_1' + \frac{1}{v} f_2' \tag{14.21}$$

$$\frac{\partial^2 E_x}{\partial z^2} = \frac{1}{v^2} f_1'' + \frac{1}{v^2} f_2'' \tag{14.22}$$

Wave velocity

A point on the wavefront is defined by $(t - z/v) = k$, where k is a constant. The velocity of this point is therefore v

$$1 - \frac{1}{v}\frac{\partial z}{\partial t} = 0 \tag{14.23}$$

$$\frac{\partial z}{\partial t} = v \tag{14.24}$$

We have thus shown that this wave moves at velocity

$$v = c = \frac{1}{\sqrt{\mu\epsilon}} \tag{14.25}$$

In free space, $c \approx 3 \times 10^8 \text{m/s}$, the measured speed of light. In a medium with relative permittivity ϵ_r and relative permeability μ_r, the wave moves with effective velocity

$$v = \frac{c}{\sqrt{\mu_r \epsilon_r}} \tag{14.26}$$

This fact alone convinced Maxwell that light is an electromagnetic wave.

Sinusoidal plane waves

For time-harmonic fields, the equations simplify to

$$\frac{dE_x}{dz} = -j\omega\mu H_y \tag{14.27}$$

$$\frac{dH_y}{dz} = j\omega\epsilon E_x \tag{14.28}$$

This gives a one-dimensional Helmholtz equation

$$\frac{d^2 E_x}{d^2 z} = -\omega^2 \mu\epsilon E_x \tag{14.29}$$

The solution is now a simple exponential

$$E_x = C_1 e^{-jkz} + C_2 e^{jkz} \tag{14.30}$$

The wave number is given by $k = \omega\sqrt{\mu\epsilon} = \frac{\omega}{v}$. We can recover a traveling wave solution

$$E_x(z, t) = \Re(E_x e^{j\omega t}) \tag{14.31}$$

$$E_x(z, t) = \Re\left(C_1 e^{j(\omega t - kz)} + C_2 e^{j(\omega t + kz)}\right) \tag{14.32}$$

$$E_x(z, t) = C_1 \cos(\omega t - kz) + C_2 \cos(\omega t + kz) \tag{14.33}$$

The wave has spatial variation $\lambda = \frac{2\pi}{k} = \frac{2\pi v}{\omega} = \frac{v}{f}$.

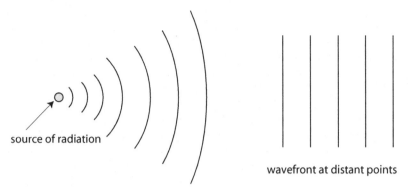

Figure 14.1 The waves radiated from a source appear as plane waves for distant points.

Magnetic field of a plane wave

We have the following relation

$$H_y = -\frac{1}{j\omega\mu}\frac{dE_x}{dz} = -\frac{1}{j\omega\mu}(-jkC_1e^{-jkz} + C_2jke^{jkz}) \tag{14.34}$$

$$H_y = \frac{k}{\mu\omega}(C_1e^{-jkz} - C_2jke^{jkz}) \tag{14.35}$$

By definition, $k = \omega\sqrt{\mu\epsilon}$

$$H_y = \sqrt{\frac{\epsilon}{\mu}}(C_1e^{-jkz} - C_2e^{jkz}) \tag{14.36}$$

The ratio E_x^+ and H_y^+ has units of impedance and is given by the constant $\eta = \sqrt{\mu/\epsilon}$. η is known as the impedance of free space.

The plane waves are the simplest wave solution of Maxwell's equations. They seem to be a gross oversimplification but, as shown in Fig. 14.1, they nicely approximate real waves that are distant from their source.

Wave equation in three dimensions

We can derive the wave equation directly in a coordinate free manner using vector analysis

$$\nabla \times \nabla \times \mathbf{E} = \nabla \times -\mu\frac{\partial \mathbf{H}}{\partial t} = \mu\frac{\partial(\nabla \times \mathbf{H})}{\partial t} \tag{14.37}$$

Substitution from Maxwell's equation yields

$$\nabla \times \mathbf{H} = \frac{\partial \mathbf{D}}{\partial t} = \epsilon\frac{\partial \mathbf{E}}{\partial t} \tag{14.38}$$

$$\nabla \times \nabla \times \mathbf{E} = -\mu\epsilon\frac{\partial^2 E}{\partial t^2} \tag{14.39}$$

Using the identity

$$\nabla \times \nabla \times \mathbf{E} = -\nabla^2\mathbf{E} + \nabla(\nabla \cdot \mathbf{E}) \tag{14.40}$$

Since $\nabla \cdot \mathbf{E} = 0$ in charge-free regions

$$\nabla^2 \mathbf{E} = \mu\epsilon \frac{\partial^2 \mathbf{E}}{\partial t^2} \tag{14.41}$$

In phasor form we have $k^2 = \omega^2 \mu\epsilon$

$$\nabla^2 \mathbf{E} = -k^2 \mathbf{E} \tag{14.42}$$

Now it is trivial to get a 1-D version of this equation

$$\nabla^2 E_x = \mu\epsilon \frac{\partial^2 E_x}{\partial t^2} \tag{14.43}$$

$$\frac{\partial^2 E_x}{\partial x^2} = \mu\epsilon \frac{\partial^2 E_x}{\partial t^2} \tag{14.44}$$

14.2 Penetration of waves into conductors

Inside a good conductor $\mathbf{J} = \sigma \mathbf{E}$. In the time-harmonic case, this implies the lack of free charges

$$\nabla \times \mathbf{H} = \mathbf{J} + \epsilon \frac{\partial \mathbf{E}}{\partial t} = (\sigma + j\omega\epsilon)\mathbf{E} \tag{14.45}$$

Since $\nabla \cdot \nabla \times \mathbf{H} \equiv 0$, we have

$$(\sigma + j\omega\epsilon)\nabla \cdot \mathbf{E} \equiv 0 \tag{14.46}$$

which in turn implies that $\rho = 0$. For a good conductor the conductive currents completely outweighs the displacement current, e.g. $\sigma \gg \omega\epsilon$. To see this, consider a good conductor with $\sigma \sim 10^7 \mathrm{S/m}$ up to very high mm-wave frequencies $f \sim 100\mathrm{GHz}$. The displacement current is still only

$$\omega\epsilon \sim 10^{11}10^{-11} \sim 1 \tag{14.47}$$

which is seven orders of magnitude smaller than the conductive current. For all practical purposes, therefore, we drop the displacement current in the volume of good conductors.

Inside of good conductors, therefore, we have

$$\nabla \times \mathbf{H} = \sigma \mathbf{E} \tag{14.48}$$

$$\nabla \times \nabla \times \mathbf{E} = \nabla(\nabla \cdot \mathbf{E}) - \nabla^2 \mathbf{E} = -\nabla^2 \mathbf{E} = -j\omega\mu\nabla \times \mathbf{H} \tag{14.49}$$

$$\nabla^2 \mathbf{E} = j\omega\mu\sigma \mathbf{E} \tag{14.50}$$

One can immediately conclude that $\mathbf{J}$ satisfies the same equation

$$\nabla^2 \mathbf{J} = j\omega\mu\sigma \mathbf{J} \tag{14.51}$$

Applying the same logic to $\mathbf{H}$, we have

$$\nabla \times \nabla \times \mathbf{H} = \nabla(\nabla \cdot \mathbf{H}) - \nabla^2 \mathbf{H} = -\nabla^2 \mathbf{H} = (j\omega\epsilon + \sigma)\nabla \times \mathbf{E} \tag{14.52}$$

$$\nabla^2 \mathbf{H} = j\omega\mu\sigma \mathbf{H} \tag{14.53}$$

Let's solve the 1D Helmholtz equation once again for the conductor

$$\frac{d^2 E_z}{d^2 x} = j\omega\mu\sigma E_z = \tau^2 E_z \tag{14.54}$$

We define $\tau^2 = j\omega\mu\sigma$ so that

$$\tau = \frac{1+j}{\sqrt{2}}\sqrt{\omega\mu\sigma} \tag{14.55}$$

Or more simply, $\tau = (1+j)\sqrt{\pi f \mu\sigma} = \frac{1+j}{\delta}$. The quantity $\delta = \frac{1}{\sqrt{\pi f \mu\sigma}}$ has units of meters and is an important number.

The general solution for the plane wave is given by

$$E_z = C_1 e^{-\tau x} + C_2 e^{\tau x} \tag{14.56}$$

Since E_z must remain bounded, $C_2 \equiv 0$

$$E_z = E_0 e^{-\tau x} = \underbrace{E_x e^{-x/\delta}}_{\text{mag}} \underbrace{e^{-jx/\delta}}_{\text{phase}} \tag{14.57}$$

Similarly, the solution for the magnetic field and current follow the same form

$$H_y = H_0 e^{-x/\delta} e^{-jx/\delta} \tag{14.58}$$
$$J_z = J_0 e^{-x/\delta} e^{-jx/\delta} \tag{14.59}$$

Penetration depth

The wave decays exponentially into the conductor. For this reason, δ is called the *penetration depth*, or more commonly, the *skin depth*. The fields drop to $1/e$ of their values after traveling one skin depth into the conductor. After several skin depths, the fields are essentially zero. You may also say that the wave exists only on the "skin" of the conductor. For a good conductor at $f = 1\text{GHz}$

$$\delta = \frac{1}{\sqrt{\mu\sigma f \pi}} \sim 10^{-6}\text{m} \tag{14.60}$$

As the frequency is increased, $\delta \to 0$, or the fields completely vanish in the volume of the conductor.

Why do fields decay in the volume of conductors? The induced fields cancel the incoming fields. As $\sigma \to \infty$, the fields decay to zero inside the conductor. The total surface current flowing in the conductor volume is given by

$$J_{sz} = \int_0^\infty J_z dx = \int_0^\infty J_0 e^{-(1+j)x/\delta} dx \tag{14.61}$$

$$J_{sz} = \frac{J_0 \delta}{1+j} \tag{14.62}$$

At the surface of the conductor, $E_{z0} = \frac{J_0}{\sigma}$

Thus we can define a surface impedance

$$Z_s = \frac{E_{z0}}{J_{sz}} = \frac{1+j}{\sigma\delta} \tag{14.63}$$

$$Z_s = R_s + j\omega L_i \tag{14.64}$$

The real part of the impedance is a resistance

$$R_s = \frac{1}{\delta\sigma} = \sqrt{\frac{\pi f \mu}{\sigma}}$$

(14.65)

The imaginary part is inductive

$$\omega L_i = R_s$$

(14.66)

The phase of this impedance is always $\pi/4$.

Interpretation of surface impedance

The resistance term is equivalent to the resistance of a conductor of thickness δ. The inductance of the surface impedance represents the "internal" inductance for a large plane conductor. Note that as $\omega \to \infty$, $L_i \to 0$. The fields disappear from the volume of the conductor and the internal inductance is zero, whereas the internal resistance equals $\omega L_i = R_s = \sqrt{\pi f \mu/\sigma}$. We commonly apply this surface impedance to conductors of finite width or even coaxial lines. It is usually a pretty good approximation to make as long as the conductor width and thickness is much larger than δ (see Chapter 5 on inductance).

Time-harmonic wave equation

Start by taking the curl of Faraday's equation

$$\nabla \times (\nabla \times \mathbf{E}) = -j\omega\nabla \times \mathbf{B}$$

(14.67)

$$\nabla \times \mathbf{H} = \sigma\mathbf{E} + j\omega\epsilon\mathbf{E}$$

(14.68)

$$\nabla \times (\nabla \times \mathbf{E}) = -j\omega\mu(\sigma\mathbf{E} + j\omega\epsilon\mathbf{E})$$

(14.69)

In a source free region, $\nabla \cdot \mathbf{E} = 0$, and thus

$$\nabla \times (\nabla \times \mathbf{E}) = \nabla(\nabla \cdot \mathbf{E}) - \nabla^2\mathbf{E} = -\nabla^2\mathbf{E}$$

(14.70)

We thus have Helmholtz' equation

$$\nabla^2\mathbf{E} - \gamma^2\mathbf{E} = 0$$

(14.71)

where $\gamma^2 = j\omega\mu(\sigma + j\omega\epsilon) = \alpha + j\beta$

Lossy materials

In addition to conductive losses σ, materials can also have dielectric and magnetic losses. A lossy dielectric is characterized by a complex permittivity $\epsilon = \epsilon_r + j\epsilon_i$, where ϵ_i arises due to the phase lag between the field and the polarization. Likewise $\mu = \mu_r + j\mu_i$. Most materials we study are weakly magnetic and thus $\mu \approx \mu_r$. For now assume that ϵ, μ, and σ are real scalar quantities.

$$\gamma = \sqrt{(-\omega^2\epsilon\mu)\left(1 + \frac{\sigma}{j\omega\epsilon}\right)}$$

(14.72)

Propagation constant and loss

Let's compute the real and imaginary part of γ

$$\gamma = j\omega\sqrt{\epsilon\mu}\left(1 - j\frac{\sigma}{\omega\epsilon}\right)^{\frac{1}{2}} \tag{14.73}$$

Consider $(1 - jh) = re^{-j\theta}$, so that

$$y = \sqrt{1 - jh} = \sqrt{r}e^{-j\theta/2} \tag{14.74}$$

Note that $\tan\theta = -h$, and $r = \sqrt{1 + h^2}$. Finally

$$\cos\frac{\theta}{2} = \sqrt{\frac{1 + \cos\theta}{2}} = \sqrt{\frac{1 + \frac{1}{r}}{2}} = \sqrt{\frac{r + 1}{2r}} \tag{14.75}$$

Similarly

$$\sin\frac{\theta}{2} = \sqrt{\frac{1 - \cos\theta}{2}} = \sqrt{\frac{r - 1}{2r}} \tag{14.76}$$

$$y = \sqrt{r}e^{-j\theta/2} = \sqrt{\frac{r + 1}{2}} - j\sqrt{\frac{r - 1}{2}} = a + jb \tag{14.77}$$

Using the above manipulations, we can now break γ into real and imaginary components

$$\gamma = j\omega\sqrt{\mu\epsilon}(a + jb) = -\omega\sqrt{\mu\epsilon}b + j\omega\sqrt{\mu\epsilon}a = \alpha + j\beta \tag{14.78}$$

where

$$\alpha = -\omega\sqrt{\mu\epsilon}\left(-\frac{\sqrt{r - 1}}{\sqrt{2}}\right) \tag{14.79}$$

and

$$\beta = \omega\sqrt{\mu\epsilon}\sqrt{\frac{r + 1}{2}} \tag{14.80}$$

for $r = \sqrt{1 + \left(\frac{\sigma}{\omega\epsilon}\right)^2}$. We have now finally shown that

$$\alpha = \omega\sqrt{\frac{\mu\epsilon}{2}}\left[\sqrt{1 + \left(\frac{\sigma}{\omega\epsilon}\right)^2} - 1\right]^{1/2} \tag{14.81}$$

$$\beta = \omega\sqrt{\frac{\mu\epsilon}{2}}\left[\sqrt{1 + \left(\frac{\sigma}{\omega\epsilon}\right)^2} + 1\right]^{1/2} \tag{14.82}$$

It's easy to show that the imaginary part of ϵ can be lumped into an effective conductivity term. In practice, most materials are either *low loss*, such that $\frac{\sigma_{\text{eff}}}{\omega\epsilon} \ll 1$, or *good conductors*, such that $\frac{\sigma_{\text{eff}}}{\omega\epsilon} \gg 1$. In these extreme cases, simplified versions of the above equations are applicable. For the low loss materials, we have

$$\alpha = \omega\sqrt{\frac{\mu\epsilon}{2}}\sqrt{1 - 1} \approx 0 \tag{14.83}$$

$$\beta = \omega\sqrt{\frac{\mu\epsilon}{2}}\sqrt{2} = \omega\sqrt{\mu\epsilon} \tag{14.84}$$

For a good conductor, we have

$$\alpha = \omega\sqrt{\frac{\mu\epsilon}{2}}\sqrt{\frac{\sigma}{\omega\epsilon}} = \sqrt{\pi f\mu\sigma} \tag{14.85}$$

$$\beta = \omega\sqrt{\frac{\mu\epsilon}{2}}\sqrt{\frac{\sigma}{\omega\epsilon}} = \sqrt{\pi f\mu\sigma} = \alpha \tag{14.86}$$

Since $\beta = \alpha$, the propagation constant has a phase of $45°$.

Effective dielectric constant

We can also lump the conductivity into an effective dielectric constant

$$\nabla \times \mathbf{H} = \sigma\mathbf{E} + j\omega\epsilon\mathbf{E} = j\omega\epsilon_{\text{eff}}\mathbf{E} \tag{14.87}$$

where $\epsilon_{\text{eff}} = \epsilon - j\sigma/\omega$. In the *low-loss* case, this is a good way to include the losses. When ϵ or μ become complex, the wave impedance is no longer real and the electric and magnetic fields fall out of phase. Since $H = E/\eta_c$

$$\eta_c = \sqrt{\frac{\mu}{\epsilon_{\text{eff}}}} = \sqrt{\frac{\mu}{\epsilon - j\sigma/\omega}} = \frac{\sqrt{\frac{\mu}{\epsilon}}}{\sqrt{1 - j\frac{\sigma}{\omega\epsilon}}} \tag{14.88}$$

Propagation in low-loss materials

If $\frac{\sigma}{\omega\epsilon} \ll 1$, then our equations simplify

$$\gamma = j\omega\sqrt{\mu\epsilon}\left(1 - j\frac{1}{2}\frac{\sigma}{\omega\epsilon}\right) \tag{14.89}$$

To first order, the propagation constant is unchanged by the losses ($\sigma_{\text{eff}} = \sigma - \omega\epsilon''$)

$$\beta = \omega\sqrt{\mu\epsilon} \tag{14.90}$$

$$\alpha = \frac{1}{2}\sigma_{\text{eff}}\sqrt{\frac{\mu}{\epsilon}} \tag{14.91}$$

A more accurate expression can be obtained with a first-order expansion of $(1 + x)^n$

$$\beta = \omega\sqrt{\mu\epsilon}\left(1 + \frac{1}{8}\left(\frac{\sigma_{\text{eff}}}{\omega\epsilon'}\right)^2\right) \tag{14.92}$$

Waves in conductors

As we saw in the previous section, this approximation is valid when $\frac{\sigma}{\omega\epsilon} \gg 1$

$$\gamma = \alpha + j\beta = \sqrt{j\omega\mu\sigma} = \omega\mu\sigma\, e^{j45°} \tag{14.93}$$

$$\alpha = \beta = \sqrt{\frac{\omega\mu\sigma}{2}} \tag{14.94}$$

The phase velocity is given by $v_p = \omega/\beta$

$$v_p = \sqrt{\frac{2\omega}{\mu\sigma}} \tag{14.95}$$

This is a function of frequency! This is a very dispersive medium. The wavelength is given by

$$\lambda = \frac{v_p}{f} = 2\sqrt{\frac{\pi}{f\mu\sigma}} \tag{14.96}$$

Example: Take $\sigma = 10^7$ S/m and $f = 100$ MHz. Using the above equations

$$\lambda = 10^{-4} \text{ m} \tag{14.97}$$

$$v_p = 10^4 \text{ m/s} \tag{14.98}$$

Note that $\lambda_0 = 3$ m in free space, and thus the wave is very much smaller and much slower moving in the conductor.

14.3 Poynting vector

Energy storage and loss in fields

We have learned that the power density of electric and magnetic fields is given by

$$w_e = \frac{1}{2}\mathbf{E} \cdot \mathbf{D} = \frac{1}{2}\epsilon E^2 \tag{14.99}$$

$$w_m = \frac{1}{2}\mathbf{H} \cdot \mathbf{B} = \frac{1}{2}\mu H^2 \tag{14.100}$$

Also, the power loss per unit volume due to Joule heating in a conductor is given by

$$p_{\text{loss}} = \mathbf{E} \cdot \mathbf{J} \tag{14.101}$$

Using $\mathbf{J} = \nabla \times \mathbf{H} - \frac{\partial \mathbf{D}}{\partial t}$, this can be expressed as

$$\mathbf{E} \cdot \mathbf{J} = \mathbf{E} \cdot \nabla \times \mathbf{H} - \frac{\partial \mathbf{D}}{\partial t} \tag{14.102}$$

We will demonstrate that the Poynting vector $\mathbf{S} = \mathbf{E} \times \mathbf{H}$ plays an important role in the energy of an EM field. To see this, use the following vector identity

$$\nabla \cdot (\mathbf{E} \times \mathbf{H}) = \mathbf{H} \cdot (\nabla \times \mathbf{E}) - \mathbf{E} \cdot (\nabla \times \mathbf{H}) \tag{14.103}$$

and substitute for $\mathbf{E} \cdot \nabla \times \mathbf{H}$ in Eq. (14.102) to obtain

$$\mathbf{E} \cdot \mathbf{J} = \mathbf{H} \cdot (\nabla \times \mathbf{E}) - \nabla \cdot (\mathbf{E} \times \mathbf{H}) - \mathbf{E} \cdot \frac{\partial \mathbf{D}}{\partial t} \tag{14.104}$$

Simplifying the above equation we have the following equation

$$= \mathbf{H} \cdot \left(-\frac{\partial \mathbf{B}}{\partial t}\right) - \mathbf{E} \cdot \frac{\partial \mathbf{D}}{\partial t} - \nabla \cdot (\mathbf{E} \times \mathbf{H}) \tag{14.105}$$

The first and second term can be written in the following form

$$\mathbf{H} \cdot \frac{\partial \mathbf{B}}{\partial t} = \mathbf{H} \cdot \left(\frac{\partial \mu \mathbf{H}}{\partial t} \right) = \frac{1}{2} \frac{\partial \mu \mathbf{H} \cdot \mathbf{H}}{\partial t} = \frac{1}{2} \frac{\partial \mu |\mathbf{H}|^2}{\partial t} \tag{14.106}$$

$$\mathbf{E} \cdot \frac{\partial \mathbf{D}}{\partial t} = \mathbf{E} \cdot \left(\frac{\partial \epsilon \mathbf{E}}{\partial t} \right) = \frac{1}{2} \frac{\partial \epsilon \mathbf{E} \cdot \mathbf{E}}{\partial t} = \frac{1}{2} \frac{\partial \mu |\mathbf{E}|^2}{\partial t} \tag{14.107}$$

Poynting's Theorem

We are now in a position to prove Poynting's Theorem. Collecting terms in Eq. (14.105)

$$\mathbf{E} \cdot \mathbf{J} = -\frac{\partial}{\partial t} \left(\frac{1}{2} \mu |\mathbf{H}|^2 \right) - \frac{\partial}{\partial t} \left(\frac{1}{2} \epsilon |\mathbf{E}|^2 \right) - \nabla \cdot (\mathbf{E} \times \mathbf{H}) \tag{14.108}$$

Applying the Divergence Theorem

$$\int_V \mathbf{E} \cdot \mathbf{J} dV = -\frac{\partial}{\partial t} \int_V \left(\frac{1}{2} \mu |\mathbf{H}|^2 + \frac{1}{2} \epsilon |\mathbf{E}|^2 \right) dV - \int_S \mathbf{E} \times \mathbf{H} dV \tag{14.109}$$

The above equation can be re-stated as

power dissipated		rate of change of		a surface integral
in volume V	$= -$	energy storage in	$-$	over the volume
(heat)		volume V		of $\mathbf{E} \times \mathbf{H}$

Interpretation of the Poynting vector

We now have a physical interpretation of the last term in the above equation. By the conservation of energy, it must be equal to the energy flow out of the volume. We may be so bold, then, to interpret the vector $\mathbf{S} = \mathbf{E} \times \mathbf{H}$ as the energy flow density of the field. While this seems reasonable, it is important to note that the physical meaning is only attached to the integral of $\mathbf{S}$ and not to discrete points in space.

Example 24

Current-Carrying Wire

Consider the above wire carrying a uniform current I (Fig. 14.2). From circuit theory we know that the power loss in the wire is simply $I^2 R$. This is easily confirmed

$$P_L = \int_V \mathbf{E} \cdot \mathbf{J} dV = \int_V \frac{1}{\sigma} \mathbf{J} \cdot \mathbf{J} dV = \frac{1}{A^2 \sigma} \int_V I^2 dV \tag{14.110}$$

$$P_L = \frac{A \cdot \ell}{A^2 \sigma} I^2 = \frac{\ell}{A \sigma} I^2 \tag{14.111}$$

Let's now apply Poynting's Theorem. Since the current is DC, we can neglect all time variation $\frac{\partial}{\partial t} = 0$ and thus the energy storage of the system is fixed in time. The magnetic field around the wire is simply given by

$$\mathbf{H} = \hat{\phi} \frac{I}{2\pi r} \tag{14.112}$$

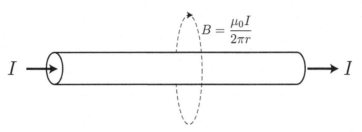

Figure 14.2 A wire carrying uniform current.

Figure 14.3 The electromagnetic fields and the Poynting vector around a current-carrying wire.

The electric field is proportional to the current density. At the surface of the wire

$$\mathbf{E} = \frac{1}{\sigma}\mathbf{J} = \frac{I}{\sigma A}\hat{\mathbf{z}}$$ (14.113)

As shown in Fig. 14.3, the Poynting vector at the surface thus points into the conductor

$$\mathbf{S} = \mathbf{E} \times \mathbf{H} = \frac{I}{\sigma A}\hat{\mathbf{z}} \times \hat{\boldsymbol{\phi}}\frac{I}{2\pi r} = \frac{-\hat{\mathbf{r}}I^2}{2\pi r\sigma A}$$ (14.114)

The power flow into the wire is thus given by

$$\int_s \mathbf{S}\cdot ds = \int_0^\ell \int_0^{2\pi} \frac{I^2}{2\pi r\sigma A}r\,d\theta\,dz = I^2 R$$ (14.115)

This result confirms that the energy flowing into the wire from the field is heating up the wire.

This result is surprising because it hints that the signal in a wire is carried by the fields, and not by the charges. In other words, if a signal propagates down a wire, the information is carried by the fields, and the current flow is impressed upon the conductor from the fields. We know that the sources of EM fields are charges and currents. But we also know that if the configuration of charges changes, the fields "carry" this information.

14.4 EM power carried by a plane wave

In a lossless medium, we have found that

$$E_x = E_0 \cos(\omega t - \beta z)$$ (14.116)

$$H_y = \frac{E_0}{\eta_0}\cos(\omega t - \beta z)$$ (14.117)

where $\beta = \omega\sqrt{\mu\epsilon}$ and $\eta = \sqrt{\mu/\epsilon}$. The Poynting vector $\mathbf{S}$ is easily calculated

$$\mathbf{S} = \mathbf{E} \times \mathbf{H} = \hat{\mathbf{z}}\frac{E_0^2}{\eta_0}\cos^2(\omega t - \beta z) \tag{14.118}$$

$$\mathbf{S} = \hat{\mathbf{z}}\frac{E_0^2}{2\eta_0}(1 + \cos(2(\omega t - \beta z))) \tag{14.119}$$

Average power of a plane wave

If we average the Poynting vector over time, the magnitude is

$$S_{av} = \frac{E_0^2}{2\eta_0} \tag{14.120}$$

This simple equation is very useful for estimating the electric field strength of an electromagnetic wave far from its source (where it can be approximated as a plane wave). The energy stored in the electric and magnetic fields are

$$w_e = \frac{1}{2}\epsilon|E_x|^2 = \frac{1}{2}\epsilon E_0^2 \cos^2(\omega t - \beta z) \tag{14.121}$$

$$w_m = \frac{1}{2}\mu|H_y|^2 = \frac{1}{2}\mu\frac{E_0^2}{\eta_0^2}\cos^2(\omega t - \beta z) \tag{14.122}$$

Plane wave "resonance"

It's now clear that

$$w_m = \frac{1}{2}\mu\frac{E_0^2}{\frac{\mu}{\epsilon}}\cos^2(\omega t - \beta z) = w_e \tag{14.123}$$

In other words, the stored magnetic energy is equal to the stored electric energy. In analogy with an LC circuit, we say that the wave is in resonance. We can also show that

$$\frac{\partial}{\partial t}\int_V (w_m + w_e)dV = -\oint_S \mathbf{S}\cdot d\mathbf{S} \tag{14.124}$$

Example 25

Cell phone base station
A cell phone base station transmits 10kW of power. Estimate the average electric field at a distance of 1m from the antenna.

Assuming that the medium around the antenna is lossless, the energy transmitted by the source at any given location from the source must be given by

$$P_t = \oint_{\text{Surf}} \mathbf{S}\cdot d\mathbf{S} \tag{14.125}$$

where Surf is a surface covering the source of radiation. Since we do not know the antenna radiation pattern, let's assume an isotropic source (equal radiation in all directions). In that case, the average Poynting vector at a distance r from the source is given by

$$S = \frac{P_t}{4\pi r^2} = \frac{10^4}{4\pi} \frac{W}{m^2} \tag{14.126}$$

This equation is simply derived by observing that the surface area of a sphere of radius r is given by $4\pi r^2$. Using $S = \frac{1}{2}\frac{E_0^2}{\eta_0}$, we have

$$E_0 = \sqrt{2\eta_0 S} = \sqrt{2 \times 377 \times \frac{10^4}{4\pi}} = 775 \frac{V}{m} \tag{14.127}$$

Example 26

Cell phone handset
A cell phone handset transmits 1W of power. What is the average electric field at a distance of 10cm from the handset?

$$S = \frac{P_t}{4\pi r^2} = \frac{1}{4\pi.1^2} \frac{W}{m^2} = 7.96 \frac{W}{m^2} \tag{14.128}$$

and so $E_0 = 77.5$ V/m. We can see that the predicted electric field near a handset is at a much lower level. What's a safe level? In reality, for such a close distance from the antenna, we cannot make a far field (plane wave) approximation. Calculation of the field requires a more detailed computation involving the antenna pattern and the shape of the head.

14.5 Complex Poynting Theorem

In the previous sections we derived the Poynting Theorem for general electric/magnetic fields. Now we would like to derive the Poynting Theorem for time-harmonic fields. We can't simply take our previous results and simply transform $\frac{\partial}{\partial t} \rightarrow j\omega$. This is because the Poynting vector is a non-linear function of the fields. Unfortunately, we must start from the beginning

$$\nabla \times \mathbf{E} = -j\omega\mathbf{B} \tag{14.129}$$

$$\nabla \times \mathbf{H} = j\omega\mathbf{D} + \mathbf{J} = (j\omega\epsilon + \sigma)\mathbf{E} \tag{14.130}$$

Using our knowledge of circuit theory, $P = V \times I^*$, we compute the following quantity

$$\nabla \cdot (\mathbf{E} \times \mathbf{H}^*) = \mathbf{H}^* \cdot \nabla \times \mathbf{E} - \mathbf{E} \cdot \nabla \times \mathbf{H}^* \tag{14.131}$$

$$\nabla \cdot (\mathbf{E} \times \mathbf{H}^*) = \mathbf{H}^* \cdot (-j\omega\mathbf{B}) - \mathbf{E} \cdot (j\omega\mathbf{D}^* + \mathbf{J}^*) \tag{14.132}$$

Applying the Divergence Theorem

$$\int_V \nabla \cdot (\mathbf{E} \times \mathbf{H}^*) dV = \oint_S (\mathbf{E} \times \mathbf{H}^*) \cdot d\mathbf{S} \tag{14.133}$$

$$\oint_S (\mathbf{E} \times \mathbf{H}^*) \cdot d\mathbf{S} = -\int_V \mathbf{E} \cdot \mathbf{J}^* dV - \int_V j\omega(\mathbf{E} \cdot \mathbf{D}^* + \mathbf{H}^* \cdot \mathbf{B}) dV \tag{14.134}$$

Let's define $\sigma_{\text{eff}} = \omega\epsilon'' + \sigma$, and $\epsilon = \epsilon'$. Since most materials are non-magnetic, we can ignore magnetic losses

$$\int_S (\mathbf{E} \times \mathbf{H}^*) \cdot d\mathbf{S} = -\int_V \sigma \mathbf{E} \cdot \mathbf{E}^* dV - j\omega \int_V (\mu \mathbf{H}^* \cdot \mathbf{H} + \epsilon \mathbf{E} \cdot \mathbf{E}^*) dV \tag{14.135}$$

Notice that the first volume integral is a real number, whereas the second volume integral is imaginary

$$\Re\left(\oint_S \mathbf{E} \times \mathbf{H}^* \cdot d\mathbf{S}\right) = -2\int_V P_c dV \tag{14.136}$$

$$\Im\left(\oint_S \mathbf{E} \times \mathbf{H}^* \cdot d\mathbf{S}\right) = -4\omega\int_V (w_m + w_e) dV \tag{14.137}$$

Let's compute the average vector $\mathbf{S}$

$$\mathbf{S} = \Re\left(\mathbf{E}e^{j\omega t}\right) \times \Re\left(\mathbf{H}e^{j\omega t}\right) \tag{14.138}$$

First observe that $\Re(\mathbf{A}) = \frac{1}{2}(\mathbf{A} + \mathbf{A}^*)$, so that

$$\Re(\mathbf{G}) \times \Re(\mathbf{F}) = \frac{1}{2}(\mathbf{G} + \mathbf{G}^*) \times \frac{1}{2}(\mathbf{F} + \mathbf{F}^*) \tag{14.139}$$

$$= \frac{1}{4}(\mathbf{G} \times \mathbf{F} + \mathbf{G} \times \mathbf{F}^* + \mathbf{G}^* \times \mathbf{F} + \mathbf{G}^* \times \mathbf{F}^*) \tag{14.140}$$

$$= \frac{1}{4}[(\mathbf{G} \times \mathbf{F}^* + \mathbf{G}^* \times \mathbf{F}) + (\mathbf{G} \times \mathbf{F} + \mathbf{G}^* \times \mathbf{F}^*)] \tag{14.141}$$

$$= \frac{1}{2}\Re(\mathbf{G} \times \mathbf{F}^* + \mathbf{G} \times \mathbf{F}) \tag{14.142}$$

Average complex Poynting vector

Finally, we have computed the complex Poynting vector with the time dependence

$$\mathbf{S} = \frac{1}{2}\Re(\mathbf{E} \times \mathbf{H}^* + \mathbf{E} \times \mathbf{H}e^{2j\omega t}) \tag{14.143}$$

Taking the average value, the complex exponential vanishes, so that

$$\mathbf{S}_{\text{av}} = \frac{1}{2}\Re(\mathbf{E} \times \mathbf{H}^*) \tag{14.144}$$

We have thus justified that the quantity $\mathbf{S} = \mathbf{E} \times \mathbf{H}^*$ represents the complex power stored in the field.

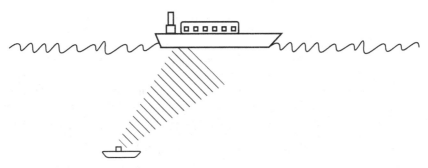

Figure 14.4 A submarine to ship communication link.

Example 27

Submarine Communication

Consider a submarine at a depth of $z = 100$ m (Fig. 14.4). We would like to communicate with this submarine using a VLF $f = 3$ kHz. The conductivity of sea water is
$\sigma = 4$ Sm^{-1}, $\epsilon_r = 81$, and $\mu \approx 1$.

Note that we are forced to use low frequencies due to the conductivity of the ocean
water. The loss conductive tangent

$$\tan \delta_c = \frac{\sigma}{\omega \epsilon} \sim 10^5 \gg 1 \tag{14.145}$$

Thus the ocean is a *good* conductor even at this low frequency of 3 kHz. The propagation loss and constant are thus equal

$$\alpha = \beta = \sqrt{\frac{\omega \mu \sigma}{2}} \approx 0.2 \tag{14.146}$$

The wavelength in sea water is much smaller than in air ($\lambda_0 = 100$ km in air)

$$\lambda = \frac{2\pi}{\beta} = 29 \text{ m} \tag{14.147}$$

Thus the phase velocity of the wave is also much smaller

$$v_p = f\lambda \approx 9 \times 10^4 \text{ m/s} \tag{14.148}$$

The skin depth, or the depth at which the wave is attenuated to about 37% of its value,
is given by

$$\delta = \frac{1}{\alpha} = 4.6 \text{ m} \tag{14.149}$$

The wave impedance is complex with a phase of $45°$

$$|\eta_c| = \sqrt{\frac{\mu \omega}{\sigma}} \approx 8 \times 10^{-2} \ \Omega \tag{14.150}$$

$$\angle \eta_c = 45° \tag{14.151}$$

Notice that $\eta_c \ll \eta_0$, the ocean water thus generates a very large magnetic field for wave propagation

$$H = \frac{E_0}{\eta_c} e^{-\alpha z} \cos(6\pi \times 10^3 t - \beta z - \phi_\eta) \tag{14.152}$$

where ϕ_η is the angle of the complex wave impedance, $45°$ in this particular case.

Now let's compute the required transmission power if the receiver at the depth of $z = 100$ m is capable of receiving a signal of at least $1~\mu V/m$.[1]

$$E_0 e^{-\alpha z} \geq E_{min} = 1~\mu V/m \tag{14.153}$$

This requires $E_0 = 2.8$ kV/m, and a corresponding magnetic field of $H_0 = E_0/\eta_c = 37$ kA/m.

This is a very large amount of power to generate at the source. The power density at the source is

$$S_{av} = \frac{1}{2} \Re(\mathbf{E} \times \mathbf{H}^*) \tag{14.154}$$

$$S_{av} = \frac{1}{2} (2.84 \times 37 \cos(45°)) \times 10^{-3} = 37~MW/m^2 \tag{14.155}$$

At a depth of 100 m, the power density drops to extremely small levels

$$S_{av}(100m) = 4.6 \times 10^{-12}~MW/m^2 \tag{14.156}$$

14.6 Reflections from a perfect conductor

Consider a plane wave incident *normally* on to a conducting surface, as shown in Fig. 14.5

$$E_i = \hat{\mathbf{x}} E_{i0} e^{-j\beta_1 z} \tag{14.157}$$

$$H_i = \hat{\mathbf{y}} \frac{E_{i0}}{\eta_0} e^{-j\beta_1 z} \tag{14.158}$$

The reflected wave (if any) has the following form

$$E_r = \hat{\mathbf{x}} E_{r0} e^{j\beta_1 z} \tag{14.159}$$

$$H_r = -\hat{\mathbf{y}} \frac{E_{r0}}{\eta_0} e^{j\beta_1 z} \tag{14.160}$$

The conductor forces the tangential electric field to vanish at the surface $z = 0$

$$\mathbf{E}(z = 0) = 0 = \hat{\mathbf{x}}(E_{i0} + E_{r0}) \tag{14.161}$$

This implies that the reflected wave has equal and opposite magnitude and phase

$$E_{r0} = -E_{i0} \tag{14.162}$$

This is similar to wave reflection from a transmission line short-circuit load. We now write

---

[1] The receiver sensitivity is set by the noise power at the input of the receiver. If the signal is too small, it is swamped by the noise.

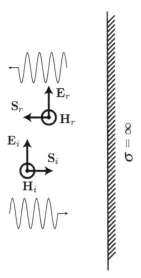

Figure 14.5 Normal reflection from a conductive boundary.

the total electric and magnetic field in region 1

$$\mathbf{E}(z) = \hat{\mathbf{x}} E_{i0}(e^{-j\beta_1 z} - e^{j\beta_1 z}) = -\hat{\mathbf{x}} E_{i0} j2 \sin(\beta_1 z) \tag{14.163}$$

$$\mathbf{H}(z) = \hat{\mathbf{y}} \frac{E_{i0}}{\eta_0}(e^{-j\beta_1 z} + e^{j\beta_1 z}) = \hat{\mathbf{y}} \frac{E_{i0}}{\eta_0} 2 \cos(\beta_1 z) \tag{14.164}$$

The net complex power carried by the wave

$$\mathbf{E} \times \mathbf{H}^* = -\hat{\mathbf{z}} \frac{E_{i0}^2}{\eta_0} 4j \sin(\beta_1 z) \cos(\beta_1 z) \tag{14.165}$$

is reactive. That means that the average power is zero

$$\mathbf{S}_{av} = \frac{1}{2} \Re(\mathbf{E} \times \mathbf{H}^*) = 0 \tag{14.166}$$

The reflected wave interferes with the incident wave to create a standing wave

$$E(z, t) = \Im(E(z)e^{j\omega t}) = \Im(E_{i0} j2 \sin(\beta_1 z)e^{j\omega t}) \tag{14.167}$$

$$E(z, t) = 2E_{i0} \sin(\beta_1 z) \cos(\omega t) \tag{14.168}$$

$$H(z, t) = \frac{2E_{i0}}{\eta_1} \cos(\beta_1 z) \sin(\omega t) \tag{14.169}$$

Note that the E and H fields are in time quadrature (90° phase difference). The instantaneous power is given by

$$S = \frac{4E_{i0}^2}{\eta_1} \underbrace{\sin(\beta_1 z) \cos(\beta_1 z)}_{\frac{1}{2}\sin(2\beta_1 z)} \underbrace{\cos(\omega t) \sin(\omega t)}_{\frac{1}{2}\sin(2\omega t)} \tag{14.170}$$

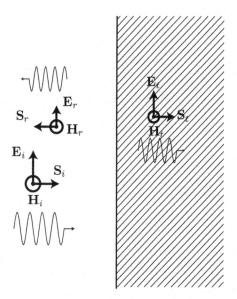

Figure 14.6 The reflection and transmission of a plane wave into a dielectric boundary.

The electric and magnetic powers are readily calculated

$$w_e = \frac{1}{2}\epsilon_1|E_1|^2 = 2\epsilon_1|E_{i0}|^2 \sin^2(\beta_1 z)\cos^2(\omega t) \tag{14.171}$$

$$w_m = \frac{1}{2}\mu_1|H_1|^2 = 2\epsilon_1|E_{i0}|^2 \cos^2(\beta_1 z)\sin^2(\omega t) \tag{14.172}$$

Note that the magnetic field at the boundary of the conductor is supported (or equivalently *induces*) a surface current

$$J_s = \hat{n} \times H = -\hat{x}\frac{2E_{i0}}{\eta_1} \text{ A/m} \tag{14.173}$$

If the material is a good conductor, but lossy, then this causes power loss at the conductor surface.

14.7 Normal incidence on a dielectric

Consider an incident wave on to a dielectric region as shown in Fig. 14.6. We have the incident and possibly reflected waves

$$E_i = \hat{x}E_{i0}e^{-j\beta_1 z} \tag{14.174}$$

$$H_i = \hat{y}\frac{E_{i0}}{\eta_1}e^{-j\beta_1 z} \tag{14.175}$$

$$E_r = \hat{x}E_{r0}e^{j\beta_1 z} \tag{14.176}$$

$$H_r = -\hat{y}\frac{E_{r0}}{\eta_1}e^{j\beta_1 z} \tag{14.177}$$

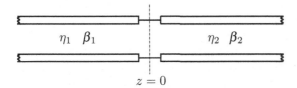

$z = 0$

Figure 14.7 An equivalent transmission line representing normal incidence through a dielectric interface.

But we must also allow the possibility of a transmitted wave into region 2

$$E_t = \hat{x} E_{t0} e^{-j\beta_2 z} \qquad (14.178)$$

$$H_t = \hat{y} \frac{E_{t0}}{\eta_2} e^{-j\beta_2 z} \qquad (14.179)$$

At the interface of the two dielectrics, assuming no interface charge (Fig. 14.7), we have

$$E_{t1} = E_{t2} \qquad (14.180)$$

$$E_{i0} + E_{r0} = E_{t0} \qquad (14.181)$$

$$H_{t1} = H_{t2} \qquad (14.182)$$

$$\frac{E_{i0}}{\eta_1} - \frac{E_{r0}}{\eta_1} = \frac{E_{t0}}{\eta_2} \qquad (14.183)$$

We have met these equations before. The solution is

$$E_{r0} = \frac{\eta_2 - \eta_1}{\eta_2 + \eta_1} E_{i0} \qquad (14.184)$$

$$E_{t0} = \frac{2\eta_2}{\eta_2 + \eta_1} E_{i0} \qquad (14.185)$$

Transmission line analogy

These equations are identical to the case of the interface of two transmission lines. The reflection and transmission coefficients are thus identical

$$\rho = \frac{E_{r0}}{E_{i0}} = \frac{\eta_2 - \eta_1}{\eta_2 + \eta_1} \qquad (14.186)$$

$$\tau = \frac{E_{t0}}{E_{i0}} = 1 + \rho \qquad (14.187)$$

Reflection and transmission with three materials

Consider three dielectric materials, as shown in Fig. 14.8. Instead of solving the problem the *long* way, let's use the transmission line analogy. First solve the problem at the interface of region 2 and 3. Region 3 acts like a load to region 2. Now transform this load impedance by the length of region 2 to present an equivalent load to region 1.

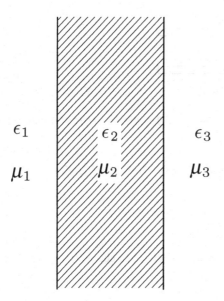

Figure 14.8 Reflection and transmission with three materials.

Example 28

Glass Coating
A very practical example is the case of minimizing reflections for eyeglasses. Due to the impedance mismatch, light normally reflects at the interface of air and glass. One method to reduce this reflection is to coat the glass with a material to eliminate the reflections. From our transmission line analogy, we know that this coating is acting like a quarter wave transmission line with

$$\eta = \sqrt{\eta_0 \, \eta_{\text{glass}}} \tag{14.188}$$

14.8   References

This chapter draws from many sources, particularly Feynman [18], Cheng [6], and Inan and Inan [62].

15 Microwave circuits

Microwave circuit theory is a powerful set of tools that allows us to treat microwave devices as circuit elements. Implicitly, we have been using microwave theory throughout this book, in particular our treatment of distributed systems. In this chapter, though, we establish the concepts of network theory on a firm electromagnetics foundation. This leads to a powerful and general set of tools for evaluating the properties of N-ports. In particular, we shall develop the concept of the scattering matrix, or S parameters, and relate it to the more familiar concept of network impedance and admittance matrices. The properties of three- and four-port devices, in particular lossless reciprocal devices, will be studied in depth.

15.1 What are microwave circuits?

To define microwave circuits, we must first understand where circuit theory comes from. Crudely speaking, circuit theory is an approximation to Maxwell's equations valid when structure dimensions are small relative to the wavelength (at the highest frequency of interest). Alternatively, circuit theory is valid when the speed of light is infinite $c \to \infty$.

For example, at $f = 60\,\text{Hz}$, we have

$$\lambda = \frac{c}{f} = \frac{3 \times 10^8}{60} = 0.5 \times 10^7 \tag{15.1}$$

If we arbitrarily require that the dimension be a factor of a thousand smaller than the wavelength, we have

$$\frac{\ell}{\lambda} = 10^{-3} \to \ell = 5\,\text{km} \tag{15.2}$$

Now let's consider $f = 1\,\text{GHz}$. This corresponds to the popular cellular bands. Now $\lambda = c/f = 30\,\text{cm}$, so using the same requirement we have

$$\frac{\ell}{\lambda} = 10^{-3} \to \ell = 0.3\,\text{mm} \tag{15.3}$$

This is a lot more restrictive! We see that this is strictly valid for relatively small structures on the Si chip. So inside a small transistor with a dimension of tens of microns, certainly circuit theory is valid.[1]

---

[1] Recall that $\lambda = v/f = c/\sqrt{\epsilon\mu}\,f$, so inside the Si substrate the wavelength for TEM waves drops by roughly $\sqrt{\epsilon_r} = \sqrt{12}$. Inside the oxide, where most of the energy flows for on-chip transmission lines, the drop is much smaller as $\sqrt{\epsilon_r} = \sqrt{4} = 2$.

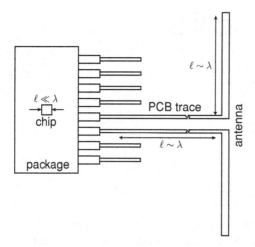

Figure 15.1 The chip, package, and board interface is often the regime where lumped circuits turn into distributed circuits.

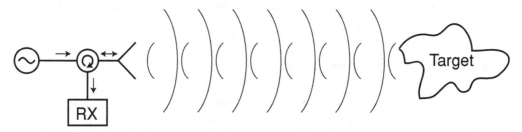

Figure 15.2 A typical radar system employs a circulator to isolate the transmitted signal from the received signal.

Microwave circuit theory is an extension of circuit theory to higher frequencies where the circuit dimensions approach the wavelength, $\ell \sim \lambda$. We need this theory in order to avoid solving Maxwell's equations! We can also use our intuition and experience from circuit theory (e.g. lumped filter design) and apply it to higher frequencies. We have to be careful in applying our intuition. For instance, as we have seen, a transmission line of length $\lambda/4$ converts an open circuit termination into a short circuit! This behavior is very counterintuitive from a lumped circuit theory perspective.

Example 29

Chip/Package/Board
In Fig. 15.1, the structures "on-chip" may behave like lumped elements (transistors, inductors, capacitors, etc.). The leads, board traces, and radiation structures, though, are "large" relative to the wavelength and require microwave theory.

Example 30

Radar, invented by Sir Robert Watson-Watt in 1935, and developed at MRL during World War II (1939–1945), allows us to detect distant objects by observing the microwave scattering from a target. In this chapter we will learn to build the basic passive building blocks, such as a coupler and circulator.

The circulator, shown schematically in Fig. 15.2, only allows a signal to travel in the direction shown. Thus the receive channel is isolated from the transmit section. This is important because the receiver is very sensitive and is designed to work with very weak signals.

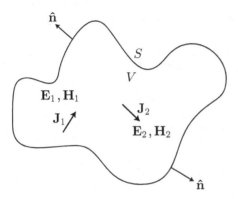

Figure 15.3 The electromagnetic theory of reciprocity interrelates the fields and sources for two separate regions.

15.2 Microwave networks

To develop microwave network theory, we must invoke some important theorems from electromagnetics. The most important theorem involves the concept of reciprocity.

15.3 Lorentz reciprocity theorem

Consider two sources $\mathbf{J}_1$ and $\mathbf{J}_2$ inside S. Inside S and on the boundary the fields satisfy Maxwell's equations

$$\nabla \times \mathbf{E}_1 = -j\omega \mathbf{H}_1 \tag{15.4}$$

$$\nabla \times \mathbf{H}_1 = \mathbf{J}_1 + j\omega\epsilon_1 \mathbf{E}_1 \tag{15.5}$$

Now use the following vector identity

$$\nabla \cdot (\mathbf{E}_1 \times \mathbf{H}_2 - \mathbf{E}_2 \times \mathbf{H}_1) = \tag{15.6}$$

$$= (\nabla \times \mathbf{E}_1) \cdot \mathbf{H}_2 - (\nabla \times \mathbf{H}_2) \cdot \mathbf{E}_1 - (\nabla \times \mathbf{E}_2) \cdot \mathbf{H}_1 + (\nabla \times \mathbf{H}_1) \cdot \mathbf{E}_2 \tag{15.7}$$

$$= -j\omega \mathbf{H}_1 \cdot \mathbf{H}_2 - (\mathbf{J}_2 + j\omega\epsilon \mathbf{E}_2) \cdot \mathbf{E}_1 + j\omega \mathbf{H}_2 \cdot \mathbf{H}_1 + (\mathbf{J}_1 + j\omega\epsilon \mathbf{E}_1) \cdot \mathbf{E}_2 \tag{15.8}$$

After the massacre, only a few terms survive. Apply the Divergence Theorem (you saw it coming)

$$\int_V \nabla \cdot (\mathbf{E}_1 \times \mathbf{H}_2 - \mathbf{E}_2 \times \mathbf{H}_1) dV = \oint_S (\mathbf{E}_1 \times \mathbf{H}_2 - \mathbf{E}_2 \times \mathbf{H}_1) \cdot d\mathbf{S} = \quad (15.9)$$

$$= \int_V (\mathbf{J}_1 \cdot \mathbf{E}_2 - \mathbf{J}_2 \cdot \mathbf{E}_1) dV \quad (15.10)$$

For a sourceless region, the RHS is identically zero and we have one form of reciprocity

$$\oint_S \mathbf{E}_1 \times \mathbf{H}_2 \cdot d\mathbf{S} = \oint_S \mathbf{E}_2 \times \mathbf{H}_1 \cdot d\mathbf{S} \quad (15.11)$$

On the other hand, if the integral encloses all of the sources, we can show that the surface integral term is zero. Let's take a few cases. Say S is a perfectly conducting surface so that $E_t = 0$. Then $\mathbf{n} \times \mathbf{E} = 0$ and

$$(\mathbf{E}_1 \times \mathbf{H}_2) \cdot \hat{\mathbf{n}} = (\hat{\mathbf{n}} \times \mathbf{E}_1) \cdot \mathbf{H}_2 = 0 \quad (15.12)$$

Conductive surface

It takes a bit more work, but we can also show that the above holds if the surface S has surface impedance Z_s

$$\mathbf{E}_t = Z_s \mathbf{J}_s = -Z_s \hat{\mathbf{n}} \times \mathbf{H} \quad (15.13)$$

$$\hat{\mathbf{n}} \times \mathbf{E} = -Z_s \hat{\mathbf{n}} \times (\hat{\mathbf{n}} \times \mathbf{H}) \quad (15.14)$$

$$(\hat{\mathbf{n}} \times \mathbf{E}_1) \cdot \mathbf{H}_2 - (\hat{\mathbf{n}} \times \mathbf{E}_2) \cdot \mathbf{H}_1 = -Z_s(\hat{\mathbf{n}} \times \hat{\mathbf{n}} \times \mathbf{H}_1) \cdot \mathbf{H}_2 + Z_s(\hat{\mathbf{n}} \times \hat{\mathbf{n}} \times \mathbf{H}_2) \cdot \mathbf{H}_1$$
$$(15.15)$$

$$= -Z_s(\hat{\mathbf{n}} \times \mathbf{H}_2) \cdot (\hat{\mathbf{n}} \cdot \mathbf{H}_1) + Z_s(\hat{\mathbf{n}} \times \mathbf{H}_1) \cdot (\hat{\mathbf{n}} \cdot \mathbf{H}_2) = 0 \quad (15.16)$$

Radiation boundary

Consider a sphere at infinity. At infinity the fields become TEM

$$\mathbf{H} = \sqrt{\frac{\epsilon}{\mu}} \hat{\mathbf{a}}_r \times \mathbf{E} \quad (15.17)$$

$$(\hat{\mathbf{n}} \times \mathbf{E}_1) \cdot \mathbf{H}_2 - (\hat{\mathbf{n}} \times \mathbf{E}_2) \cdot \mathbf{H}_1 = \quad (15.18)$$

$$= \sqrt{\frac{\epsilon}{\mu}} ((\hat{\mathbf{a}}_r \times \mathbf{E}_1) \cdot (\hat{\mathbf{a}}_r \times \mathbf{E}_2) - (\hat{\mathbf{a}}_r \times \mathbf{E}_2) \cdot (\hat{\mathbf{a}}_r \times \mathbf{E}_1)) = 0 \quad (15.19)$$

we can actually show that for any surface enclosing all the sources, the integral vanishes so that

$$\int_V \mathbf{E}_1 \cdot \mathbf{J}_2 dV = \int_V \mathbf{E}_2 \cdot \mathbf{J}_1 dV \quad (15.20)$$

For point sources

$$E_1 J_2 = E_2 J_1 \quad (15.21)$$

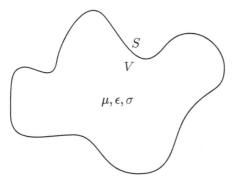

Figure 15.4 The uniqueness theorem for a region V proves that the electromagnetic fields are only a function of the initial fields in the region and either the tangential electric field or the normal magnetic field on the boundary.

Uniqueness theorem

An electromagnetic field is uniquely determined within a bounded region V at all times $t > 0$ by the initial values of the electric and magnetic vectors throughout V and the values of the tangential component of the electric vector or the magnetic vector over the boundaries for $t \geq 0$. To prove this, assume the solution is not unique and two distinct solutions are $\mathbf{E}_1/\mathbf{H}_1$ and $\mathbf{E}_2/\mathbf{H}_2$.

$$\mathbf{E}_1(t = 0) = \mathbf{E}_2(t = 0) \tag{15.22}$$
$$\mathbf{H}_1(t = 0) = \mathbf{H}_2(t = 0) \tag{15.23}$$

Assume linear field equations (exclude ferromagnetic materials) so that the difference fields are also a solution.

$$\mathbf{E} \triangleq \mathbf{E}_1 - \mathbf{E}_2 \tag{15.24}$$
$$\mathbf{H} \triangleq \mathbf{H}_1 - \mathbf{H}_2 \tag{15.25}$$

Assume sources are outside of V so that within V Poynting's Theorem is satisfied

$$\frac{\partial}{\partial t} \int_V \frac{1}{2}(\epsilon|E|^2 + \mu|H|^2)dV + \int_V \rho|J|^2 dV = -\oint_S (\mathbf{E} \times \mathbf{H}) \cdot \mathbf{dS} \tag{15.26}$$

Since both solutions satisfy the boundary conditions, $\mathbf{n} \times \mathbf{E} = 0$ or $\mathbf{n} \times \mathbf{H} = 0$, the RHS is zero. We have the following

$$\frac{\partial}{\partial t} \int_V \frac{1}{2} \underbrace{(\epsilon|E|^2 + \mu|H|^2)}_{\geq 0} dV = -\underbrace{\int_V \rho|J|^2 dV}_{\leq 0} \tag{15.27}$$

Since integrand is zero at time $t = 0$ and non-zero for $t > 0$, the only consistent solution is $\mathbf{E} = 0$ and $\mathbf{H} = 0$ for $t \geq 0$.

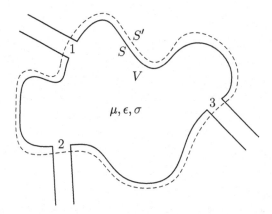

Figure 15.5 The region V can be excited at N ports. The steady-state fields inside are determined uniquely by either the currents or voltages at the ports.

15.4   The network formulation

In Fig. 15.5, imagine that S is a perfect conductor. Then $E_t \equiv 0$ on the surface of S except at the reference planes. Thus if the voltage or current is given at each reference plane, the fields are uniquely determined inside V. For instance if voltages V_1, V_2, $\cdots$ are specified, then the currents into each port are a linear combination of the voltages (if the materials are linear)

$$I_1 = Y_{11}V_1 + Y_{12}V_2 + Y_{13}V_3 + \cdots \tag{15.28}$$
$$I_2 = Y_{21}V_1 + Y_{22}V_2 + Y_{23}V_3 + \cdots \tag{15.29}$$
$$I_3 = Y_{31}V_1 + Y_{32}V_2 + Y_{33}V_3 + \cdots \tag{15.30}$$
$$\vdots$$

Or, in general, we can define an $N \times N$ complex matrix Y such that

$$i = Yv \tag{15.31}$$

Similarly if the currents are at the reference planes, then tangential magnetic fields are also known and a unique solution of Maxwell's equations inside V follows. The tangential E-fields, or the voltages, can then be computed as a linear combination of the voltages

$$V_1 = Z_{11}I_1 + Z_{12}I_2 + Z_{13}I_3 + \cdots \tag{15.32}$$
$$V_2 = Z_{21}I_1 + Z_{22}I_2 + Z_{23}I_3 + \cdots \tag{15.33}$$
$$V_3 = Z_{31}I_1 + Z_{32}I_2 + Z_{33}I_3 + \cdots \tag{15.34}$$
$$\vdots$$

What if the boundary is not a perfect conductor? Then introduce a new surface S' several skin depths within the conductor so that the tangential fields are essentially zero. Then the same argument as above applies except now the conductive portion will lead to loss and contribute to the real part of Y_{ij} or Z_{ij}.

Symmetry of impedance matrix

Suppose all terminals (reference planes, ports) are shorted except at the ith plane and denote the solution to Maxwell's equations as $\mathbf{E}_i, \mathbf{H}_i$. Similarly $\mathbf{E}_j, \mathbf{H}_j$ corresponds to the case when all terminals except the jth plane are shorted. By the Lorentz Reciprocity Theorem we have

$$\oint_S (\mathbf{E}_i \times \mathbf{H}_j - \mathbf{E}_j \times \mathbf{H}_i) \cdot d\mathbf{S} = 0 \tag{15.35}$$

for a sourceless region bounded by S. Let S consist of conducting walls bounding the junction and the N terminal planes. The integral over the walls vanishes if the walls are perfectly conducting or if the walls exhibit a surface impedance Z_m. So the above reduces to

$$\sum_{n=1}^{N} \int_{t_n} (\mathbf{E}_i \times \mathbf{H}_j - \mathbf{E}_j \times \mathbf{H}_i) \cdot d\mathbf{S} = 0 \tag{15.36}$$

But $\mathbf{n} \times \mathbf{E}_i$ and $\mathbf{n} \times \mathbf{E}_j$ are zero at all terminal planes except i and j

$$\int_{t_i} \mathbf{E}_i \times \mathbf{H}_j \cdot d\mathbf{S} = \int_{t_j} \mathbf{E}_j \times \mathbf{H}_i \cdot d\mathbf{S} \tag{15.37}$$

or

$$V_i(I_i)_j = V_j(I_j)_i \tag{15.38}$$

$(I_i)_j$ is the current at the terminal plane i arising from an applied voltage at plane j

$$I_i = (I_i)_j = Y_{ij} V_j \tag{15.39}$$
$$I_j = (I_j)_i = Y_{ji} V_i \tag{15.40}$$

Therefore

$$V_i V_j Y_{ij} = V_j V_i Y_{ji} \tag{15.41}$$

or

$$Y_{ij} = Y_{ji} \tag{15.42}$$

Since $Z = Y^{-1}$, the inverse of a symmetric matrix is also symmetric

$$A A^{-1} = I \tag{15.43}$$
$$I^t = I = A^t (A^{-1})^t \tag{15.44}$$
$$(A^{-1})^t = (A^t)^{-1} \tag{15.45}$$
$$A^t = A \rightarrow A^{-1} = (A^{-1})^t \tag{15.46}$$

Loss-free networks

For any network we have

$$-\oint(\mathbf{E} \times \mathbf{H}^*) \cdot d\mathbf{S} = \sum_{m=1}^{N} V_m I_m^* = 2W_L + 4j\omega(W_m - W_e) \qquad (15.47)$$

Since $V_m = \sum_{n=1}^{N} Z_{mn} I_n$ and $W_L \equiv 0$

$$\sum_{m=1}^{N}\sum_{n=1}^{N} Z_{mn} I_n I_m^* = 4j\omega(W_m - W_e) \qquad (15.48)$$

Let all ports be open except port i

$$Z_{ii} I_i I_i^* = 4j\omega(W_m - W_e) \qquad (15.49)$$

Thus the diagonal terms are imaginary. Now let all ports be open circuited except port i and j

$$Z_{ij} I_i I_j^* + Z_{ji} I_j I_i^* + Z_{ii}|I_i|^2 + Z_{jj}|I_j|^2 = 4j\omega(W_m - W_e) \qquad (15.50)$$

We thus have

$$\Re\left(Z_{ij} I_i I_j^* + Z_{ji} I_j I_i^*\right) = 0 \qquad (15.51)$$

Since the network is reciprocal, $Z_{ij} = Z_{ji}$, so

$$\Re\left(Z_{ij} \underbrace{(I_i I_j^* + I_j I_i^*)}_{\text{real}} \right) = 0 \qquad (15.52)$$

That means that Z_{ij} has to be imaginary. In conclusion, for a *lossless reciprocal* network, Z is imaginary. Since $Y = Z^{-1}$, Y is also imaginary.

15.5 Scattering matrix

Voltages and currents are difficult to measure directly at microwave frequencies. The Z matrix requires "opens," and it is hard to create an ideal open due to parasitic capacitance and radiation. Likewise, a Y matrix requires "shorts," again ideal shorts are impossible at high frequency due to the finite inductance.[2] Scattering parameters, or S parameters, are easier to measure at high frequency. The measurement is direct and only involves measurement of relative quantities (such as the SWR or the location of the first minima relative to the load). It is important to realize that, although we associate S parameters with high-frequency and wave propagation, the concept is valid for any frequency.

---

[2]   Many active devices could oscillate under the open or short termination.

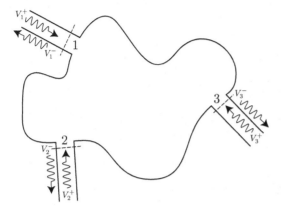

Figure 15.6 An N-port network can be characterized by incident and scattered waves.

Incident and scattering waves

With reference to Fig. 15.6, let's define the vector v^+ as the incident "forward" waves on each transmission line connected to the N port. Define the reference plane as the point where the transmission line terminates onto the N port. The vector v^- is then the reflected or "scattered" waveform at the location of the port.

$$v^+ = \begin{pmatrix} V_1^+ \\ V_2^+ \\ V_3^+ \\ \vdots \end{pmatrix} \tag{15.53}$$

$$v^- = \begin{pmatrix} V_1^- \\ V_2^- \\ V_3^- \\ \vdots \end{pmatrix} \tag{15.54}$$

Because the N port is linear, we expect that scattered field to be a linear function of the incident field

$$v^- = Sv^+ \tag{15.55}$$

S is the scattering matrix

$$S = \begin{pmatrix} S_{11} & S_{12} & \cdots \\ S_{21} & \ddots & \\ \vdots & & \end{pmatrix} \tag{15.56}$$

The fact that the S matrix exists can be easily proved if we recall that the voltage and current on each transmission line termination can be written as

$$V_i = V_i^+ + V_i^- \tag{15.57}$$

$$I_i = Y_0(V_i^+ - V_i^-) \tag{15.58}$$

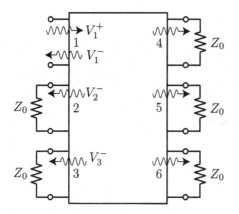

Figure 15.7 To calculate S_{ij}, terminate all ports k, $k \neq j$ with load terminations and drive port j with a unit source and observe the voltage waveform at port i.

Inverting these equations

$$V_i + Z_0 I_i = V_i^+ + V_i^- + V_i^+ - V_i^- = 2V_i^+ \tag{15.59}$$

$$V_i - Z_0 I_i = V_i^+ + V_i^- - V_i^+ + V_i^- = 2V_i^- \tag{15.60}$$

Thus v^+, v^- are simply linear combinations of the port voltages and currents. By the uniqueness theorem then, $v^- = S v^+$.

As shown in Fig. 15.7, the term S_{ij} can be computed directly by the following formula

$$S_{ij} = \left. \frac{V_i^-}{V_j^+} \right|_{V_k^+ = 0 \, \forall \, k \neq j} \tag{15.61}$$

In other words, to measure S_{ij}, drive port j with a wave amplitude of V_j^+ and terminate all other ports with the characteristic impedance of the lines (so that $V_k^+ = 0$ for $k \neq j$). Then observe the wave amplitude coming out of the port i.

Example 31

S matrix for a one-port capacitor
Let's calculate the S parameter for a capacitor shown in Fig. 15.8a

$$S_{11} = \frac{V_1^-}{V_1^+} \tag{15.62}$$

This is of course just the reflection coefficient for a capacitor

$$S_{11} = \rho_L = \frac{Z_C - Z_0}{Z_C + Z_0} = \frac{\frac{1}{j\omega C} - Z_0}{\frac{1}{j\omega C} + Z_0} \tag{15.63}$$

$$= \frac{1 - j\omega C Z_0}{1 + j\omega C Z_0} \tag{15.64}$$

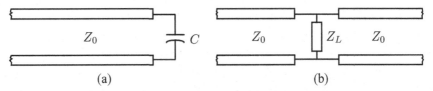

(a)   (b)

Figure 15.8 (a) The S matrix of a capacitor is simply the reflection coefficient of the capacitive termination. The two-port S matrix of a shunt admittance is computed by finding the reflection and transmission due to a lumped discontinuity.

The reflection coefficient S_{11} has a magnitude equal to 1 at all frequencies

$$|S_{11}| = \frac{|1 - j\omega C Z_0|}{|1 + j\omega C Z_0|} = 1 \tag{15.65}$$

A plot of S_{11} as a function of frequency on the Smith Chart traces a half unit arc starting at the "open" impedance and ending on the short impedance.

Example 32

S matrix for a two-port shunt element
Consider a shunt impedance connected at the junction of two transmission lines, as shown in Fig. 15.8b. The voltage at the junction is of course continuous. The currents, though, differ

$$V_1 = V_2 \tag{15.66}$$

$$I_1 + I_2 = Y_L V_2 \tag{15.67}$$

To compute S_{11}, enforce $V_2^+ = 0$ by terminating the line. Thus we can be rewrite the above equations

$$V_1^+ + V_1^- = V_2^- \tag{15.68}$$

$$Y_0(V_1^+ - V_1^-) = Y_0 V_2^- + Y_L V_2^- = (Y_L + Y_0)V_2^- \tag{15.69}$$

We can now solve the above equation for the reflected and transmitted wave

$$V_1^- = V_2^- - V_1^+ = \frac{Y_0}{Y_L + Y_0}(V_1^+ - V_1^-) - V_1^+ \tag{15.70}$$

$$V_1^-(Y_L + Y_0 + Y_0) = (Y_0 - (Y_0 + Y_L))V_1^+ \tag{15.71}$$

$$S_{11} = \frac{V_1^-}{V_1^+} = \frac{Y_0 - (Y_0 + Y_L)}{Y_0 + (Y_L + Y_0)} = \frac{Z_0||Z_L - Z_0}{Z_0||Z_L + Z_0} \tag{15.72}$$

The above equations can be written by inspection since $Z_0||Z_L$ is the effective load seen at the junction of port 1. Thus for port 2 we can write

$$S_{22} = \frac{Z_0||Z_L - Z_0}{Z_0||Z_L + Z_0} = \frac{-Z_0}{Z_0 + 2Z_L} \tag{15.73}$$

Likewise, we can solve for the transmitted wave, or the wave scattered into port 2

$$S_{21} = \frac{V_2^-}{V_1^+} \tag{15.74}$$

Since $V_2^- = V_1^+ + V_1^-$, we have

$$S_{21} = 1 + S_{11} = \frac{2Z_0||Z_L}{Z_0||Z_L + Z_0} = \frac{2Z_L}{2Z_L + Z_0} \tag{15.75}$$

By symmetry, we can deduce S_{12} as

$$S_{12} = \frac{2Z_0||Z_L}{Z_0||Z_L + Z_0} \tag{15.76}$$

Conversion formula

Since V^+ and V^- are related to V and I, it is easy to find a formula to convert for Z or Y to S

$$V_i = V_i^+ + V_i^- \quad \rightarrow \quad v = v^+ + v^- \tag{15.77}$$

$$Z_{i0}I_i = V_i^+ - V_i^- \quad \rightarrow \quad Z_0i = v^+ - v^- \tag{15.78}$$

Now starting with $v = Zi$, we have

$$v^+ + v^- = ZZ_0^{-1}(v^+ - v^-) \tag{15.79}$$

Note that Z_0 is the scalar port impedance

$$v^-(I + ZZ_0^{-1}) = (ZZ_0^{-1} - I)v^+ \tag{15.80}$$

$$v^- = (I + ZZ_0^{-1})^{-1}(ZZ_0^{-1} - I)v^+ = Sv^+ \tag{15.81}$$

We now have a formula relating the Z matrix to the S matrix

$$S = (ZZ_0^{-1} + I)^{-1}(ZZ_0^{-1} - I) = (Z + Z_0I)^{-1}(Z - Z_0I) \tag{15.82}$$

This equation should look familiar since it is a generalization of the scalar reflection coefficient for a load given by the same equation!

$$\overline{\rho} = \frac{Z/Z_0 - 1}{Z/Z_0 + 1} \tag{15.83}$$

To solve for Z in terms of S, simply invert the relation

$$Z_0^{-1}ZS + IS = Z_0^{-1}Z - I \tag{15.84}$$

$$Z_0^{-1}Z(I - S) = S + I \tag{15.85}$$

$$Z = Z_0(I + S)(I - S)^{-1} \tag{15.86}$$

As expected, these equations degenerate into the correct form for a 1×1 system

$$Z_{11} = Z_0 \frac{1 + S_{11}}{1 - S_{11}} \tag{15.87}$$

Reciprocal networks

We have found that the Z and Y matrix are symmetric. Now let's see what we can infer about the S matrix.

$$v^+ = \frac{1}{2}(v + Z_0 i) \tag{15.88}$$

$$v^- = \frac{1}{2}(v - Z_0 i) \tag{15.89}$$

Substitute $v = Zi$ in the above equations

$$v^+ = \frac{1}{2}(Zi + Z_0 i) = \frac{1}{2}(Z + Z_0)i \tag{15.90}$$

$$v^- = \frac{1}{2}(Zi - Z_0 i) = \frac{1}{2}(Z - Z_0)i \tag{15.91}$$

Since $i = i$, the above equations must result in consistent values of i. Or

$$2(Z + Z_0)^{-1} v^+ = 2(Z - Z_0)^{-1} v^- \tag{15.92}$$

Thus

$$S = (Z - Z_0)(Z + Z_0)^{-1} \tag{15.93}$$

Consider the transpose of the S matrix

$$S^t = ((Z + Z_0)^{-1})^t (Z - Z_0)^t \tag{15.94}$$

Recall that Z_0 is a diagonal matrix

$$S^t = (Z^t + Z_0)^{-1}(Z^t - Z_0) \tag{15.95}$$

If $Z^t = Z$ (reciprocal network), then we have

$$S^t = (Z + Z_0)^{-1}(Z - Z_0) \tag{15.96}$$

Previously we found that

$$S = (Z + Z_0)^{-1}(Z - Z_0) \tag{15.97}$$

So that we see that the S matrix is also symmetric (under reciprocity)

$$S^t = S \tag{15.98}$$

Another proof

Note that in effect we have shown that

$$(Z + I)^{-1}(Z - I) = (Z - I)(Z + I)^{-1} \tag{15.99}$$

This is easy to demonstrate if we note that

$$Z^2 - I = Z^2 - I^2 = (Z + I)(Z - I) = (Z - I)(Z + I) \tag{15.100}$$

In general matrix multiplication does not commute, but here it does

$$(Z - I) = (Z + I)(Z - I)(Z + I)^{-1} \tag{15.101}$$
$$(Z + I)^{-1}(Z - I) = (Z - I)(Z + I)^{-1} \tag{15.102}$$

Thus we see that $S^t = S$.

Scattering parameters of a lossless network

Consider the total power dissipated by a network (must sum to zero)

$$P_{av} = \frac{1}{2}\Re(v^t i^*) = 0 \tag{15.103}$$

Expanding in terms of the wave amplitudes

$$= \frac{1}{2}\Re\left((v^+ + v^-)^t Z_0^{-1}(v^+ - v^-)^*\right) \tag{15.104}$$

Where we assume that Z_0 are real numbers and equal. The notation is about to get ugly

$$= \frac{1}{2Z_0}\Re\left(v^{+t}v^{+*} - v^{+t}v^{-*} + v^{-t}v^{+*} - v^{-t}v^{-*}\right) \tag{15.105}$$

Notice that the middle terms sum to a purely imaginary number. Let $x = v^+$ and $y = v^-$

$$y^t x^* - x^t y^* = y_1 x_1^* + y_2 x_2^* + \cdots - x_1 y_1^* + x_2 y_2^* + \cdots = a - a^* \tag{15.106}$$

We have shown that

$$P_{av} = \frac{1}{2Z_0}\left(\underbrace{v^{+t}v^+}_{\substack{\text{total incident} \\ \text{power}}} - \underbrace{v^{-t}v^{-*}}_{\substack{\text{total reflected} \\ \text{power}}} \right) = 0 \tag{15.107}$$

This is a rather obvious result. It simply says that the incident power is equal to the reflected power (because the N port is lossless). Since $v^- = Sv^+$

$$v^{-t}v^- = (Sv^+)^t(Sv^+)^* = v^{+t} S^t S^* v^{+*} \tag{15.108}$$

This can only be true if S is a unitary matrix

$$S^t S^* = I \tag{15.109}$$
$$S^* = (S^t)^{-1} \tag{15.110}$$

Orthogonal properties of S

Expanding out the matrix product

$$\delta_{ij} = \sum_k (S^T)_{ik} S_{kj}^* = \sum_k S_{ki} S_{kj}^* \tag{15.111}$$

For $i = j$ we have

$$\sum_k S_{ki} S_{ki}^* = 1 \qquad (15.112)$$

For $i \neq j$ we have

$$\sum_k S_{ki} S_{kj}^* = 0 \qquad (15.113)$$

The dot product of any column of S with the conjugate of that column is unity, while the dot product of any column with the conjugate of a different column is zero. If the network is reciprocal, then $S^t = S$ and the same applies to the rows of S. Note also that $|S_{ij}| \leq 1$.

Shift in reference planes

A very nice property of the S matrix is that if we move the reference planes, we can easily recalculate the S parameters. We'll derive a new matrix S' related to S. Let us call the waves at the new reference w

$$v^- = S v^+ \qquad (15.114)$$
$$w^- = S' w^+ \qquad (15.115)$$

Since the waves on the lossless transmission lines only experience a phase shift, we have a phase shift of $\theta_i = \beta_i \ell_i$

$$w_i^- = v^- e^{-j\theta_i} \qquad (15.116)$$
$$w_i^+ = v^+ e^{j\theta_i} \qquad (15.117)$$

Or we have

$$
\begin{bmatrix}
e^{j\theta_1} & 0 & \cdots \\
0 & e^{j\theta_2} & \cdots \\
0 & 0 & e^{j\theta_3} & \cdots \\
& \vdots & &
\end{bmatrix}
w^- = S
\begin{bmatrix}
e^{-j\theta_1} & 0 & \cdots \\
0 & e^{-j\theta_2} & \cdots \\
0 & 0 & e^{-j\theta_3} & \cdots \\
& \vdots & &
\end{bmatrix}
w^+ \qquad (15.118)
$$

So we see that the new S matrix is simply

$$
S' =
\begin{bmatrix}
e^{-j\theta_1} & 0 & \cdots \\
0 & e^{-j\theta_2} & 0 & \cdots \\
0 & 0 & e^{-j\theta_3} & \cdots \\
\vdots & \vdots & & \ddots
\end{bmatrix}
S
\begin{bmatrix}
e^{-j\theta_1} & 0 & \cdots \\
0 & e^{-j\theta_2} & 0 \\
0 & 0 & e^{-j\theta_3} & 0 \\
\vdots & \vdots & & \ddots
\end{bmatrix}
\qquad (15.119)
$$

15.6 Properties of three-ports

Let us begin this section with a simple question. Is it possible to construct a lossless three-port device that is matched in all three-ports

$$S_{11} = S_{22} = S_{33} = 0 \qquad (15.120)$$

such that the power flowing into port 1 is divided into ports 2 and 3, or in terms of the S matrix

$$S = \begin{bmatrix} 0 & S_{12} & S_{13} \\ S_{21} & 0 & S_{23} \\ S_{31} & S_{32} & 0 \end{bmatrix} \tag{15.121}$$

If we assume that the device is constructed with reciprocal materials, then $S^T = S$, or

$$S = \begin{bmatrix} 0 & S_{12} & S_{13} \\ S_{12} & 0 & S_{23} \\ S_{13} & S_{23} & 0 \end{bmatrix} \tag{15.122}$$

By the unitary property of the matrix (lossless), we have that the columns of the matrix are orthogonal and of unit magnitude. Writing this out we have

$$S_{13} S_{23}^* = 0 \tag{15.123}$$
$$S_{12} S_{13}^* = 0 \tag{15.124}$$
$$S_{12} S_{23}^* = 0 \tag{15.125}$$

and

$$|S_{12}|^2 + |S_{13}|^2 = 1 \tag{15.126}$$
$$|S_{12}|^2 + |S_{23}|^2 = 1 \tag{15.127}$$
$$|S_{13}|^2 + |S_{23}|^2 = 1 \tag{15.128}$$

The above system represents six equations and six unknowns. But no matter how we try to solve the equations, we find that the system is inconsistent. For instance, let's try to satisfy Eqs. (15.123) and (15.124) by setting $S_{13} = S_{12} = 0$. Then Eq. 15.126 is violated. In a like manner, we can try $S_{23} = S_{12} = 0$, which violates Eq. 15.127. Finally, if we try $S_{23} = S_{13} = 0$, we violate Eq. (15.128). So in conclusion, we cannot build a reciprocal lossless matched three-port!

Non-reciprocal three-port circulator

If we relax the conditions by allowing a non-reciprocal element, then the S matrix is given by Eq. (15.121). Applying the unitary (lossless) condition we have

$$S_{31}^* S_{32} = 0 \tag{15.129}$$
$$S_{21}^* S_{23} = 0 \tag{15.130}$$
$$S_{12}^* S_{13} = 0 \tag{15.131}$$

and

$$|S_{12}|^2 + |S_{13}|^2 = 1 \tag{15.132}$$
$$|S_{21}|^2 + |S_{23}|^2 = 1 \tag{15.133}$$
$$|S_{31}|^2 + |S_{32}|^2 = 1 \tag{15.134}$$

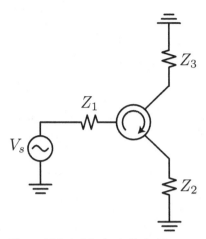

Figure 15.9 A "clockwise" circulator couples power from one-port to the next port in the direction shown.

Let's take a particular solution that satisfies Eqs. (15.129)–(15.131), or $S_{12} = S_{23} = S_{31} = 0$. Then by Eqs. (15.132)–(15.134) we have $|S_{13}| = |S_{21}| = |S_{32}| = 1$, or in matrix form

$$S = \begin{bmatrix} 0 & 0 & 1 \\ 1 & 0 & 0 \\ 0 & 1 & 0 \end{bmatrix} \tag{15.135}$$

Such a device is called a circulator, shown in Fig. 15.9. The origin of the name is clear from the figure. Under matched conditions, if power is injected into any port, it circulates in the direction of the arrow. For instance, if power is injected into port 1, all of it ends up in the port 3 termination.

It's also possible to build a counterclockwise circulator. This is the second possible solution of the above equations. The circulator is a very nice element and it has many applications, for instance in a radar shown in the earlier example. We now clearly see the function of the device. If the power is transmitted into port 1, and an antenna is placed at port 3, then the received power will be incident on port 2, in effect isolating the transmitter signal from the received signal. This is important because the transmitted signal is orders of magnitude stronger than the weak received signal. Thus a very sensitive receiver circuit can be designed to operate with weak signals without having to worry about the non-linearity induced by the strong transmitter interfering signal.

Three-port with single port mismatch

As a final example, let's consider a reciprocal lossless three-port device matched only at two ports. Or in terms of the S matrix

$$S = \begin{bmatrix} 0 & S_{12} & S_{13} \\ S_{12} & 0 & S_{23} \\ S_{13} & S_{23} & S_{33} \end{bmatrix} \tag{15.136}$$

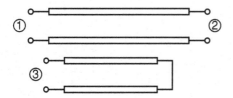

Figure 15.10 A three-port with two matched ports degenerates into a trivial isolated structure with a delay from port 1 to 2 and an isolated mismatched third port.

Applying the unitary conditions, we have

$$S_{13}S_{23}^* = 0 \tag{15.137}$$
$$S_{12}S_{13}^* + S_{23}S_{33}^* = 0 \tag{15.138}$$
$$S_{12}S_{23}^* + S_{13}S_{33}^* = 0 \tag{15.139}$$

and

$$|S_{12}|^2 + |S_{13}|^2 = 1 \tag{15.140}$$
$$|S_{12}|^2 + |S_{23}|^2 = 1 \tag{15.141}$$
$$|S_{13}|^2 + |S_{23}|^2 + |S_{33}|^2 = 1 \tag{15.142}$$

Subtracting Eqs. (15.137) and (15.138) we have

$$|S_{13}|^2 - |S_{23}|^2 = 0 \tag{15.143}$$

or more simply $|S_{13}| = |S_{23}|$. Then Eq. (15.140) implies that $S_{13} = S_{23} = 0$. Now Eq. (15.137) implies that $|S_{12}| = 1$ and Eq. (15.139) implies that $|S_{33}| = 1$. After the onslaught the resulting S matrix takes the following form

$$S = \begin{bmatrix} 0 & e^{j\phi} & 0 \\ e^{j\phi} & 0 & 0 \\ 0 & 0 & e^{j\theta} \end{bmatrix} \tag{15.144}$$

But that means that ports 1 and 2 are completely isolated from port 3! In other words this three-port is really just a trivial case of a transmission line connecting ports 1 and 2 (to introduce the delay) and a transmission line into a lossless termination at port 3 (such as a short or an open). One such circuit is shown in Fig. 15.10.

The important conclusion is that any lossless three-port must employ non-reciprocal elements in order to do something "interesting"!

Power dividers and combiners

Consider the simplest power divider, the T junction of three transmission lines, as shown in Fig. 15.11. We already know that we cannot match more than one port! Let's match port 1, or let

$$Y_{in} = \frac{1}{Z_2} + \frac{1}{Z_3} = \frac{1}{Z_0} \tag{15.145}$$

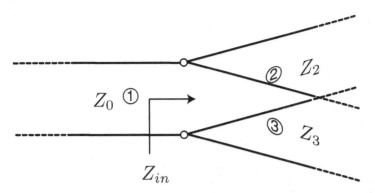

Figure 15.11 A simple power splitter formed by the junction of three transmission lines can only match one port. Furthermore, ports 2 and 3 are not isolated.

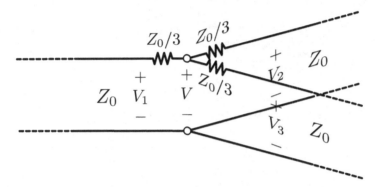

Figure 15.12 A lossy power splitter formed by the junction of three transmission with three resistors.

For an equal power splitter with $Z_1 = Z_0 = 50\Omega$, we make $Z_2 = Z_3 = 100\Omega$. If needed, we can use a broadband transformer to synthesize 100Ω from 50Ω transmission lines. Unfortunately, the output ports do not see a match. From port 2 we see

$$Y_{in,2} = \frac{1}{Z_3} + \frac{1}{Z_0} = \frac{1}{100} + \frac{1}{50} \tag{15.146}$$

or $Z_{in,2} = 33.3\Omega$. Further, the output ports are not isolated.

If we introduce loss into the circuit, as shown in Fig. 15.12a, then we can in fact match all three ports. By symmetry, the impedance looking into each port 2 or 3, represented by Fig. 15.12b, is given by

$$Z = \frac{Z_0}{3} + Z_0 = \frac{4}{3}Z_0 \tag{15.147}$$

So the impedance from port 1 is given by $Z/2$ in series with $Z_0/3$, or

$$Z_{in} = \frac{Z_0}{3} + \frac{2}{3}Z_0 = Z_0 \tag{15.148}$$

Since the network is symmetric, all the ports are matched, so $S_{11} = S_{22} = S_{33} = 0$. To find the coupling to ports 2 and 3, simply calculate the transfer function based on the voltage

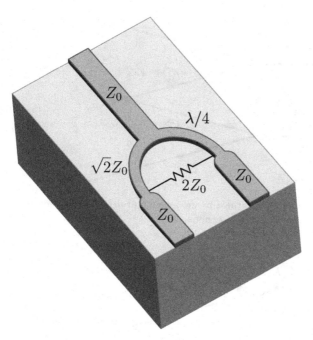

Figure 15.13 A microstrip three-port Wilkinson power divider/combiner.

division. At the center of the structure the voltage V is simply

$$V = V_1 \frac{2Z_0/3}{Z_0/3 + 2Z_0/3} = \frac{2}{3}V_1 \tag{15.149}$$

so that the voltage at ports 2 or 3 is given by

$$V_2 = V_3 = V \frac{Z_0}{Z_0 + Z_0/3} = \frac{3}{4}V = \frac{3}{4} \times \frac{2}{3}V_1 = \frac{1}{2}V_1 \tag{15.150}$$

which means that $S_{21} = S_{31} = S_{23} = 1/2$. The power drops by 6 dB. An ideal power divider would have an insertion loss of 3 dB, so half the power is consumed by the resistors.

Wilkinson divider

The Wilkinson power divider (or combiner), shown in microstrip form in Fig. 15.13, employs quarter wave lines to split the power. The resistor $R = 2Z_0$, which introduces loss, is also a critical part of the structure. We shall find that the performance of the divider, under matched conditions, is close to an ideal divider or combiner. An equivalent circuit for the divider is shown in Fig. 15.14. The calculation of the S matrix is simplified considerably by using an "even" and "odd" mode analysis technique.

Even mode

In the even mode, we excite ports 2 and 3 with an equal voltage of $+2V$ and so no current flows through the resistor $R = 2Z_0$. In fact, the circuit can be simplified further as shown

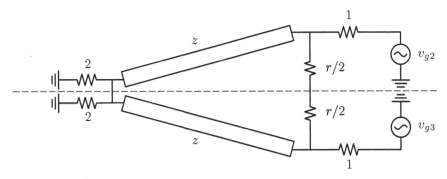

Figure 15.14 The equivalent circuit for a Wilkinson power divider.

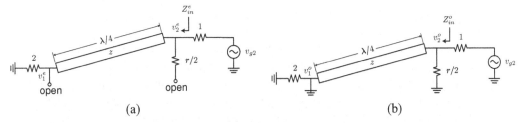

Figure 15.15 Equivalent half circuit under (a) even and (b) odd mode excitation.

in Fig. 15.15a, where symmetric points are converted to open circuit connections. In the equivalent even mode half circuit, the termination at port 1 is $2Z_0$ since it is in parallel with another identical circuit. So the input impedance from port 2 is simply the quarter wave transformed impedance

$$Z_{in}^e = \frac{Z^2}{2Z_0} \tag{15.151}$$

To match the even mode we should make $Z = \sqrt{2}Z_0$. The wave on the transmission line consists of a forward and backward wave

$$v(z) = V^+(e^{-j\beta x} + \Gamma e^{j\beta x}) \tag{15.152}$$

At port 2, or $x = -\lambda/4$, we have

$$v_2^e = v(-\lambda/4) = jV^+(1 - \Gamma) = v_{g2}/2 \tag{15.153}$$

and at port 1 we have

$$v_1^e = v(0) = V^+(1 + \Gamma) \tag{15.154}$$

If we substitute for V^+ from Eq. (15.153), we have

$$v_1^e = \frac{v_{g2}}{2j} \frac{1 + \Gamma}{1 - \Gamma} \tag{15.155}$$

since $\Gamma = (2 - \sqrt{2})/(2 + \sqrt{2})$, the above simplifies to

$$v_1^e = -j\sqrt{2}\frac{v_{g2}}{2} \tag{15.156}$$

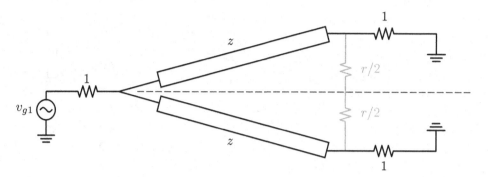

Figure 15.16 Driving port 1 excites the even mode of the Wilkinson structure.

Odd mode

In the odd mode, we excite port 2 with $+1V$ and port 3 with $-1V$. The equivalent half circuit from port 2 to port 1 is shown in Fig. 15.15b. Note that since the circuit is excited with a balanced signal, any symmetric point, such as the middle of the resistor, can be replaced with a short circuit. This is similar to the analysis of a differential pair half circuit. Since the quarter wave transformer converts the short circuit at port 1 to an open at port 2, the input impedance is simply given by

$$Z_{in}^2 = R/2 = Z_0 \tag{15.157}$$

Likewise $V_1^o = 0$ due to the short, and $V_2^o = \frac{1}{2}v_{gs2}$.

Port impedance

The input impedance at port 1 can be calculated from the equivalent circuit shown in Fig. 15.16. Due to the balanced excitation, we can ignore the resistor r and calculate the input impedance by inspection

$$Z_{in,1} = Z_{in,even} = \frac{1}{2}\frac{(\sqrt{2}Z_0)^2}{Z_0} = Z_0 \tag{15.158}$$

which shows that port 1 is matched.

Since we have shown that all the ports are matched (for all modes independently), we conclude that

$$S_{11} = S_{22} = S_{33} = 0 \tag{15.159}$$

To calculate S_{12} (and by symmetry S_{21}) we simply sum the even and odd modes which excites port 2 with a voltage of $2v_{gs2}$ and port 3 with zero volts.

$$S_{12} = S_{21} = \frac{V_1^e + V_1^o}{V_2^e + V_2^o} = -j/\sqrt{2} \tag{15.160}$$

There is no need to calculate $S_{13} = S_{31}$ since by symmetry it is the same as port 2. So we see that the Wilkinson divider (combiner) has 3 dB of insertion loss, just like an ideal

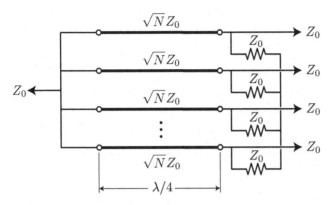

Figure 15.17 An N-way Wilkinson divider/combiner.

power divider. Since the even and odd mode circuits decouple ports 2 and 3, these ports are isolated and so $S_{23} = S_{32} = 0$.

Somehow it seems like we cheated nature! Our previous analysis implied that it was impossible to build a lossless reciprocal three-port, but it seems that we have done just that. Even though we used a resistor R, no power is dissipated in this resistor under matched terminations. But the above analysis only applies under matched conditions. Under a mismatch, the reflected power is dissipated through the resistor R.

The Wilkinson divider/combiner can be generalized in two ways [47]. We can divide or combine power unevenly if we terminate $Z_2 = kZ_0$ and $Z_3 = Z_0/k$ and satisfy the following equations

$$Z_{03} = Z_0\sqrt{\frac{1 + k^2}{k^3}} \tag{15.161}$$

$$Z_{02} = k^2 Z_{03} \tag{15.162}$$

$$R = Z_0 \left(k + \frac{1}{k} \right) \tag{15.163}$$

The Wilkinson divider/combiner can also become an N-way (or $N + 1$ port) by employing the structure shown in Fig. 15.17.

15.7 Properties of four-ports

If the analysis of a three-port left you feeling queasy, then the four-port will definitely make you sick. Because a matched lossless reciprocal four port has 12 non-zero elements in the S matrix

$$S = \begin{bmatrix} 0 & S_{12} & S_{13} & S_{14} \\ S_{12} & 0 & S_{23} & S_{24} \\ S_{13} & S_{23} & 0 & S_{34} \\ S_{14} & S_{24} & S_{34} & 0 \end{bmatrix} \tag{15.164}$$

Let C_i denote the ith column. Then applying the unitary conditions to the matrix we have the following ten equations

$$
\begin{array}{cccc}
C_1 \cdot C_1^* & C_1 \cdot C_2^* & C_2 \cdot C_3^* & C_3 \cdot C_4^* \\
C_2 \cdot C_2^* & C_1 \cdot C_3^* & C_2 \cdot C_4^* & \\
C_3 \cdot C_3^* & C_1 \cdot C_4^* & & \\
C_4 \cdot C_4^* & & &
\end{array}
\tag{15.165}
$$

which means that we can solve for two independent parameters.

Let's start with row 1 and row 2 product times S_{24}^* and row 3 and row 4 product times S_{13}^*

$$
\begin{aligned}
(S_{13}^* S_{23} + S_{14}^* S_{24} = 0) \times S_{24}^* \\
(S_{13} S_{14}^* + S_{23} S_{24}^* = 0) \times S_{13}^*
\end{aligned}
\tag{15.166}
$$

If we subtract the above two equations, we come up with the following equation

$$
S_{14}^*(|S_{24}|^2 - |S_{13}|^2) = 0
\tag{15.167}
$$

Let's do the same thing again, now with rows 1 and 3 and rows 2 and 4

$$
\begin{aligned}
(S_{12}^* S_{23} + S_{14}^* S_{34} = 0) \times S_{12} \\
(S_{14}^* S_{12} + S_{34}^* S_{23} = 0) \times S_{34}
\end{aligned}
\tag{15.168}
$$

and again subtracting the above equations, we have

$$
S_{23}(|S_{12}|^2 - |S_{34}|^2) = 0
\tag{15.169}
$$

These two equations, (15.167) and (15.169), can be satisfied if $S_{14} = 0$ and $S_{23} = 0$. This leads to a *directional coupler*. Our four column magnitude equations lead to

$$
|S_{12}|^2 + |S_{13}|^2 = 1 \tag{15.170}
$$
$$
|S_{13}|^2 + |S_{34}|^2 = 1 \tag{15.171}
$$
$$
|S_{12}|^2 + |S_{24}|^2 = 1 \tag{15.172}
$$
$$
|S_{24}|^2 + |S_{34}|^2 = 1 \tag{15.173}
$$

If we subtract the first two of the above equations, we have $|S_{12}| = |S_{34}|$. Likewise, subtracting the bottom two equations we are led to $|S_{13}| = |S_{24}|$.

We will now make some additional assumptions about the four-port element to put the structure into a canonical form. We are free to adjust the terminal reference planes, since this only adds a phase shift to our S matrix. In particular, we can add transmission line sections to ensure that the phase from port 1 to port 2 matches the phase shift from port 3 to port 4. In fact, we can make this phase shift a multiple of 360°. Thus we can make

$$
S_{12} = S_{34} = \alpha
\tag{15.174}
$$

Then our equations imply that

$$S_{13} = \beta e^{j\theta} \tag{15.175}$$

$$S_{24} = \beta e^{j\phi} \tag{15.176}$$

Because row 2 and 3 are orthogonal, we have

$$S_{12}^* S_{13} + S_{24}^* S_{34} = 0 \tag{15.177}$$

or

$$\alpha^* \beta e^{j\theta} + \beta^* e^{-j\phi} \alpha = 0 \tag{15.178}$$

since α and β are real numbers, this simplifies to

$$e^{j\theta} + e^{-j\phi} = 0 \tag{15.179}$$

or

$$\theta + \phi = \pi \pm 2n\pi \tag{15.180}$$

There are two important canonical cases to consider. First we can have a symmetric coupler such that $\theta = \phi = \pi/2$, so the S matrix is given by

$$S = \begin{bmatrix} 0 & \alpha & j\beta & 0 \\ \alpha & 0 & 0 & j\beta \\ j\beta & 0 & 0 & \alpha \\ 0 & j\beta & \alpha & 0 \end{bmatrix} \tag{15.181}$$

We see that the power into port 1 is split into ports 2 and 3 with power gain α^2 and β^2 and a phase shift difference of 90°. The anti-symmetric coupler has $\theta = 0$ and $\phi = \pi$, which leads to

$$S = \begin{bmatrix} 0 & \alpha & \beta & 0 \\ \alpha & 0 & 0 & -\beta \\ \beta & 0 & 0 & \alpha \\ 0 & -\beta & \alpha & 0 \end{bmatrix} \tag{15.182}$$

Now we see that α and β are related by conservation of energy. Recall that this is a lossless four port and so naturally

$$|\alpha|^2 + |\beta|^2 = 1 \tag{15.183}$$

so an ideal canonical coupler has only one degree of freedom!

If we had chosen other possible solutions to the equations, we would have found other directional couplers or isolated two-ports. In fact, we can show that any non-trivial lossless, matched four port is a directional coupler.

Directional coupler

A schematic symbol for a directional coupler is shown in Fig. 15.18. Ports 1–4 are given the following descriptive names. Port 1 is the input port, port 2 is the "through" port, port 3 is the "coupled" port, and finally port 4 is the "isolated" port.

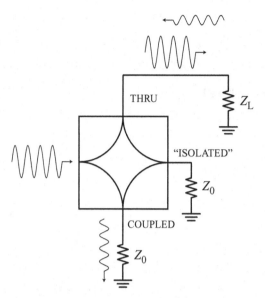

Figure 15.18 A directional coupler.

Imagine we connect a source to port 1 and terminate all the other ports in Z_0. Then the wave coming out of port 2 is given by

$$v_2^- = S_{21}v_1^+ + S_{24}v_4^+ \tag{15.184}$$

Note that by construction $S_{23} = 0$, which shows that ports 2 and 3 are isolated from one another. Since port 4 is terminated $v_4^+ = 0$. Substituting for a symmetric coupler we have

$$v_2^- = \alpha v_1^+ \tag{15.185}$$

so that port 2 receives α^2 fraction of the power. In many applications we make $\alpha \approx 1$ so this is the "through" port, or the port that receives most of the power. On the other hand, necessarily $\beta \approx 0$ so a tiny fraction of the power is then "coupled" to port 3 since

$$v_3^- = j\beta v_1^+ \tag{15.186}$$

Of course there is no restriction in the values of α but the convention is to call this the through port. The coupled port has a phase shift of 90° for a symmetric coupler and a phase shift of 180° for an anti-symmetric coupler.

Port 4 is completely isolated from port 1 since

$$v_4^- = S_{23}v_2^+ + S_{23}v_3^+ = 0 \tag{15.187}$$

and thus both ports 2 and 3 are matched.

Now by symmetry, we see that if we put our source on port 4, then port 3 is the through port, port 2 is the coupled port, and port 1 is isolated. And, likewise, if we put our source at port 2, then port 3 is isolated, port 1 is the through port, and port 4 is coupled! I won't repeat myself by putting the source at port 3! The diagram of the directional coupler gives this image.

A non-ideal coupler has finite *directivity* and *isolation*. For such a coupler we define

$$D = \text{Directivity} = 10 \log \frac{P_3}{P_4} = 20 \log \frac{\beta}{|S_{14}|} \tag{15.188}$$

$$I = \text{Isolation} = 10 \log \frac{P_1}{P_4} = -20 \log |S_{14}| \tag{15.189}$$

The coupling is naturally defined by

$$C = \text{Coupling} = 10 \log \frac{P_3}{P_1} = -20 \log \beta \tag{15.190}$$

An important anti-symmetric coupler is the hybrid coupler with $\alpha = \beta = 1/\sqrt{2}$. This coupler has a phase shift of $90°$

$$S = \frac{1}{\sqrt{2}} \begin{bmatrix} 0 & 1 & j & 0 \\ 1 & 0 & 0 & j \\ j & 0 & 0 & 1 \\ 0 & j & 1 & 0 \end{bmatrix} \tag{15.191}$$

An important symmetric coupler is called a $180°$ hybrid. Based on the implementation, other names include the *Magic-T* and *Rat-Race* coupler. It also splits power equally so its S matrix is given by

$$S = \frac{1}{\sqrt{2}} \begin{bmatrix} 0 & 1 & 1 & 0 \\ 1 & 0 & 0 & -1 \\ 1 & 0 & 0 & 1 \\ 0 & -1 & 1 & 0 \end{bmatrix} \tag{15.192}$$

Often a $180°$ hybrid is drawn with port 1 labeled the *summation* port, or Σ port, and port 4 labeled the *difference* port, or the Δ port. This is because the power coming out of port 1 is the *sum* of the incident voltages of ports 2 and 3

$$v_\Sigma^- = v_1^- = \frac{1}{\sqrt{2}}(v_2^+ + v_3^+) \tag{15.193}$$

whereas the power on port 4 is the difference voltage

$$v_\Delta^- = v_4^- = \frac{1}{\sqrt{2}}(v_3^+ - v_2^+) \tag{15.194}$$

Example 33

Cell phone power measurement

Nearly every cell phone employs a directional coupler, as shown in Fig. 15.19, to measure the power delivered to the load antenna. A simple capacitive divider could be used to sample the load voltage, but a directional coupler can distinguish between the forward power and the reflected power due to a load mismatch. The antenna impedance changes based on operating conditions, such as proximity to the human head and especially to nearby metal objects. It is therefore important to measure the power under widely varying loads.

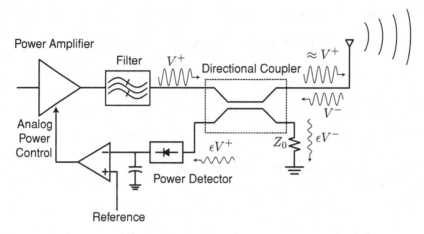

Figure 15.19 A directional coupler can detect the radiated power regardless of the load mismatch.

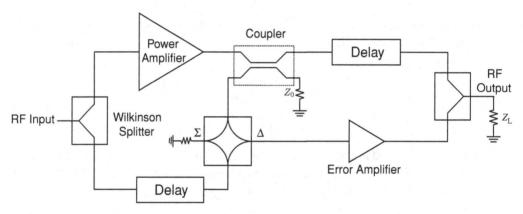

Figure 15.20 A feedforward linearization scheme.

Example 34

A feedforward distortion cancellation scheme

In Fig. 15.20, several 3 and 4 port building blocks are employed in the design of a linear power amplifier (PA). In this feedforward linearization scheme, the distortion produced by the power amplifier is detected and canceled from the output. The input signal is split over two parallel paths, the amplifying path and the correction path. In the correction path a delay line is employed to match the delay of the power amplifier. Next, a 180° hybrid is employed to take the difference of the coupled PA signal and the delayed input signal. If the coupling coefficient matches the PA gain, then the difference signal is simply the error, or distortion, produced by the PA. This signal is amplified by the error amplifier and again combined with the power signal. Note the power signal is likewise delayed to match the delay of the error amplifier. Since the Δ terminal of the hybrid is used, the error signal has a 180° phase shift relative to the main path. Again the gain of the error amplifier should match the gain of the coupler so that the final combined signals produce a cancellation of the distortion terms. The error amplifier does not produce any significant distortion since it operates with small distortion signals. This scheme works well and can reduce the distortion products bounded by the matching of the gain and phase in the two paths.

Branch line coupler

A branch line coupler, also called a quadrature coupler, shown in Fig. 15.21, is composed of a ring of quarter-wave transmission lines. The analysis of this structure can be performed with the even and odd mode analysis introduced in the previous section. The reader can quickly draw the equivalent even and odd mode circuits, shown in Fig. 15.22. In the odd mode, the equivalent circuit has two short circuited sections of $\lambda/8$ lines at the input and output. In the even mode, though, the lines are open circuited. For each case the equivalent circuit of Fig. 15.23 is applicable, with $Y = \pm j Y_{02}$ for the open and short circuit $\lambda/8$ reactance.

If we inject power into port 1, this is equivalent to a sum of even and odd mode excitation, since the even and odd powers add at port 1 and subtract at port 2. The input impedance is given by

$$Y_{in} = Y + \frac{Y_{01}^2}{Y_0 + Y} = Y_0 \frac{(Y_{01}/Y_0)^2 + (Y/Y_0) + (Y/Y_0)^2}{1 + (Y/Y_0)} \qquad (15.195)$$

where the second term accounts for the action of the quarter wave line. Since the input impedance is known, we can readily calculate the reflection coefficient for each mode

$$\rho = \frac{1 - Y_{in}/Y_0}{1 + Y_{in}/Y_0} \qquad (15.196)$$

If we calculate this expression for the even mode, we obtain

$$\rho_{1,e} = \frac{1 - (Y_{01}/Y_0)^2 + (Y_{02}/Y_0)^2}{1 + 2j(Y_{02}/Y_0) + (Y_{01}/Y_0)^2 - (Y_{02}/Y_0)^2} \qquad (15.197)$$

and similarly, in odd mode

$$\rho_{1,o} = \frac{1 - (Y_{01}/Y_0)^2 + (Y_{02}/Y_0)^2}{1 - 2j(Y_{02}/Y_0) + (Y_{01}/Y_0)^2 - (Y_{02}/Y_0)^2} \qquad (15.198)$$

The reflection coefficient for port 4 can be deduced without any further calculation. For the even mode it is clearly the same, or $\rho_{4,e} = \rho_{1,e}$. By anti-symmetry, the reflection for the odd mode is simply $\rho_{4,o} = -\rho_{4,o}$.

If we excite port 1 with a signal (sum of even and odd modes) we desire a match at port 1 and no signal at port 4, since it is the isolated port. To obtain an input match at port 1, we require the sum of the even mode and odd mode reflections to cancel, or

$$\rho_{1,e} + \rho_{1,o} = 0 \qquad (15.199)$$

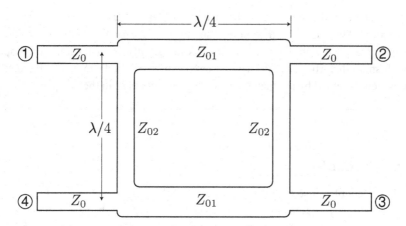

Figure 15.21 A branch line four-port directional coupler.

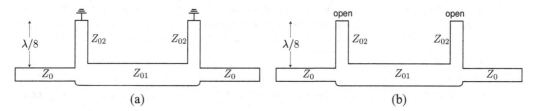

Figure 15.22 Equivalent half geometry for branch line coupler under (a) odd and (b) even mode excitation.

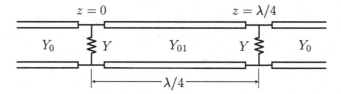

Figure 15.23 The equivalent circuit for a branch line coupler excited in even or odd mode. The admittance Y represents the $\lambda/8$ open or short stub.

To isolate port 4, we require

$$\rho_{1,e} - \rho_{1,o} = 0 \tag{15.200}$$

These two equations are satisfied if and only if each reflection (even and odd) is zero independently, $\rho_{1,e} = \rho_{1,o} \equiv 0$. Or

$$1 - (Y_{01}/Y_0)^2 + (Y_{02}/Y_0)^2 = 0 \tag{15.201}$$

Solving this equation, we arrive at the following condition for isolation and match

$$Z_{02}/Z_0 = \frac{Z_{01}/Z_0}{\sqrt{1 - (Z_{01}/Z_0)^2}} \tag{15.202}$$

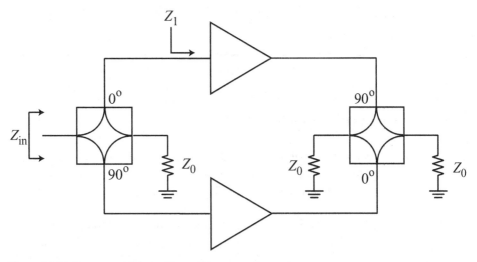

Figure 15.24 Power combining with quadrature couplers.

The quadrature coupler is characterized by a coupling factor determined by the ratio of Z_{01} and Z_0

$$C = 10\log\left(\frac{1}{1 - (Z_{01}/Z_0)^2}\right) \tag{15.203}$$

For a 3 dB coupler, we need $Z_{01} = Z_0/\sqrt{2}$. The operation of a branch line coupler can be understood intuitively in the following way. The phase delay from port 1 to port 2 and 3 differ by $90°$ due to the physical size of the structure. The two paths to port 4, on the other hand, differ by $180°$, and thus this port is isolated.

Example 35

Power combining with quadrature couplers

In Fig. 15.24 we see that an input signal is split into two paths by a hybrid quadrature coupler. Each path then travels through a power amplifier, and the output of the amplifiers is combined using a second quadrature coupler.

The net phase shift through each path is equalized, since at the input the top amplifier uses the $0°$ port but at the output it uses the quadrature, or $90°$ port. One advantage of power combining is to increase the output power delivered to the load by a factor of nearly 2 (assuming low loss couplers). In addition, if one amplifier fails, the system will continue to function, albeit at reduced output power. The isolation in the coupler isolates each power amplifier and simplifies the design. A less obvious advantage is that the quadrature coupler at the input and output generates a matched input impedance, regardless of the input impedance of each power amplifier, $Z_1 \neq Z_0$. Observe that the reflected power due to the input mismatch at the top power amplifier has precisely $180°$ phase difference with the bottom path. If the amplifiers are matched, the reflected powers cancel. The same argument holds at the output of the power amplifier.

15.8 Two conductor coupler

To build a coupler of small coupling ratios, we can simply take two transmission lines in parallel, as shown in Fig. 15.25. For small coupling, the lateral coupling is enough, but for larger ratios a broadside coupler configuration may be needed. A top view for a lateral coupler is shown in Fig. 15.25.

It is important to realize that this coupler supports two modes, an even mode and an odd mode, each with its own characteristic impedance and propagation constant. In Fig. 15.26 we sketch the fields for the odd and even modes. In the even mode, a magnetic wall can be placed between the structures, whereas in odd mode an electric wall can be placed at the point of symmetry.[3] The distributed equivalent circuit for these modes is shown in Fig. 15.27. In the odd mode, the anti-parallel currents lower the inductance per unit length to $L - M$, whereas in even mode the magnetic fields adds to produce $L + M$ inductance per unit length. Likewise, in even mode the capacitance of the mode is given by the C_0, whereas in the odd mode the coupling capacitance produces a capacitance per unit length $C_0 + 2C_c$ (Miller effect). The odd and even mode impedances are given by

$$Z_{0o} = \sqrt{\frac{L - M}{C_0 + 2C_c}} \qquad (15.204)$$

and

$$Z_{0e} = \sqrt{\frac{L + M}{C_0}} \qquad (15.205)$$

A detailed analysis [47] yields the following for the coupling factor of a $\lambda/4$ length coupler

$$C = 20 \log \left(\frac{Z_{0e} + Z_{0o}}{Z_{0e} - Z_{0o}} \right) \qquad (15.206)$$

In general, for a shorter coupler, the coupling drops and it is given by

$$c = \frac{jC \tan \theta}{\sqrt{1 - C^2} + j \tan \theta} \qquad (15.207)$$

---

[3] A magnetic wall is the dual of an electric wall. An electric wall is a perfect conductor forcing the electric field to be incident at a normal angle and the magnetic field is tangential to the surface. A magnetic wall, on the other hands, forces the electric field to be incident tangentially whereas the magnetic field must be incident normally.

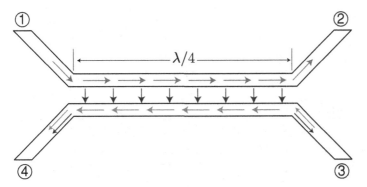

Figure 15.25 Two microstrip transmission lines placed in close proximity form a directional coupler. The electrical and magnetic coupling is designed to add in phase at port 4 but cancel at port 3.

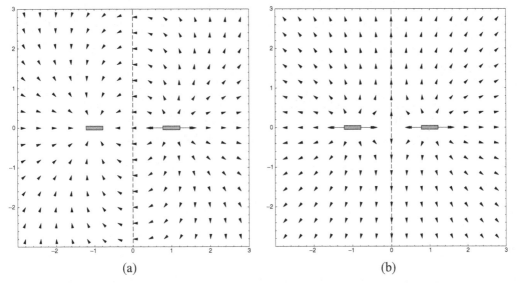

(a)   (b)

Figure 15.26 (a) Two conductors excited in the odd mode. The line of symmetry is an electrical wall. (b) Two conductors excited in the even mode. The line of symmetry is a magnetic wall.

where θ is the electrical length of the line. The "through" power is given by

$$c = \frac{\sqrt{1 - C^2}}{\sqrt{1 - C^2}\cos\theta + j\sin\theta} \qquad (15.208)$$

In this structure the operation of the coupler is intuitively easy to understand as well. Note that a voltage at port 1 couples to ports 3 and 4 with equal phase, whereas a current from port 1 to 2 couples to port 3 and 4 with opposite phase, as shown in Fig. 15.25. In general, then, the magnetic and electrical coupling interact differently at each port. If we design the structure properly, then, the coupling cancels at one port (isolated port) and adds at the other port [64].

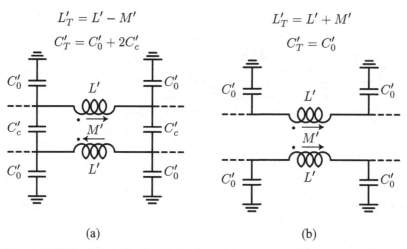

(a)   (b)

Figure 15.27 Equivalent distributed circuit model for two parallel lines excited in the (a) odd and (b) even modes.

15.9 References

This chapter is adapted from my lecture notes on Microwave Circuits, a course taught at UC Berkeley. In preparing the notes I have relied on classic books by Collin [9] and Smythe [59]. Other important sources includes Pozar [47] and Vendelin [64].

References

[1]   I. Aoki, S.D. Kee, D.B. Rutledge, and A. Hajimiri. Distributed active transformer-a new power-combining and impedance-transformation technique. *IEEE MTT-S International Microwave Symposium Digest*, **50**(1): 316–331, 2002.

[2]   A. Bevilacqua and A. M. Niknejad. An ultra wideband cmos low noise amplifier for 3.1-10.6 ghz wireless receivers. In *IEEE International Solid-State Circuits Conference*, pages 382–383, 2004.

[3]   Cao Yu, R.A. Groves, Huang Xuejue, N.D. Zamdmer, J.-O. Plouchart, R.A. Wachnik, Tsu-Jae King, and Chenming Hu. Frequency-independent equivalent-circuit model for on-chip spiral inductors. *IEEE Journal of Solid-State Circuits*, **38**: 419–426, March 2003.

[4]   R. Carson. *High-Frequency Amplifiers*. New York: John Wiley, 1982.

[5]   Wei-Kai Chen. *Active Network Analysis*. World Scientific Publishing, 1991.

[6]   D. Cheng. *Field and Wave Electromagnetics*. Prentice Hall, 1989.

[7]   G. Chien, F. Weishi, Y.A. Hsu, and L. Tse. A 2.4ghz cmos transceiver and baseband processor chipset for 802.11b wireless lan application. In *IEEE International Solid-State Circuits Conference*, pages 358–499, 2003.

[8]   Kenneth K. Clarke and Donald T. Hess. *Communication Circuits: Analysis and Design*. Addison-Wesley, 1971.

[9]   R. E. Collin. *Foundations for Microwave Engineering*. McGraw-Hill, 1966.

[10] R. E. Collin. *Field Theory of Guided Waves*. New York: IEEE Press, 2nd edition, 1990.

[11] D. Ham and W. Andress. A circular standing wave oscillator. In *IEEE International Solid-State Circuits Conference*, pages 380–381, 533, 2004.

[12] Charles A. Desoer and Ernest S. Kuh. *Basic Circuit Theory*. New York: McGraw-Hill, 1969.

[13] C. H. Doan, S. Emami, A. M. Niknejad, and R. W. Brodersen. Design of cmos for 60ghz applications. In *IEEE International Solid-State Circuits Conference*, pages 440–538, 2004.

[14] Chinh Doan. Ph.D. Thesis in preperation. University of California, Berkeley, 2006.

[15] Mohan Vamsi Dunga. A scalable MOS device substrate resistance model for RF and microwave circuit simulation: research project. University of California, Berkeley, 2004.

[16] E. Abou-Allam and T. Manku. An improved transmission-line model for mos transistors. *IEEE Transactions on Circuits and Systems II: Analog and Digital Signal Processing*, **46**(11): 1380–1387, 1999.

[17] B.-E. Kim *et al.* A 9dbm iip3 direct-conversion satellite broadband tuner-demodulator soc. In *IEEE International Solid-State Circuits Conference*, pages 446–507, 2003.

[18]   Richard Phillips Feynman, Robert B. Leighton, and Matthew L. Sands. *The Feynman Lectures on Physics, Vol II*. Reading, MA: Addison-Wesley, 1963.

[19]   Ranjit Gharpurey. Analysis and simulation of substrate coupling in integrated circuits. University of California, Berkeley, 1995.

[20]   G. Gonzalez. *Microwave Transistor Amplifiers*. Prentice Hall, 1984.

[21]   H. M. Greenhouse. Design of planar rectangular microelectronic inductors. *IEEE Trans. Parts, Hybrids and Packaging*, PHP-10:101–9, June 1974.

[22]   F. W. Grover. *Inductance Calculations*. Princeton, NJ: Van Nostrand, 1946.

[23]   M. S. Gupta. Power gain in feedback amplifiers: a classic revisited. *IEEE Transactions on Microwave Theory and Techniques*, 40(5): 864–879, May 1992.

[24]   H. Hashemi, X. Guan, and A. Hajimiri. A fully integrated 24 ghz 8-path phased-array receiver in silicon. In *IEEE International Solid-State Circuits Conference*, pages 390–391, 534, 2004.

[25]   S. J. Haefner. Alternating current resistance of rectangular conductors. *Proc. IRE*, pages 434–447, 1937.

[26]   J. Hagen. *Radio-Frequency Electronics: Circuits and Applications*. Cambridge: Cambridge University Press, 1996.

[27]   R. Howe and C. Sodini. *Microelectronics: An Integrated Approach*. Prentice Hall, 1996.

[28]   J. Craninckx and M.S.J. Steyaert. A 1.8-ghz low-phase-noise CMOS VCO using optimized hollow spiral inductors. *IEEE Journal of Solid-State Circuits*, 32(5): 736–744, 1997.

[29]   Howard Johnson and Martin Graham. *High-Speed Digital Design: A Handbook of Black Magic*. Prentice Hall, 1993.

[30]   Charles Kittel. *Introduction to Solid State Physics*. New York: John Wiley, 7th edition, 1996.

[31]   Herbert L. Krauss, Charles W. Bostian, and Frederick H. Raab. *Solid State Radio Engineering*. New York: John Wiley, 1980.

[32]   Ken Kundert. Power supply noise reduction. www.designers-guide.org, January 2004.

[33]   K. L. Scott, T. H. Hirano, H. Yang, H. Singh, R. T. Howe, and A. M. Niknejad. High-performance inductors using capillary based fluidic self-assembly. *Journal of Microelectromechanical Systems*, 13: 300–309, April 2004.

[34]   Thomas H. Lee. *The Design of CMOS Radio-Frequency Integrated Circuits*. Cambridge: Cambridge University Press, 1998.

[35]   Gang Liu. Ph.D. thesis in preperation. University of California, Berkeley, 2006.

[36]   John R. Long. Monolithic transformers for silicon RF IC design. *IEEE Journal of Solid-State Circuits*, 9: 1368–1382, September 2000.

[37]   A. M. Niknejad and R. G. Meyer. Analysis and optimization of monolithic inductors and transformers for RF ICS. In *IEEE International Solid-State Circuits Conference*, pages 375–378, 1997.

[38]   M. Zannoth, B. Kolb, J. Fenk, and R. Weigel. A fully integrated VCO at 2 ghz. *IEEE Journal of Solid-State Circuits*, 33(12): 1987–1991, 1998.

[39]   S. J. Mason. Power gain in feedback amplifiers. *Transactions of the IRE Professional Group on Circuit Theory*, CT-1(2): 20–25, June 1954.

[40]   M.W. Pospieszalski. On the measurement of noise parameters of microwave two-ports. *IEEE MTT-S International Microwave Symposium Digest*, 34(4): 456–458, April 1986.

[41]   Paul J. Nahin. *The Science of Radio*. New York: Springer-Verlag, 2nd edition, 2001.

[42]   A. M. Niknejad, C. Hu M. Chan, X. Xi, J. He, P. Su, Y. Cao, H. Wan, M. Dunga, C. Doan, S. Emami, and C.-H. Lin. Compact modeling for RF and microwave integrated circuits. In *Workshop on Compact Modeling*, 2003.

[43]   A. M. Niknejad and R G. Meyer. *Design, Simulation and Applications of Inductors and Transformers for Si RF ICs*. Boston: Kluwer Academic Publishers, 2000.

[44]   N.T. Tchamov, T. Niemi, and N. Mikkola. High-performance differential VCO based on armstrong oscillator topology. *IEEE Journal of Solid-State Circuits*, **36**(1): 139–141, 2001.

[45]   Henry W. Ott. *Noise Reduction Techniques in Electronic Systems*. New York: John Wiley, 2nd edition, 1988.

[46]   E. Pettenpaul and H. Kapusta *et al.* Cad models of lumped elements on gaas up to 18 ghz. *IEEE Transactions on Microwave Theory and Techniques*, **36**: 294–304, February 1988.

[47]   D. M. Pozar. *Microwave Egineering*. New York: John Wiley, 2nd edition, 1997.

[48]   John G. Proakis. *Digital communications*. New York: McGraw-Hill, 3rd edition, 1995.

[49]   E. Purcell. *Electricity and Magnetism, Vol. II*. McGraw-Hill Science/Engineering/Math, 1984.

[50]   Paul R. gray and Robert G. Meyer. *Analysis and Design of Analog Integrated Circuits*. New York: John Wiley, 3rd edition, 1993.

[51]   R. Lin, Lu Qiang, P. Ranade, Tsu-Jae King, and Chenming Hu. An adjustable work function technology using mo gate for cmos devices. *IEEE Electron Device Letters*, **23**: 49–51, January 2002.

[52]   S. Ramo, J. R. Whinnery, and T. Van Duzer. *Fields and Waves in Communication Electronics*. New York: John Wiley, 3rd edition, 1994.

[53]   John R. Reitz. *Foundations of Electromagnetic Theory*. Addison-Wesley, 1979.

[54]   A. E. Ruehli and H. Heeb. Circuit models for three-dimensional geometries including dielectrics. *IEEE Transactions on Microwave Theory and Techniques*, **40**: 1507–1516, July 1992.

[55]   S. Reynolds, B. Floyd, U. Pfeiffer, and T. Zwick. 60ghz transceiver circuits in sige bipolar technology. In *IEEE International Solid-State Circuits Conference*, pages 442–443, 538, 2004.

[56]   D. K. Shaeffer, T. Hai, L. Qinghung, A. Ong, V. Condito, S. Benyamin, W. Wong, and S. Xiaomin. A 40/43 gb/s sonet oc-768 sige 4:1 mux/cmu. In *IEEE International Solid-State Circuits Conference*, pages 236–237, 2003.

[57]   J. Singh. *Electronic and Optoelectronic Properties of Semiconductor Structures*. Cambridge: Cambridge University Press, 2003.

[58]   Jack R. Smith. *Modern Communication Circuits*. McGraw-Hill Science/Engineering/Math, 2nd edition, 1997.

[59]   William B. Smythe. *Static and Dynamic Electricity*. New York: McGraw-Hill, 3rd edition, 1967.

[60]   F. E. Terman. *Radio Engineers' Handbook*. New York: McGraw-Hill, 1943.

[61]   Y. Tsividis. *Operation and Modeling of the MOS Transistor*. New York: McGraw-Hill, 1987.

[62]   A. S. Inan and U. S. Inan. *Engineering Electromagnetics*. Prentice Hall, 1999.

[63]   G. D. Vendelin. *Design of Amplifiers and Oscillators by the S-Parameter Method*. New York: John Wiley, 1982.

[64]   G. D. Vendelin, U. L. Rohde, and A. M. Pavio. *Microwave Circuit Design Using Linear and Nonlinear Techniques*. New York: John Wiley, 1990.

[65]   W. Simburger, H.-D. Wohlmuth, P. Weger, and A. Heinz. A monolithic transformer coupled 5-w silicon power amplifier with 59% pae at 0.9 ghz. *IEEE Journal of Solid-State Circuits*, **34**(12): 1881–1892, 1999.

[66]   W. T. Weeks, L. L. Wu, M. F. McAllister, and A. Singh. Resistive and inductive skin effect in rectangular conductors. *IBM Journal of Research and Development*, **23**: 652–660, November 1979.

[67]   W.Y. Liu, J. Suryanarayanan, J. Nath, S. Mohammadi, L.P.B. Katehi, and M.B. Steer. Toroidal inductors for radio-frequency integrated circuits. *IEEE Transactions on Microwave Theory and Techniques*, **52**(2):646–654, 2004.

[68]   Xuemei Xi, Mohan Dunga, Jin He, Weidong Liu, Kanyu M. Cao, Xiaodong Jin, Jeff J. Ou, Mansun Chan, Ali M. Niknejad, and Chenming Hu. *BSIM4.3.0 MOSFET Model – User's Manual*. Berkeley, CA: University of California, 2003.

[69]   A. I. Zverev. *Handbook of Filter Synthesis*. New York: John Wiley, 1967.

Index